Rubber Toughened Engineering Plastics

Rubber Toughened Engineering Plastics

Edited by

A. A. Collyer

Formerly at the Division of Applied Physics
School of Science
Sheffield Hallam University, UK

CHAPMAN & HALL

London · Glasgow · Weinheim · New York · Tokyo · Melbourne · Madras

Published by Chapman & Hall, 2–6 Boundary Row, London SE1 8HN, UK

Chapman & Hall, 2–6 Boundary Row, London SE1 8HN, UK

Blackie Academic & Professional, Wester Cleddens Road, Bishopbriggs, Glasgow G64 2NZ, UK

Chapman & Hall GmbH, Pappelallee 3, 69469 Weinheim, Germany

Chapman & Hall Japan, Thomson Publishing Japan, Hirakawacho Nemoto Building, 6F, 1-7-11 Hirakawa-cho, Chiyoda-ku, Tokyo 102, Japan

Chapman & Hall Australia, Thomas Nelson Australia, 102 Dodds Street, South Melbourne, Victoria 3205, Australia

Chapman & Hall India, R. Seshadri, 32 Second Main Road, CIT East, Madras 600 035, India

First edition 1994

Typeset in Times 10/12pt by Thomson Press (India) Ltd, New Delhi
Printed in Great Britain at the University Press, Cambridge

ISBN 0 412 58380 1

A catalogue record for this book is available from the British Library

Library of Congress Catalog Card Number: 93-74886

∞ Printed on acid-free text paper, manufactured in accordance with ANSI/NISO Z39. 48-1992 (Permanence of Paper).

Contents

List of contributors

A. A. Collyer
Flat 2, 9 Elmhyrst Road
Weston Super Mare BS23 2SJ
UK

A. M. Donald
Department of Physics
University of Cambridge
Cavendish Laboratory
Madingley Road
Cambridge CB3 0HE
UK

R. J. Gaymans
University of Twente
PO Box 217
7500 AE Enschede
The Netherlands

D. J. Hourston
The Polymer Centre
Lancaster University
Lancaster LA1 4YA
UK

H. Keskkula
Department of Chemical Engineering
University of Texas
Austin
TX 78712
USA

S. Lane
The Polymer Centre
Lancaster University
Lancaster LA1 4YA
UK

G. C. McGrath
Engineering Department
TWI
Abington Hall
Abington
Cambridge CB1 6AL
UK

D. Parker
ICI Advanced Materials
PO Box 90
Wilton
Middlesbrough TS6 8JE
UK

D. R. Paul
Department of Chemical Engineering
University of Texas
Austin
TX 78712
USA

A. Savadori
EniChem
Casella Postale 12120
20120 Milano
Italy

S. J. Shaw
Materials and Structures Department, Defence Research Establishment, Building R178, Farnborough, Hampshire GU14 6TD, UK

I. Walker
International Paints Ltd
18 Hanover Square
London W1A 1AD
UK

G. W. Wheatley
3 Hermes Close
Hull HU9 4DS
UK

Preface

The rubber toughening of polymers such as polystyrene has been successfully carried out for many years, and has led to many diverse engineering materials such as high impact polystyrene (HIPS), acrylonitrile–butadiene–styrene (ABS) and styrene–butadiene–styrene copolymers (SBS). The synthetic routes to manufacture are well understood and most adequately documented. For the high temperature engineering and speciality plastics this is certainly not the case; difficulties are encountered both in the choice of rubber for the dispersed phase and in the synthetic routes involved for obtaining the optimum particle size for toughening and in obtaining adequate interfacial adhesion. This book is intended to bring out the main physical principles involved in optimum toughening and to describe the synthetic strategies used to obtain satisfactorily toughened grades in these materials.

The book may be divided into two parts: in the first section Chapters 1, 2 and 3 deal with failure mechanisms and toughening mechanisms in pure polymer matrices and in fibre reinforced composites, with Chapter 4 discussing the numerous methods available for the evaluation of toughened plastics materials. Chapter 5 reviews the wide spectrum of toughening agents available for engineering polymers. The second section of the book is devoted to describing the synthetic routes and toughening strategies involved for various polymer matrices, namely epoxies, polyamides, polyesters and polycarbonates, polysulphones and polyaryletherketones, and polyimides.

This work is intended for research and development workers in universities and industry with an interest in polymeric materials and polymer chemistry, and it is hoped that the book acts as a satisfactory focus for the current thought on rubber toughening principles and the methods employed for the rubber toughening of major engineering and speciality plastics.

1

Failure mechanisms in polymeric materials

A. M. Donald

1.1 INTRODUCTION

Polymers differ from other molecules by virtue of their size. They are macromolecules, whose total molecular weight (or relative molar mass) may reach millions. As we shall see, this has important consequences for their response to mechanical stress or strain. In particular, whereas most materials (including polymers) which exhibit any degree of ductility may show a shear response, an alternative mode of deformation is also open to polymeric materials and to them alone, the mechanism known as crazing. In order to understand toughening and failure mechanisms in polymers, one therefore has first to look at the nature of the polymer chains. Only a brief overview of salient facts will be presented here to introduce the key factors and terminology. The interested reader is referred to other texts (e.g. Refs 1–3) for a broader picture.

Many of the recent fundamental studies of micromechanisms of deformation have been carried out on the vinyl polymer polystyrene (PS), which forms the basis of the rubber toughened material high impact polystyrene (HIPS), and this will be used as a model to introduce some basic concepts. A polymer consists of a series of monomer units joined together to form a long chain. For most of the materials to be discussed in this book the polymer will consist of only one type of unit, and for PS this monomer is

```
  H  H
  |  |
—C—C—
  |  |
  H
```

This unit is repeated n times, where n is known as the degree of polymerization.

One of the attractions of polystyrene as a model material is that it can be obtained in monodisperse form. This means that a polymerization route exists (so-called anionic polymerization) enabling control over the molecular weight to be achieved, so that the chains are all essentially the same length (same *n*). This is in contrast to normal commercial materials which will be polydisperse. The degree of polydispersity is normally characterized by considering different moments of the molecular weight distribution, the most normal ratio chosen being that of $\bar{M}_w/\bar{M}_n$. In this expression $\bar{M}_w$, the weight-average molecular weight, and $\bar{M}_n$, the number average, are defined by

$$\bar{M}_w = \frac{\sum(N_i M_i)M_i}{\sum N_i M_i} \quad \text{and} \quad \bar{M}_n = \frac{\sum N_i M_i}{\sum N_i}$$

where N_i is the number of molecules of molecular weight M_i, and $\sum$ implies summation over all *i* molecular weights. If all the chains are the same length, then the ratio $\bar{M}_w/\bar{M}_n$ will be unity. For monodisperse PS the best achievable samples will have $\bar{M}_w/\bar{M}_n \approx 1.03$, whereas commercial samples of most polymers are likely to have $\bar{M}_w/\bar{M}_n \approx 2$. In practice, as will be seen below, the presence of a low molecular weight tail to the molecular weight distribution may have a significant effect on the deformation. Useful samples of PS are likely to have $\bar{M}_w > 200\,000$.

The conformation of a PS chain in both the melt and the glass has been shown to be that of a random or Gaussian coil [4]. This means that the chain follows a random walk, whose root mean square (r.m.s.) end-to-end distance is proportional to the square root of the number of monomers in the chain. Each chain can therefore be thought of as a loose coil which is penetrated by its neighbours, and this will be true both above and below the glass transition temperature T_g.

In the melt it has long been recognized that the presence of entanglements plays a key role in determining the viscoelastic response of the polymer (see, for example, Ref. 5). These entanglements were originally thought of as simple topological knots analogous to crosslinks in a rubber. This idea was based on the similarity of rheological data of linear (non-crystalline) polymers above T_g and crosslinked rubbers, as shown schematically in Fig. 1.1. For a rubber the shear modulus *G* in the so-called plateau region (Fig. 1.1) above T_g is related to the molecular weight between crosslinks M_c by

$$G = \frac{\rho RT}{M_c}$$

where ρ is the density, *R* the gas constant and *T* the absolute temperature. A similar expression is then used to relate the shear modulus in the plateau region of an uncrosslinked polymer to the molecular weight between entanglements, M_e. For PS this leads to a value for M_e of $\sim 19\,000$, this value being independent of total chain length for high molecular weight polymers. It will become clear that this quantity, which can only be established unequivocally above T_g, appears to play a crucial role in determining the response of the polymer to mechanical stresses below T_g.

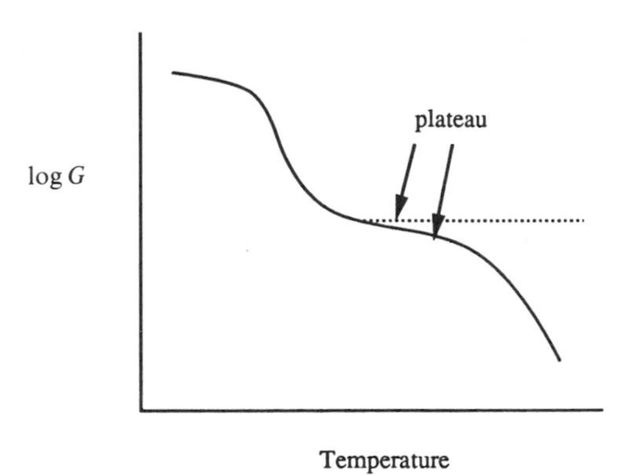

Fig. 1.1 Schematic representation of the variation of shear modulus with temperature for crosslinked (·····) and uncrosslinked (——) polymers.

Recent ideas no longer necessarily envisage the entanglements as being localized point constraints. The approaches of De Gennes [6] and Doi and Edwards [7] suggest instead that the constraints imposed on one chain by its neighbours can be represented by a tube, of diameter equal to the distance between entanglements, within which the particular chain is confined (Fig. 1.2). When the chain moves it can only do so along the tube, a process known as **reptation**; it cannot cross the boundaries of the tube. As the chain moves and its end comes out of one end of the tube, then the memory of that part of the tube is lost. Ultimately this means that the whole chain in a melt can diffuse across a sample, but the process is obviously far slower than it would be if the tube were not present.

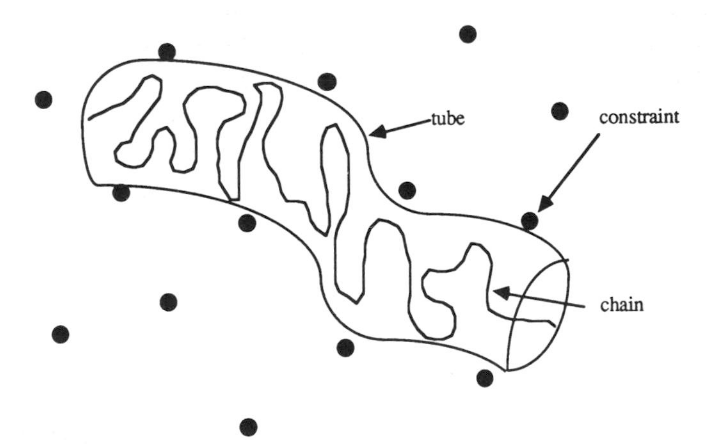

Fig. 1.2 Schematic representation of the tube model of De Gennes [6] and Doi and Edwards [7]. The chain is confined to a tube by entanglements arising from the presence of other chains.

Reference has been made above to the glass transition temperature T_g. This is frequently taken as the temperature below which long-range motions of the chain are frozen out. PS is usually obtained in atactic form, i.e. the distribution of pendent benzene rings is stereochemically irregular along the chains and thus they cannot crystallize. The material is amorphous and in engineering applications will be used below T_g in the glassy state. Although many of the important polymers used in toughened plastics are non-crystalline, this is not the case for all of them, polyetheretherketone being one example of a matrix that is highly crystalline. For crystallizable polymers one may expect an additional complexity of deformation behaviour since there is a temperature regime between T_g and the melting temperature T_m in which one population of chains – those in the amorphous regions – is mobile, while other chains (or indeed other parts of the same chains) are pinned in the crystals. At the fundamental level comparatively little is known about the behaviour in this regime (for a description of deformation in crystalline polymers the reader should consult the recent reviews by Friedrich [8] and by Narisawa and Ishikawa [9]). This chapter will confine itself to amorphous polymers for which the picture is becoming rather clearer.

1.2 MECHANICAL PROPERTIES – SOME DEFINITIONS

Whereas, to the layman, stress and strain are often used synonymously, in the context of mechanical properties they have very distinct meanings. **Stress** is a force per unit area. Since during deformation the cross-sectional area of a sample usually changes, it is important to distinguish between **true** stress – the force per instantaneous cross-sectional area – and **nominal** (or engineering) stress which is the force divided by the initial (undeformed) cross-sectional area. In general, for a solid, the stress can be represented by a stress tensor σ_{ij} containing nine terms:

$$\sigma_{ij} = \begin{bmatrix} \sigma_{11} & \sigma_{12} & \sigma_{13} \\ \sigma_{21} & \sigma_{22} & \sigma_{23} \\ \sigma_{31} & \sigma_{32} & \sigma_{33} \end{bmatrix}.$$

Of the two suffixes, the first describes the direction of the normal to the plane on which the stress acts, whereas the second describes the direction of the stress. Figure 1.3 shows the nine components of a general stress acting on a cube. The three components of stress for which the two suffixes are equal, σ_{ii}, correspond to normal stresses, since they act on planes perpendicular to their direction. The usual sign convention is that positive values of σ_{ii} correspond to tensile stresses, whereas negative values are compressive. The remaining six stresses are shear stresses which tend to cause the body to rotate. In order for this motion not to occur, it is necessary for σ_{ij} to equal σ_{ji} so that there is no resultant torque. This means that in a static situation only six of the nine compo-

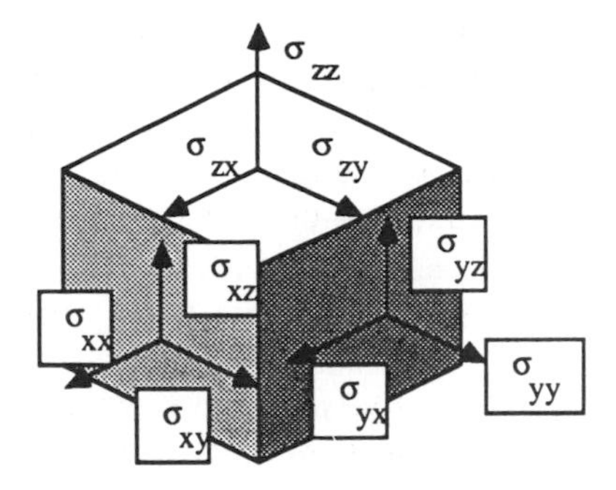

Fig. 1.3 Schematic representation of the nine components of stress acting on a cube.

nents of the stress tensor are independent. It is frequently useful to split the stress applied to a polymer into two components: a hydrostatic or dilational component, which gives rise to a volume change, and a deviatoric or pure shear component, which causes a change in shape. The hydrostatic component p can be written

$$p = \tfrac{1}{3}(\sigma_{11} + \sigma_{22} + \sigma_{33}).$$

The deviatoric component σ'_{ij} can then be written by subtracting the hydrostatic component from the original tensor to yield

$$\sigma'_{ij} = \begin{bmatrix} \sigma_{11} - p & \sigma_{12} & \sigma_{13} \\ \sigma_{21} & \sigma_{22} - p & \sigma_{23} \\ \sigma_{31} & \sigma_{32} & \sigma_{33} - p \end{bmatrix}.$$

Strain ε_{ij} is likewise a tensor, and can be written using nine components in the most general case:

$$\varepsilon_{ij} = \begin{bmatrix} \varepsilon_{11} & \varepsilon_{12} & \varepsilon_{13} \\ \varepsilon_{21} & \varepsilon_{22} & \varepsilon_{23} \\ \varepsilon_{31} & \varepsilon_{32} & \varepsilon_{33} \end{bmatrix}.$$

In a simple uniaxial test, strain is particularly easily defined. If l_0 is the original length and l_1 is the final, then the **nominal** strain ε_n is given by

$$\varepsilon_n = \frac{l_1 - l_0}{l_0}.$$

For larger strains this will diverge from the **true** strain ε_t, given by the integral of the above equation:

$$\varepsilon_t = \int_{l_1}^{l_0} \frac{dl}{l} = \ln\left(\frac{l_1}{l_0}\right).$$

Both stress and strain can be described in terms of three **principal** components acting along principal axes. In this representation the axes are so chosen that the tensor is diagonalized (in the case of stresses this means that the three principal stresses are normal stresses and the shear stresses are zero).

Although in general there are components of stress and strain acting across all faces of a cube, there are some important situations where this is not the case. One example is a thin sheet. Since the stress acting normal to a free surface is zero, for a thin sheet all the stresses acting on planes parallel to the sheet surface must be small. This means that the total stress normal to the plane of the sheet will tend to zero, giving rise to a state of **plane stress** in which σ_{11} and σ_{22} are finite but σ_{33} is zero. Similarly the state of strain in which one of the principal components vanishes is known as **plane strain**. This situation arises for instance in relatively thick samples in the vicinity of a crack tip where the material is constrained.

Stress and strain are related through a modulus. The familiar example is Young's modulus that relates stress and strain in a tensile test, but more generally, since stress and strain are both tensors, a fourth-order tensor c_{ijkl} is required:

$$\sigma_{ij} = c_{ijkl}\varepsilon_{kl}$$

where summation is implied over both *i* and *j*. Although this stiffness tensor c_{ijkl} in principle contains 81 components, symmetry reduces this to only 2 for an elastically isotropic solid such as a glassy polymer [10]. Three parameters are commonly used to characterize the polymer: Young's modulus E, the shear modulus G, and Poisson's ratio ν. (When a stress σ_{11} is applied along the 1 axis there are resulting strains of $-\nu\sigma_{11}/E$ along the 2 and 3 directions.) These three are related by the equation $E = 2(1 + \nu)G$.

It is clearly important, when attempting to characterize the response of a material to a stress or strain field, that these fields are as well characterized as possible. To this end it is necessary to work with samples of well-defined geometry. The chapter by Bowden in Ref. 1 provides an excellent review of the subject. With this information it then becomes possible to map out the different stress states that give rise to a particular type of deformation. However, many studies restrict themselves to the simplest cases of uniaxial tension or plane strain compression.

Most materials show a linear elastic response at low strains, and polymers are no exception. As the stress and strain increase, for many polymers a **yield point** is passed, as shown in Fig. 1.4. Loosely speaking, the yield point corresponds to the point after which increasing strain occurs under a lower stress, so that (as in Fig. 1.4) it corresponds to the maximum in the curve. Beyond this point, deformation is certainly occurring plastically (i.e. the deformation cannot be recovered if the stress is removed), but for many polymers the onset of plastic deformation actually precedes the yield point defined as above.

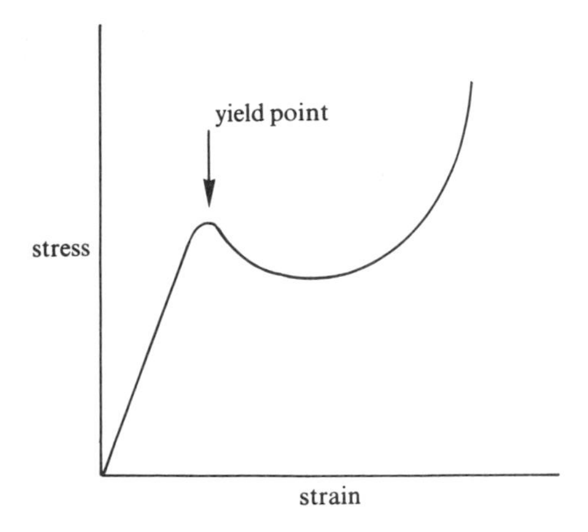

Fig. 1.4 Schematic stress–strain curve with yield point.

When identifying the yield point with the maximum in Fig.1.4 a couple of caveats are in order. Firstly, if nominal stress is plotted against strain a maximum may occur which actually reflects a geometrical instability of the sample rather than a true intrinsic yield point; to identify the intrinsic yield point requires that true stress is plotted. Secondly, not all polymers exhibit a maximum; for those that do not, an intrinsic yield point can usually be identified with a kink in the true stress–strain curve. In general, as strain continues to rise a point will be reached at which orientation hardening sets in. This means that whereas beyond the yield point strain increases under a decreasing stress, when orientation hardening sets in the stress required for further strain increases again, and often quite steeply. Although for most polymers stressed at temperatures close to T_g homogeneous plastic deformation may occur uniformly throughout a sample, in many instances of room temperature testing the deformation actually proceeds inhomogeneously, with localized regions of extensive deformation surrounded by material that has only deformed elastically. These regions may consist of either shear deformation or crazing, to be described fully below.

Since in this book we are interested in toughened polymers, we need to define what we mean by **toughness**. A tough material is one that absorbs a large amount of energy before failure, in contrast to a brittle one that does not. This means that there needs to be available to the polymer one or more deformation mechanisms which absorb energy before crack propagation occurs. Figure 1.5 shows the contrasting behaviour of PS and HIPS. PS is a brittle material which fractures before yield. Crazing does precede fracture, but only to a very limited extent. In contrast to this, in HIPS the rubber particles promote extensive craze formation throughout the sample, and since a large number are generated before any fail to give rise to crack propagation this is an

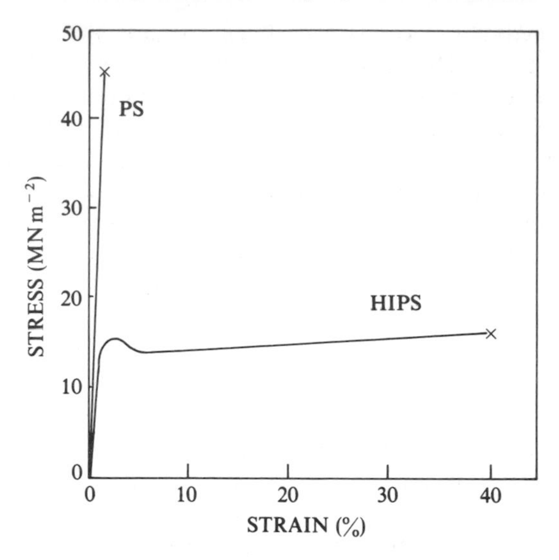

Fig. 1.5 Stress–strain curves for PS and HIPS. (Reproduced from C. B. Bucknall, *Toughened Plastics;* published by Applied Science, 1977.)

effective way of absorbing energy; fracture is delayed to much higher strains and the material is tough.

1.3 SHEAR DEFORMATION

When general yield occurs as the yield point is passed, shear is the general mode of deformation. The yield stress depends on a number of parameters due to both the state of the polymer and the nature of the applied stress. High strain rates and low temperatures will both push up the yield stress. So will annealing the glass below the glass transition temperature, a process that is known to promote molecular rearrangements leading to densification. Molecular weight is not an important parameter for yield.

At a molecular level the processes involved in yield are not well understood. The dependence of the yield stress on temperature and strain rate has led workers to use an Eyring type theory to model behaviour, as has been done for viscosity. In this type of model it is assumed that there is an activation barrier ΔE^* to be overcome for deformation to proceed, and the presence of a stress reduces the height of the barrier for jumps in the forward direction and increases it for reverse jumps. This leads to an expression for the strain rate $\dot{\varepsilon}$ in terms of the applied stress σ:

$$\dot{\varepsilon} = A \exp\left[-\frac{\Delta E^* - v^*|\sigma|}{RT} \right]$$

where v^* is the activation volume and A is a constant. The work of Bauwens-Crowet, Bauwens and Homès [12] shows that yield stress data for polycarbonate can fit this expression well. However, the physical meaning that should be attached to the parameters ΔE^* and v^* is not clear, and it cannot be said that it describes the yield process in molecular terms, although it is a convenient formulation phenomenologically. What does seem to be apparent is that the motions that are involved in yield are comparatively local, so that only short segments of a chain are involved. It is for this reason that molecular weight does not affect the yield stress. The significance of this will be seen later when contrasting shear processes with crazing.

It is often helpful to formulate a yield criterion to describe the state of stress required for yield to occur. Various different criteria have been proposed, all of which work well for some types of materials, but for polymers there is no simple criterion which can be universally applied. The simplest criterion is that known as the Tresca criterion. This states that yield will occur when the maximum shear stress on any plane achieves a critical value. Written in terms of the three principal stresses it can be expressed as

$$|\sigma_{11} - \sigma_{22}| = 2\tau_T = \sigma_y$$

where τ_T is usually taken as the yield stress of the material undergoing pure shear τ_y, and σ_y is the uniaxial tensile yield stress. This criterion was originally developed for metals, as was a second criterion known by the name of the Von Mises criterion. This expresses the yield criterion in terms of the differences of the principal stresses:

$$(\sigma_{11} - \sigma_{22})^2 + (\sigma_{22} - \sigma_{33})^2 + (\sigma_{33} - \sigma_{11})^2 = 6\tau_M$$

where τ_M is the shear stress for flow in pure shear, which is also equal to τ_y. For polymers neither of these criteria is satisfactory since they both imply that the yield stress will be independent of the hydrostatic component of the stress tensor whereas in practice this is not the case. Various modifications have been proposed to the above two criteria to take this into account (see for instance Ref. 13) and an alternative criterion is sometimes used, one that was originally developed for soils and which goes by the name of the Mohr–Coulomb criterion. This states that yield will occur when the shear stress on any plane exceeds a critical value which varies linearly with the stress normal to that particular plane. It is thus similar to the Tresca criterion, but with τ_T replaced by $\tau_C^0 - \mu_C\sigma_N$. τ_C^0 and μ_C are both material constants, the latter being equivalent to a coefficient of friction.

If the shear deformation proceeds inhomogeneously, for a bulk sample the usual manifestation is in the form of shear bands. Bowden (in Ref. 1) has discussed the geometry of shear deformation under different types of constraints. Shear banding corresponds to the case of restraint in two directions, the type of constraint one would expect in a thick sample. However, in thin samples, or in regions of a bulk sample where the constraint is less (for instance

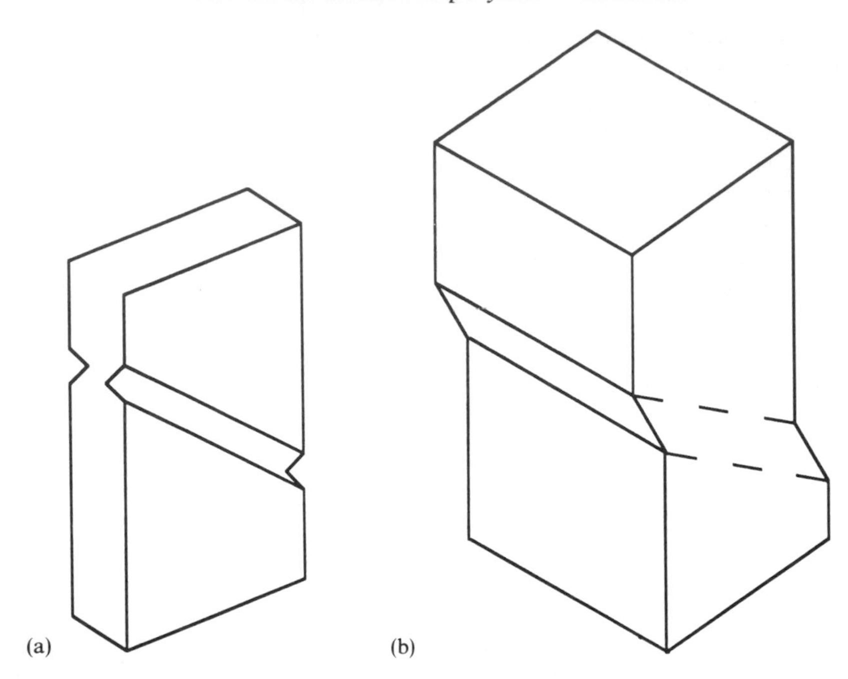

Fig. 1.6 Schematic appearance of (a) an inclined neck and (b) a shear band.

owing to local voiding of rubber particles) there is the possibility of an inclined neck. In this case local thinning occurs, with the material relaxing inwards within the neck. Figure 1.6 shows the geometry of both an inclined neck and a shear band.

If a shear band forms in an isotropic material at constant volume, it must grow at 45° to the tensile axis. For polymers there may be a substantial dilatation, and the shear band then grows at some angle greater than 45°, the deviation from 45° depending on the magnitude of the dilatation. For an inclined neck, the angle of inclination to the tensile axis is 54.7° for deformation occurring without volume change, and this angle also increases if there is a dilatation.

In general inhomogeneous deformation occurs because beyond yield the material strain softens (Fig. 1.4). This means that, once one part of the sample has yielded, it can continue to deform under a lower stress, and thus that it is easier for that part of the sample to continue to deform rather than for a different part of the sample to yield. It follows that this tendency to inhomogeneous deformation will be most pronounced under testing conditions which show a pronounced maximum in stress followed by a large yield drop. Conditions which promote this behaviour are low temperatures and/or high strain rates. Examples of the variation in true stress–strain curves for PS under different testing conditions are shown in Fig. 1.7 [13], from which it is clearly

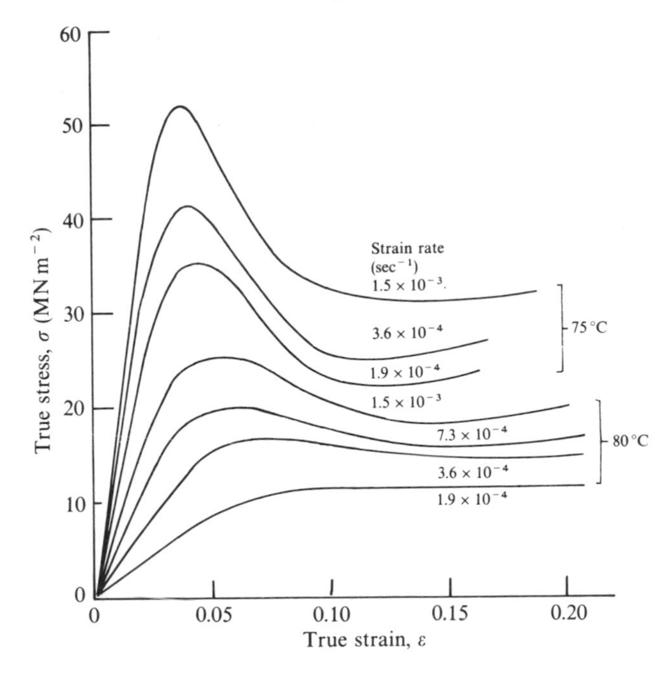

Fig. 1.7 True stress–strain curves for PS strained in plane strain compression, at various temperatures and strain rates. (Reproduced from P. B. Bowden and S. Raha, *Philosophical Magazine*, 1970.)

seen that both the yield stress and the yield drop are largest for the lower of the two temperatures studied and the fastest strain rate. One consequence of this is that the shear will become increasingly localized as the temperature drops, resulting in sharp shear bands. In contrast to this, as T_g is approached the shear becomes much more diffuse, and ultimately individual shear bands are not observed. In general, when shear bands form, molecular orientation takes place within them. This can be seen from the development of birefringence when the bands are viewed between crossed polars. As an example of sharp birefringent shear bands, Fig. 1.8 shows the appearance of bands in polyethersulphone (PES) following a uniaxial tensile test at 140 °C (some 80 °C below T_g).

In thin films subjected to uniaxial tension, the appearance of shear is rather different with so-called deformation zones (DZs) [16, 17] growing normal to the tensile axis. Once again substantial molecular alignment can be inferred from birefringence measurements [16], and the extent of localization of the strain depends on both testing conditions and the state of the polymer. Figure 1.9 shows the contrasting behaviour of annealed and unannealed samples of polycarbonate (PC), the former showing much more extreme strain localization than the latter, corresponding to the increase in yield stress and

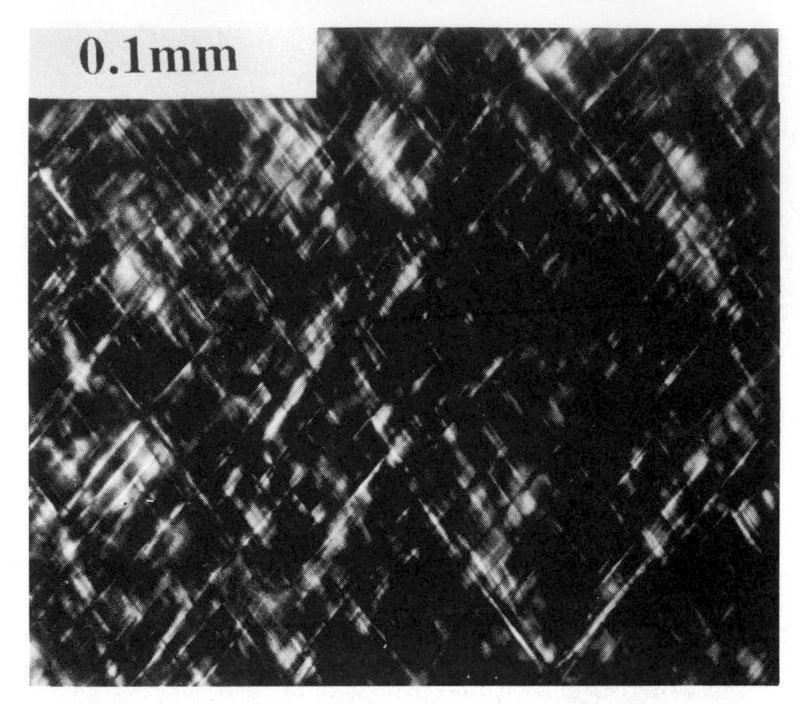

Fig. 1.8 The appearance of shear bands in PES strained at $10^{-4}\,s^{-1}$ at 140 °C viewed under crossed polars. (Reproduced from C. J. G. Plummer and A. M. Donald, *Journal of Applied Polymer Science*, 1990.)

yield drop consequent on the annealing of PC [18]. Thin films of 1 μm or less are suitable for transmission electron microscopy (TEM), and Fig. 1.9(c) shows the appearance of a DZ in annealed PC in TEM. It is clear that the contrast, and therefore the mass thickness, is uniform over the majority of the DZ and that (in contrast to the appearance of crazes to be described below) there are no voids. From the contrast of the electron image plate it is possible to measure the extension ratio λ of the material within the DZ. For PC this yields a value of 1.4 [16]. This value is similar to that measured when a uniform neck forms in a bulk sample, and can therefore be said to represent some kind of 'natural draw ratio'.

1.4 CRAZING

Whereas shear is usually associated with rather ductile behaviour, crazing is frequently associated with a brittle response, and this is particularly true of single-phase polymers (such as PS) as opposed to toughened polymers (such as HIPS). This is because crazes contain voids and easily break down to form cracks. The nature of craze morphology cannot be revealed by optical

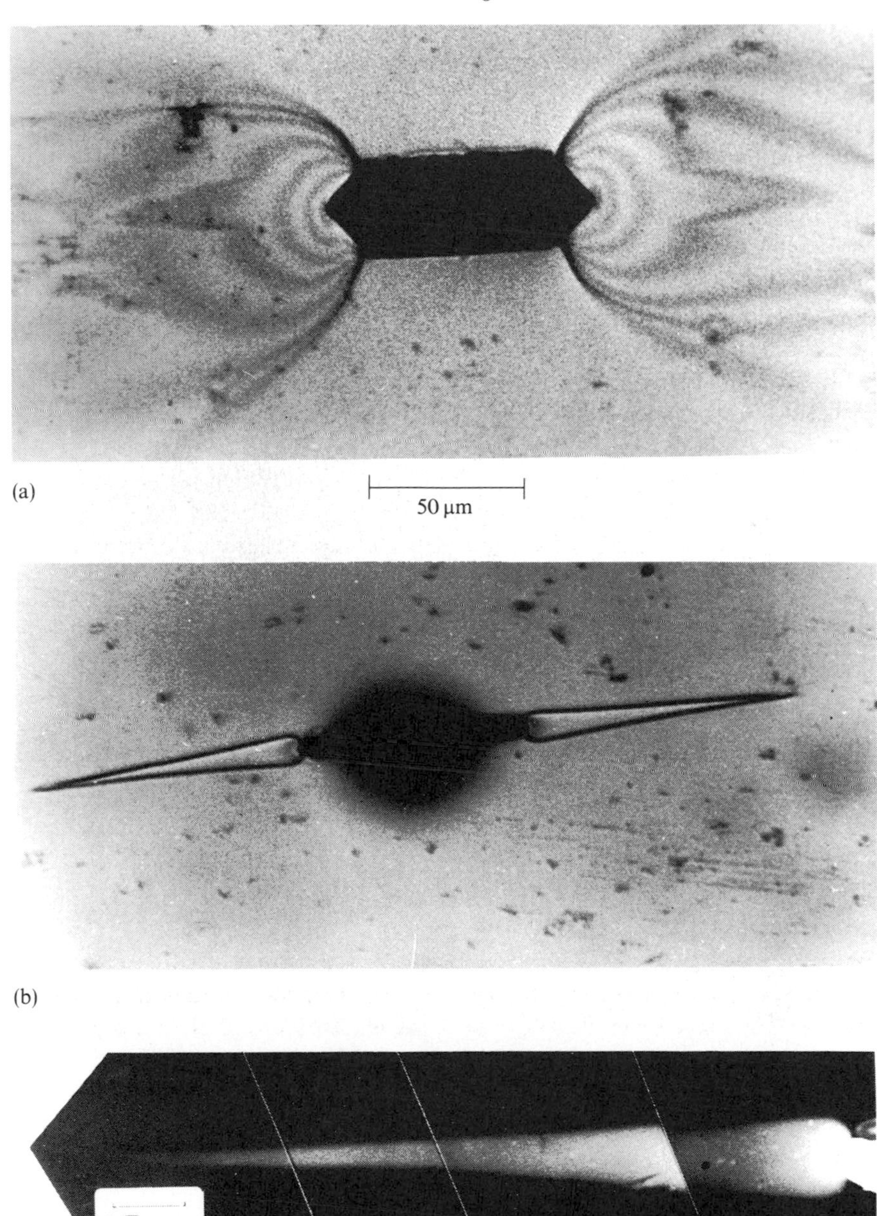

Fig. 1.9 The appearance of deformation zones in (a) unannealed PC in the optical microscope, (b) PC annealed for 1 h at 132 °C and (c) the appearance of a DZ in annealed PC in the transmission electron microscope. (Reproduced from A. M. Donald and E. J. Kramer, *Journal of Materials Science*, 1981.)

microscopy, because the scale of the structure is too fine, but the greater resolution of the TEM shows that a craze consists of an array of fibrils and voids. This void–fibril structure was first visualized by Kambour and Russell [19], and an example of a TEM image of a craze is shown in Fig. 1.10. Since a micrograph such as Fig. 1.10 consists of a projection onto a two-dimensional surface of a three-dimensional structure, all the different layers of fibrils overlapping, it is not straightforward to measure craze fibril diameters and spacing from such an image. To carry out these measurements it is preferable to use small angle X-ray scattering [18, 19]. Typical values for PS are a mean fibril diameter of 6 nm and a spacing of 20–25 nm [21].

Crazes grow normal to the principal (tensile) stress. They may grow to be millimetres or even centimetres in length and fractions of a millimetre in thickness if conditions are such as to prevent early failure and crack propagation. However, frequently they are far smaller, particularly in toughened materials. Because of their void–fibril structure and therefore their different refractive index from surrounding undeformed material, they scatter light. A stressed material that contains a high density of crazes is said to have 'stress whitened' because of its appearance as a result of this scattering.

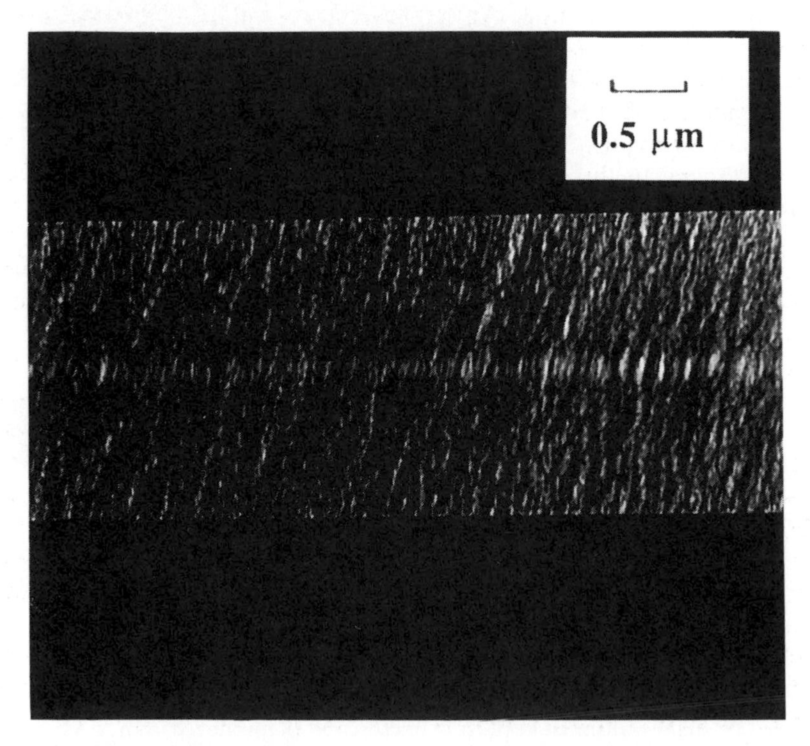

Fig. 1.10 Transmission electron micrograph of a craze in PS.

Before the internal structure of the craze had been observed, it was known that, although they resembled cracks, they were distinct because they were capable of sustaining a load. Much of the early work to characterize crazes concentrated on the stress conditions under which they grew. There is an inherent problem in such studies, because so often crazes are initiated at flaws either within or at the surface of the specimen, where of course the stress conditions are not accurately known. In addition, if optical methods for detecting the existence of crazes are used, substantial craze growth beyond the initiation stage is actually being monitored. Thus in general the stress for craze propagation is measured. Sternstein, Ongchin and Silverman [22] looked at the stress conditions for craze formation in polymethylmethacrylate (PMMA). They were able to show that crazes grow perpendicular to the major principal stress, and that this stress must exceed a critical value. However, the precise expression they derived is not now generally accepted. Bowden and Oxborough [23] formulated a criterion in terms of a critical tensile strain which depended on the hydrostatic component of the stress tensor. It can be written

$$\sigma_1 - \nu\sigma_2 - \nu\sigma_3 = Y + \frac{X}{\sigma_1 + \sigma_2 + \sigma_3}$$

where X and Y will be time–temperature dependent. That the hydrostatic component of the stress tensor is an important parameter is not surprising when it is remembered that the void–fibril structure of the craze means that there is substantial dilatation associated with crazing. Whether crazes can form at all if the hydrostatic component is zero or even negative is very hard to ascertain owing to the uncertainties in local stress fields around the flaws where crazes are likely to nucleate.

If the stress conditions for stress initiation are hard to determine, the microscopic mechanisms are even harder to establish. Plausible arguments of the various steps involved have been put forward, but concrete evidence is sparse. Kramer [24] has summarized the likely events involved in craze initiation. The basic sequence can be described as (1) local plastic deformation by shear in the vicinity of a defect, which leads to the buildup of significant lateral stresses, (2) nucleation of voids to release the triaxial constraints and (3) void growth and strain hardening of the intervening polymer ligaments as molecular orientation proceeds. The incipient craze structure is thereby stabilized and the craze can subsequently propagate under appropriate stress conditions. If the increase in stress in stage (1) promotes further shear deformation, then clearly the craze nucleus will not form. As we shall see later there are ways to rationalize which polymers tend to deform by shear and which by crazing, but these arguments apply to craze propagation explicitly. At the initiation stage there is no simple picture to determine whether or not voiding will start to occur, although one may suspect that the same parameters may well be involved.

Given this three-stage process, the question of which stage is the critical one needs also to be considered. The stage most usually considered critical is the void nucleation stage. Argon and Hanoosh [25], making this assumption, tested their ideas of a critical porosity being required on PS, using a theoretical model to relate the applied stress to the achieved porosity. One consequence of nucleation being the critical stage is that one would assume that there is a critical size associated with the critical nucleus. Evidence to support this comes from work done on HIPS [26], for which it has long been known that small rubber particles are rather inefficient at toughening. If a critical size is involved for a craze to initiate, it will be necessary for the stress level to remain sufficiently high over at least this critical distance. Since the extent of the stress concentration at a rubber particle will scale with particle size, it follows that small particles are less likely to be able to satisfy this criterion, and therefore be poor at nucleating crazes – as observed.

The mechanism for the propagation of a craze tip is rather better understood. The idea that a craze could propagate via the meniscus instability (first described for the behaviour of interfaces between liquids of different densities by Taylor [27]) was initially proposed by Argon and Salama [28]. A schematic view of how this mechanism can lead to the void–fibril structure of the craze is shown in Fig. 1.11. Experimental evidence supporting this view came from stereo images of craze tips in PS taken in the transmission electron microscope [29]. It is now thought that this same mechanism also applies to craze widening. A discussion of this will be deferred until some molecular aspects of craze formation have been introduced in section 1.6.

In order to examine the behaviour of crazes, a technique using optical interferometry has frequently been used. This is discussed in detail in Ref. 30. The technique utilizes the fact that the voided structure of the craze gives rise to a change in refractive index, and that there is a sharp interface between crazed and uncrazed material. A characteristic interference fringe pattern is set up, and analysis of the positions of the various fringes permits the craze shape to be determined after various assumptions, relating to the refractive index of the craze and its variation with strain, have been made.

In analysing the data obtained in this way, comparison of the observed craze profile with a theoretical prediction known as the Dugdale profile is frequently made. This specifically relates to the situation of an elliptical crack possessing narrow plastic zones at its tips, with a constant surface stress σ_c acting on the boundaries of the zones. The situation is shown schematically in Fig. 1.12. Applying the condition that there should be no stress singularity at the tip of the crack, Dugdale derived an expression for the length of the plastic zones, in terms of the crack half-length c and the applied stress σ [31]. If a is the total half crack plus craze length, then

$$\frac{a}{c} = \sec\left(\frac{\pi\sigma}{2\sigma_c}\right)$$

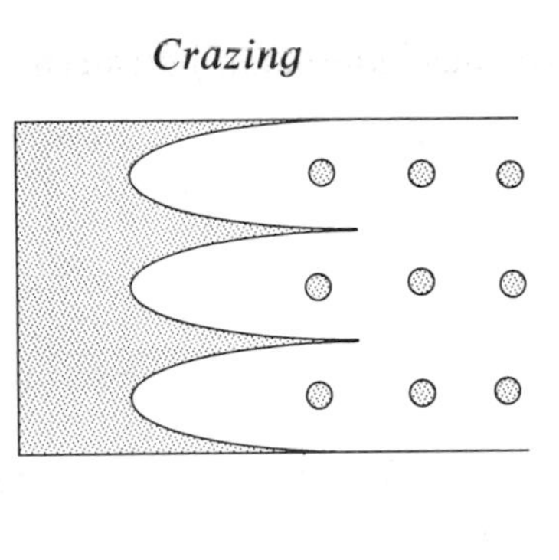

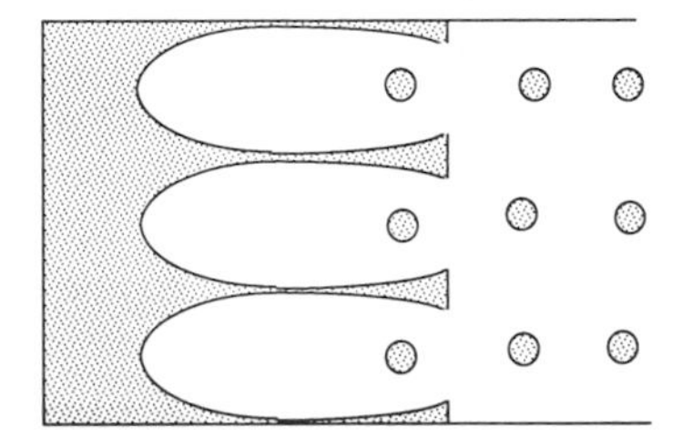

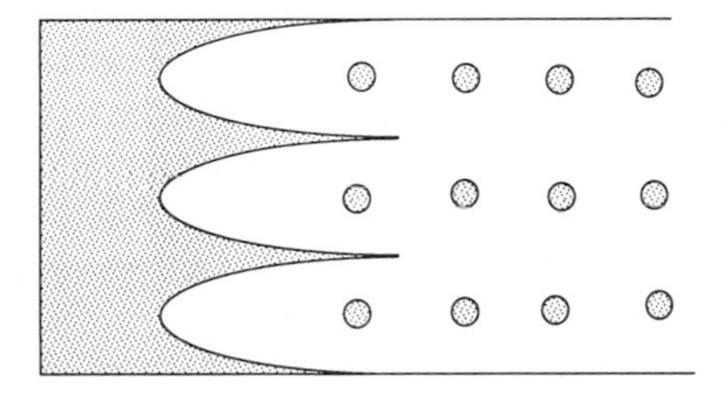

Fig. 1.11 Successive steps showing how the meniscus instability leads to the void–fibril structure of a craze. The craze tip consists of a series of 'fingers'. Deformation of these as they advance into the undeformed polymer (shown shaded) leads to fibril formation as they move on (the fibrils are also shown shaded).

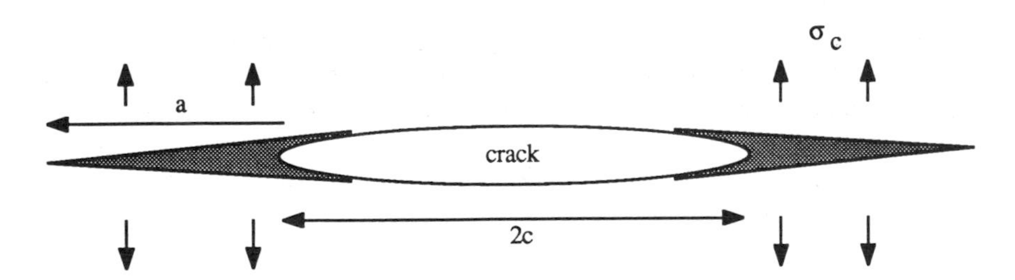

Fig. 1.12 Schematic representation of crack plus craze in the Dugdale model.

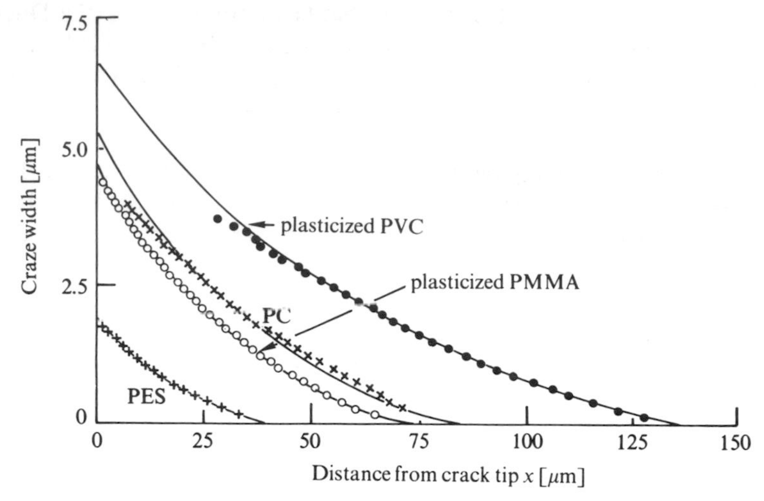

Fig. 1.13 Comparison of experimental data for various polymers with the predictions of the Dugdale model (——) (PVC, polyvinylchloride). (Redrawn from W. Döll, *Advances in Polymer Science*, 1983.)

and the displacement profile $w(x)$ over the craze length $c < x < a$ can also be evaluated in terms of these same parameters [32]. This theoretical profile can then be compared with the experimentally measured one. Figure 1.13 shows a comparison of such experimentally determined profiles with theoretical predictions for a range of polymers; it can be seen that the agreement is in general very good. In addition, the Dugdale model allows the craze width at the crack tip w_c to be related to the stress intensity factor K_I, or alternatively the strain energy release rate G (these standard terms in fracture mechanics are defined in elementary fracture mechanics texts such as Ref. 33) by the relation

$$2w_c = K_I^2/\sigma_c E^*$$

where $E^* = E$ in plane stress and $E^* = E/(1 - \nu)$ in plane strain.

One disadvantage of this optical interferometry technique is its comparatively poor spatial resolution. Techniques involving TEM have therefore been developed to overcome this problem [34–36]. Microdensitometry of the electron image plate allows the volume fraction of material in the craze v_f to be evaluated from a comparison of the mass thickness contrast of the crazed material, undeformed polymer and a hole in the film. v_f is the inverse of the extension ratio λ. Measurement of the craze width $T(x)$ from the micrographs then can yield the displacement profile $w(x)$ via the relation

$$w(x) = \tfrac{1}{2}T(x)[1 - v_f(x)].$$

This quantity $w(x)$ also then permits the stress profile along the craze to be computed following a Fourier transform method due to Sneddon [34, 37].

Thus not only can the displacement profile be compared with the Dugdale model, but the assumption of a constant stress along the craze can be directly tested. The stress profile for a PS sample is shown in Fig. 1.14(a), from which it can be seen that the stress is indeed constant everywhere except in the vicinity of the craze tip itself.

Measurement of v_f, or equivalently λ, is itself extremely illuminating. Figure 1.14(b) shows a typical craze extension ratio profile, from which it can be seen that λ, like the surface stress, is everywhere constant except at the craze tip. Two conclusions can be drawn from this: firstly, that the craze must widen by drawing in fresh material from the outside rather than by further extension by creep of existing fibrils; secondly, that the high extension ratio at the craze tip correlates with the high stress there. The remnant of this high extension ratio

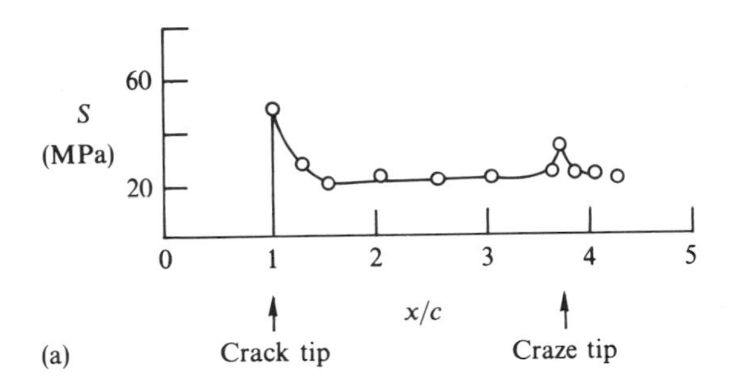

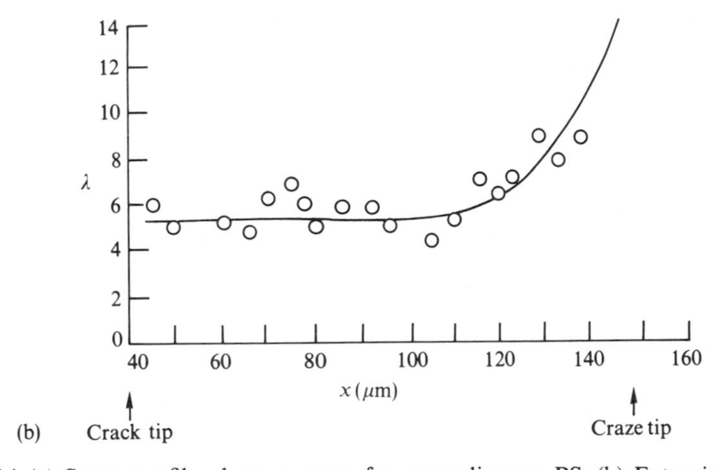

Fig. 1.14 (a) Stress profile along a craze for monodisperse PS. (b) Extension ratio profile along a craze. (Reproduced from A. M. Donald, E. J. Kramer and R. A. Bubeck, *Journal of Polymer Science, Polymer Physics Edition*, 1982.)

can be seen in Fig. 1.10 as the highly extended region running down the centre of the craze, the so-called midrib.

1.5 INTERACTIONS OF CRAZES AND SHEAR BANDS

Before turning to molecular aspects of crazing and shear deformation, one last point needs to be considered, which is of particular importance in toughened systems where there may be extensive deformation prior to failure, and that is the question of interactions between crazes and shear bands. Chau and Li have considered interactions between two shear bands in some detail [38], exploring the possibility that shear band intersections may lead to crack initiation, an idea various workers had previously considered [39–41]. They were able to observe strands of fibres pulled apart when the specimen was subjected to tensile loads. However, in other cases the microvoiding that occurs at the intersection of shear bands can actually lead to craze formation [42, 43].

Intersections involving crazes have also been considered. As might be expected, when two crazes intersect, crack initiation is also likely to occur. The nature of the intersection region between two crazes preceding crack nucleation has been followed by TEM in which it was shown that the volume fraction v_f in the intersection region is much larger than in the parent crazes, and approximately given by the product of v_f of the two contributing crazes [44]. Intersections between crazes and shear bands will not necessarily lead to fracture. When crazes meet pre-existing shear bands, they may be arrested or their path may be diverted and, in particular, the direction of propagation may rotate away from lying perpendicular to the applied stress owing to the molecular orientation in the shear band. In addition, the changed stress conditions within a shear band may lead to craze initiation within the band [45]. Likewise, a craze may be stopped when it meets a pre-existing craze.

Finally, crazes may terminate in shear bands, a process sometimes referred to as shear blunting. This appears to be a rather frequent occurrence in toughened polymers [11]. The competition between shear deformation and crazing can be interpreted in terms of the entanglement network of the glassy polymer, to be discussed in the next section, and the occurrence of shear blunting can be seen to be a common feature reflecting the changing conditions of stress and strain rate as deformation proceeds.

1.6 MOLECULAR MECHANISMS INVOLVED IN DEFORMATION

It has been mentioned in section 1.3 that shear deformation appears only to involve local segmental motion, and hence that molecular weight is not a key parameter in controlling shear response: on the other hand, the yield stress is sensitive to other factors such as aging history and strain rate, as well as temperature. In contrast to this local response, crazing must involve long-range motions, permitting the chains to be oriented and stretched out in the fibrils. The extent of the chain stretching involved is shown by the values of λ measured by TEM – for PS the value is ~ 4 [34]. In addition, for crystallizable polymers such as isotactic PS, crazing can actually lead to crystallization within the fibrils [46]. Molecular weight is known to be important for crazing, since below a critical molecular weight stable crazes do not form at all [47], although a very small unstable craze is thought to precede the crack tip [48]. In addition, it appears that there is a critical crack tip opening which is molecular weight dependent, increasing with the chain length [30]. However, early work suggested that the crazing stress was independent of molecular weight once the critical molecular weight was comfortably passed [49]. The explanation of why there is a critical molecular weight for crazing, the approximate numerical values of λ observed in a wide range of different polymers, the competition between crazing and shear and the effects of strain rate and temperature can all be rationalized within a framework based on the idea of the entanglement network.

Starting off with the simplest possible picture of entanglements being localized point constraints, one can imagine what happens when a stress is applied, causing the entanglement points to move apart. The schematic response is shown in Fig. 1.15 [50]. It can be seen that there is a critical extension ratio given by

$$\lambda = l_e/d \tag{1.1}$$

beyond which one would expect significant strain hardening to set in. The distance l_e, the chain contour length between entanglements, can be determined from knowledge of the entanglement molecular weight M_e and the average projected length of a unit along the chain l_0 via the relation

$$l_e = l_0 M_e / M_0$$

where M_0 is the molecular weight of the unit. d is the r.m.s. distance between

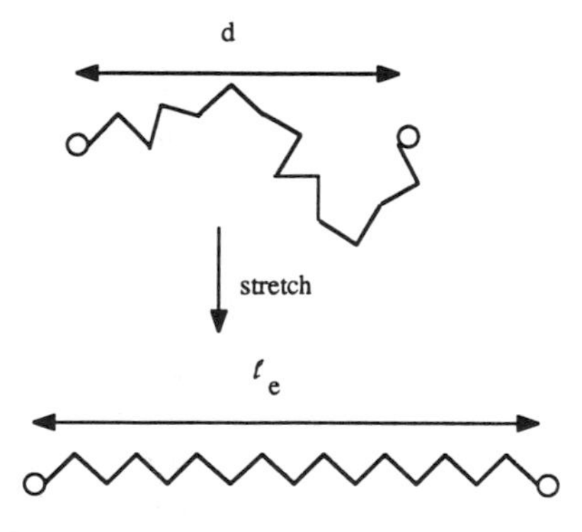

Fig. 1.15 The effect of stretching a chain between entanglement points.

entanglement points – the entanglement mesh size – and is given by

$$d = k(M_e)^{1/2}$$

where k relates the chain dimension to the chain length and can be obtained from neutron scattering [4]. Evaluation of equation (1.1) for a wide range of glassy polymers gave good agreement between the predicted values of λ and those measured via TEM [50].

However, since crazing involves voiding it is clear that this simple-minded picture must be refined, and indeed that the picture might better describe the shear response of polymers. Measurement of the extension ratio within deformation zones shows that this correlation does indeed hold, and that whereas $\lambda_{craze} \approx 0.6\lambda$, $\lambda_{DZ} \approx 0.8\lambda$ (Fig. 1.16). For the craze to form there must have been some loss of entanglements to permit the void–fibril structure to form. This can be seen by considering the schematic representation of the region at the fibril base in Fig. 1.17. The entanglement point at C must be broken for the two chains considered to be able to move freely into one or other of the fibrils. This loss of entanglement can occur either by chain **scission** – the chain literally being broken – or by **disentanglement**, which will require the relative motion of the two chains. Recent work has been able to indicate the different conditions under which the two mechanisms will be favoured.

Considering first PS, the evidence that as long as the molecular weight is sufficiently high there is no dependence of the crazing stress on M in room

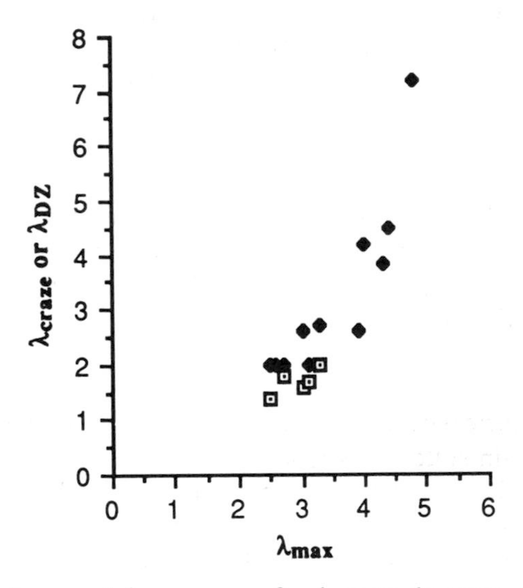

Fig. 1.16 λ_{DZ} and λ_{craze} vs λ for a range of polymers; ◆, craze; ⊡, DZ. (Compiled from A. M. Donald and E. J. Kramer, *Journal of Polymer Science, Polymer Physics Edition*, 1982, and A. M. Donald and E. J. Kramer, *Polymer*, 1982.)

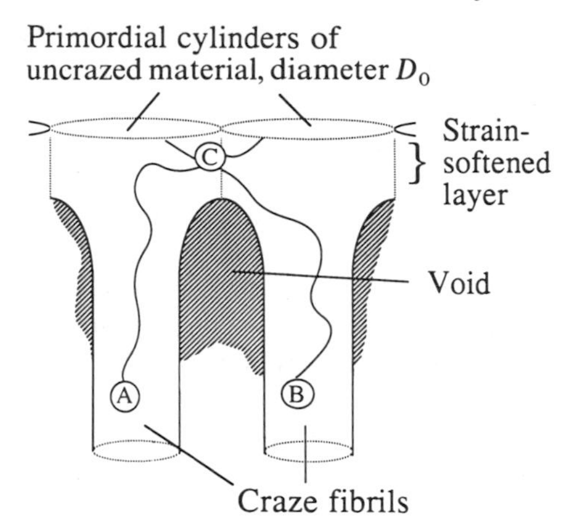

Fig. 1.17 A schematic representation of the surface drawing mechanism during craze growth. The chains marked A and B are entangled at C. This entanglement point must be broken for the chains to continue to move into the fibrils. (Reproduced from C. J. G. Plummer and A. M. Donald, *Macromolecules*, 1990.)

temperature crazing [49] suggests that scission must dominate. Assuming this to be the case, then it is possible to work out an expression for the surface energy of the voids created. There will be two contributions, the van der Waals energy term γ, and a term arising from the energy cost of breaking entanglements. A geometrical argument allows enumeration of the number of entanglements lost when creating fibrils: the number is $\frac{1}{4}dv_e$, where v_e is the entanglement density [24]. This term then leads to a contribution to the surface energy of $\frac{1}{4}dv_eU$, where U is the cost to break a bond. Hence the total energy can be written

$$\Gamma = \gamma + \tfrac{1}{4}dv_eU. \tag{1.2}$$

It can be shown [53] that the stress for craze propagation S_c by the meniscus instability (this mechanism being thought to be the operative one for craze widening as well as for tip advance [54]) is dependent on this energy term via an expression of the form

$$S_c \propto [\Gamma\sigma_y(T)]^{1/2}(\dot{\varepsilon})^{1/2n} \tag{1.3}$$

where n is the empirical constant in a simple power law constitutive equation for the flow of strain-softened polymer.

This expression immediately shows that whereas for PS, which has a rather low value of v_e, the value for S_c may be reasonable, for polymers with substantially higher values of the entanglement density v_e the surface energy Γ, and hence S_c, may become prohibitively large. These high v_e polymers are those that have low values of λ, and examination of Fig. 1.16 shows that these are the polymers, such as PC, which are conventionally regarded as tough rather than brittle. In other words, those polymers that within this framework would be expected to have high values of S_c are indeed those that appear not to deform easily by crazing. This shift from crazing to a more ductile shear response is particularly easily followed by changing the value of ν_e in a systematic way by adding the high ν_e polymer polyphenyleneoxide to PS; the values of λ for crazes change in line with expectation, and crazing is suppressed as the value of ν_e rises [55, 56].

However, scission is not the only mechanism whereby entanglements may be lost. What is the effect of disentanglement, and how does this compete with scission? For PS, the evidence is that scission is the operative mechanism at ambient temperatures, as evidenced by the lack of molecular weight dependence of the crazing stress. However, as the temperature is raised towards T_g, a dependence on M becomes apparent. This was first observed as a change in the nature of deformation with molecular weight [57]: low molecular weight samples continue to craze at all temperatures whereas long chains show a transition to a more ductile response at temperatures close to T_g. This can be understood in terms of disentanglement.

Recalling first of all the process of reptation discussed in section 1.1, it will be remembered that the entangled nature of the chains in a melt causes constraints on their motion, restricting motion to a tube. In the melt, Brownian motion allows the chain ends to move out of the tube randomly, and so eventually the whole chain can diffuse, losing all memory of its original conformation. At first sight it is perhaps surprising that large-scale motion of the chain can still occur below T_g, when such long-range motions are usually regarded as frozen out. In the case of craze growth it is not Brownian motion that permits chain motion, but the presence of the applied stress. Motion of the chains under a stress has been termed **forced reptation**, which differs from the original concept of reptation in that motion is not random but rather biased to occur in the direction of the applied stress [54, 58]. This model predicts a molecular weight dependence for the motion, albeit with a somewhat different dependence on molecular weight from that of conventional reptation. The detailed model for forced reptation leads to the conclusion that disentanglement will be easiest for short chains, at low strain rates and at high temperatures, all of which are dependences that are intuitively obvious. The increase in importance of disentanglement for short chains is one reason why the presence of a low molecular weight tail in the molecular weight distribution can significantly affect the deformation behaviour of a polymer.

This picture of forced reptation can now be used to explain the fact that low molecular weight PS crazes at all temperatures whereas high molecular weight PS shows a transition to shear. If disentanglement processes come into play then the second term in equation (1.2) no longer contributes, and correspondingly S_c will drop. This means that S_c for short chains falls below that for long. As the temperature is raised, the shear yield stress will also be dropping (section 1.3), and the possibility arises that shear will occur at lower stresses than crazing (it can be shown [53] that the stress for shear deformation zone formation has a stronger temperature dependence than S_c). This crossover between crazing and yield will occur at a lower temperature for long chains than for short, because the latter have a higher value of S_c once disentanglement becomes important. This is shown schematically in Fig. 1.18.

PS behaves as one might expect intuitively: in general it becomes less brittle as the temperature is raised. However, not all polymers behave like this. The high v_e polymers are likely to deform by shear at room temperature but become increasingly brittle as the temperature is raised. Can this also be understood within this framework? The answer is yes. The first point to note is that these are the polymers for which the high value of Γ in equation (1.2) leads to high values of S_c in equation (1.3). Thus these polymers do not deform by scission crazing at room temperature, since yielding occurs at a stress below S_c.

On the other hand, as the temperature is raised the chains become mobile and the possibility of disentanglement arises. This occurs particularly easily because the high v_e polymers are also those which are commercially available with low molecular weights, because even comparatively short chains contain a sufficient number of entanglements to give useful mechanical properties. (PC and PES typically have molecular weights of a few tens of thousands whereas commercially useful PS will probably have a molecular weight a factor of 10 higher.) As disentanglement becomes a possibility S_c will drop, as discussed above in the context of PS, and hence there can be a transition from crazing to

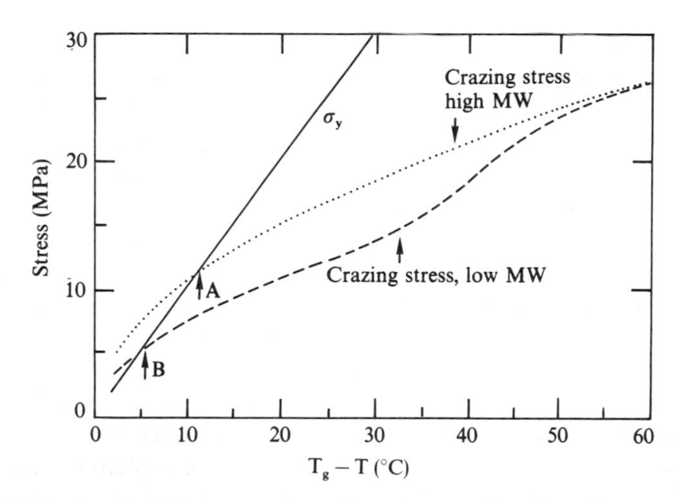

Fig. 1.18 Variation of yield stress and crazing stress as a function of temperature. For high molecular weight PS, which is unable to disentangle at any temperature up to T_g, the transition to shear (indicated by arrow A) occurs at a lower temperature than for low molecular weight PS which can disentangle. The transition temperature of the latter is indicated by arrow B. (Reproduced from E. J. Kramer and L. L. Berger, *Advances in Polymer Science*, 1990.)

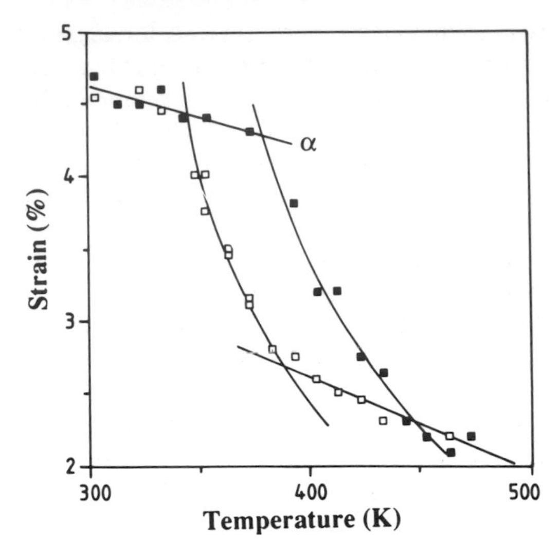

Fig. 1.19 The strain for deformation onset as a function of temperature for two different molecular weights of PES aged for 70 h at 200 °C. The curve marked α corresponds to shear deformation, and the remaining points correspond to crazing. $M_w \approx 47\,000$ and $M_w \approx 69\,000$. (Reproduced from C. J. G. Plummer and A. M. Donald, *Journal of Polymer Science, Polymer Physics Edition*, 1989.)

shear as the temperature is raised. The temperature at which this transition occurs will depend on chain length, since a higher temperature will be required for sufficient mobility to occur for long chains than for short. An example of this is shown in Fig. 1.19 for PES [59]. Other factors that will affect the transition temperature will be aging history (since aging pushes up the yield stress) and strain rate (since a low strain rate will encourage disentanglement). The net effect is that the normally ductile and tough polymers such as PC and PES will be liable to show brittle failure at temperatures approaching T_g, and this is likely to be particularly apparent under low strain rate–creep conditions [60].

Finally, the ideas of a competition between shear and disentanglement or scission driven crazing enable the different morphologies of deformation when mixed modes are present to be understood. The common case of crazes blunted by shear referred to in section 1.5 corresponds to scission crazes growing under high strain rates (which favour chain breakages) giving way to shear as the local strain rate drops. Other mixed modes may correspond to initial shear changing to disentanglement crazes at long times, with the crazes growing essentially under creep conditions [56].

1.7 CONCLUSIONS

Of the two usual modes of deformation preceding cracking in a homogeneous glassy polymer, crazing and shear, the former is the more likely to be associated with a brittle response and the latter with tougher and more ductile behaviour. If a toughened polymer contains a matrix liable to deform by crazing then one aim must be to ensure that a large number of crazes form before any single one breaks down to a crack. The extension ratio of the craze is an important parameter in this respect, since if the external stress is σ, then the true stress in the fibrils will be $\lambda\sigma$; in other words, it is enhanced by the factor λ.

It is possible to rationalize which polymers craze and which preferentially show shear deformation using the idea of the entanglement network borrowed from melt rheology. Conventionally tough polymers will be those which contain a high density of entanglements, whereas brittle polymers will have a comparatively sparse population. However, there is a subtle interplay between the two modes of deformation and there will be transitions from one to the other depending on temperature, aging history, strain rate and molecular weight.

REFERENCES

1. Haward, R. N. (1973) *Physics of Glassy Polymers*, Applied Science, London.
2. Flory, P. J. (1953) *Principles of Polymer Chemistry*, Cornell University Press, Ithaca, NY.
3. Young, R. J. (1981) *Introduction to Polymers*, Chapman & Hall, London.
4. Wignall, G. D., Ballard, D. G. H. and Schatten, J. (1976) *J. Macromol. Sci. Phys. B*, **12**, 75.
5. Ferry, J. D. (1980) *Viscoelastic Properties of Polymers*, 3rd edn, Wiley, New York.
6. De Gennes, P. G. (1971) *J. Chem. Phys.*, **55**, 572.
7. Doi, M. and Edwards, S. F. (1978) *J. Chem. Soc. Faraday Trans.*, **74**, 1789, 1802.
8. Friedrich, K. (1983) *Adv. Polym. Sci.*, **52–53**, 225.
9. Narisawa, I. and Ishikawa, M. (1990) *Adv. Polym. Sci.*, **91–92**, 353.
10. Ward, I. M. (1983) *Mechanical Properties of Solid Polymers*, Wiley, Chichester, p. 23.
11. Bucknall, C. B. (1977) *Toughened Plastics*, Applied Science, Barking.
12. Bauwens-Crowet, C., Bauwens, J. C. and Homès, G. (1969) *J. Polym. Sci. A*, **27**, 735.
13. Kinloch, A. J. and Young, R. J. (1983) *Fracture Behaviour of Polymers*, Applied Science, London, p. 116.
14. Bowden, P. B. and Raha, S. (1970) *Philos. Mag.*, **22**, 463.
15. Plummer, C. J. G. and Donald, A. M. (1990) *J. Appl. Polym. Sci.*, **41**, 1197.
16. Donald, A. M. and Kramer, E. J. (1981) *J. Mater. Sci.*, **16**, 2967.
17. Donald, A. M. and Kramer, E. J. (1981) *J. Mater. Sci.*, **16**, 2977.
18. Adam, G. A., Cross, A. and Haward, R. N. (1975) *J. Mater. Sci.*, **10**, 1582.
19. Kambour, R. P. and Russell, R. R. (1971) *Polymer*, **12**, 237.
20. Paredes, E. and Fischer, E. W. (1979) *Makromol. Chem.*, **180**, 2707.
21. Brown, H. R. and Kramer, E. J. (1981) *J. Macromol. Sci. Phys. B*, **19**, 487.
22. Sternstein, S. S., Ongchin, L. and Silverman, A. (1968) *Appl. Polym. Symp.*, **7**, 175.
23. Bowden, P. B. and Oxborough, R. J. (1973) *Philos. Mag.*, **28**, 547.
24. Kramer, E. J. (1983) *Adv. Polym. Sci.*, **52–53**, 1.
25. Argon, A. S. and Hanoosh, J. G. (1977) *Philos. Mag.*, **36**, 1195.
26. Donald, A. M. and Kramer, E. J. (1982) *J. Appl. Polym. Sci.*, **27**, 3729.
27. Taylor, G. I. (1958) *Proc. R. Soc. London, Ser. A*, **245**, 312.
28. Argon, A. S. and Salama, M. M. (1977) *Philos. Mag.*, **35**, 1217.

29. Donald, A. M. and Kramer, E. J. (1981) *Philos. Mag. A*, **43**, 857.
30. Döll, W. (1983) *Adv. Polym. Sci.*, **52–53**, 105.
31. Dugdale, D. S. (1960) *J. Mech. Phys. Solids*, **8**, 100.
32. Goodier, J. N. and Field, F. A. (1963) In *Fracture of Solids* (eds D. C. Drucker and J. J. Gilman), Wiley, New York, p. 103.
33. Knott, J. F. (1973) *Fundamentals of Fracture Mechanics*, Butterworths, London.
34. Lauterwasser, B. D. and Kramer, E. J. (1979) *Philos. Mag. A*, **39**, 469.
35. Brown, H. R. (1979) *J. Polym. Sci., Polym. Phys. Ed.*, **17**, 1431.
36. Donald, A. M., Kramer, E. J. and Bubeck, R. A. (1982) *J. Polym. Sci., Polym. Phys. Ed.*, **20**, 1129.
37. Sneddon, I. N. (1951) *Fourier Transforms*, McGraw-Hill, New York, p. 395.
38. Chau, C. C. and Li, J. C. M. (1981) *J. Mater. Sci.*, **16**, 1858.
39. Bowden, P. B. (1970) *Philos. Mag.*, **22**, 455.
40. Argon, A. S. (1973) *Pure Appl. Chem.*, **43**, 247.
41. Friedrich, K. and Schafer, J. K. (1979) *J. Mater. Sci.*, **14**, 1480.
42. Mills, N. J. (1976) *J. Mater. Sci.*, **11**, 363.
43. Narisawa, I., Ishikawa, M. and Ogawa, H. (1980) *J. Mater. Sci.*, **15**, 2059.
44. King, P. S. and Kramer, E. J. (1981) *J. Mater. Sci.*, **16**, 1843.
45. Donald, A. M., Kramer, E. J. and Kambour, R. P. (1982) *J. Mater. Sci.*, **17**, 1739.
46. Morel, D. and Grubb, D. T. (1984) *J. Mater. Sci. Lett.*, **3**, 5.
47. Wellinghoff, S. and Baer, E. (1975) *J. Macromol. Sci. Phys. B*, **11**, 367.
48. Kramer, E. J. (1978) *J. Mater. Sci.*, **14**, 1381.
49. Fellers, J. and Kee, B. F. (1974) *J. Appl. Polym. Sci.*, **18**, 2355.
50. Donald, A. M. and Kramer, E. J. (1982) *J. Polym. Sci., Polym. Phys. Ed.*, **20**, 899.
51. Plummer, C. J. G. and Donald, A. M. (1990) *Macromolecules*, **23**, 3929.
52. Donald, A. M. and Kramer, E. J. (1982) *Polymer*, **23**, 1183.
53. Berger, L. L. and Kramer, E. J. (1987) *Macromolecules*, **20**, 1980.
54. Kramer, E. J. and Berger, L. L. (1990) *Adv. Polym. Sci.*, **91–92**, 1.
55. Donald, A. M. and Kramer, E. J. (1982) *Polymer*, **23**, 461.
56. Donald, A. M. and Kramer, E. J. (1982) *J. Mater. Sci.*, **17**, 1871.
57. Donald, A. M. (1985) *J. Mater. Sci.*, **20**, 2630.
58. McLeish, T. C. B., Plummer, C. G. J. and Donald, A. M. (1989) *Polymer*, **30**, 1651.
59. Plummer, C. J. G and Donald, A. M. (1989) *J. Polym. Sci., Polym. Phys. Ed.*, **27**, 325.
60. Davies, M. and Moore, R. (1987) ICI plc Internal Report.

2

Rubber toughening mechanisms in polymeric materials

I. Walker and A. A. Collyer

2.1 INTRODUCTION

The manner in which toughened polymer matrices fail under impact depends on extrinsic factors such as rate, temperature, notch, loading type and specimen geometry, and the manner of failure (crazing, yielding or a combination), and on intrinsic variables such as the microstructure of the blend and the chain structure of the matrix [1]. In this chapter, the intrinsic variables are largely examined, although temperature effects will be mentioned.

Amorphous thermoplastic materials are used in service below their glass transition temperatures, T_g; in general, they are brittle and notch sensitive at these temperatures, but creep is kept to a minimum. Under these conditions the molecules are in a frozen state and unable to respond without rupture to rapidly applied stresses or impacts.

Partly crystalline thermoplastics may be used in service at temperatures between T_g and T_m, the melting temperature. Above T_g, these materials are generally tough, but give rise to creep under load. This is due to the mobility of the molecules in the amorphous regions surrounding the rigid crystallites, which provide the strength and rigidity.

At room temperature partly crystalline engineering thermoplastics are brittle because of their high T_g values, whereas polyethylene is tough because its T_g value is below this temperature. However, if a polymer possesses secondary transitions below T_g, it may behave in a ductile manner below T_g, because some part of the molecule may have sufficient mobility to accommodate impacts without rupture. Both polyvinylchloride and polyamide 66 are examples of this. In the latter, a secondary transition occurs at $-50\,°C$, whereas its T_g value is $60\,°C$.

The manner in which rubber toughening works depends on the way in which the matrix polymer usually fails under impact. When the craze initiation stress of the matrix is lower than the yield stress, the failure mechanism is by crazing, and rubber toughening is mainly achieved by the dispersed rubber particles acting as craze initiators. Conversely, if the craze initiation stress is higher than the yield stress, the matrix will fail by shear yielding, and toughening is usually achieved by the dispersed rubber particles acting as initiators of shear bands. Mixed crazing and yielding occurs when the craze initiation stress and the yield stress are of comparable size, or when certain interactions occur between shear bands and crazes [2–6].

Wellinghoff and Baer [7] noted that vinyl polymers failed by crazing and named them type I materials. Here chain scission is more likely than in type II materials, which often consist of main chain aromatic polymers, which tend to fail by shear yielding.

Type I materials are brittle at temperatures 10–20 °C below T_g [1, 7–9]. They have low crack initiation and low crack dissipation energies during impacts. They possess low unnotched and notched impact strengths. When rubber toughened they still tend to fail by crazing; examples of these include polystyrene (PS) and styrene–acrylonitrile (SAN).

In these materials the impact energy is most effectively dissipated by the formation of large craze envelopes at the crack tip. The dispersed impact modifier must arrest these crazes and must be sufficiently large so as not to be engulfed by the approaching craze. The rubber particle size must exceed the craze thickness and the interfacial adhesion must be sufficient to permit the effective transfer of stress to the rubber inclusion to blunt the craze. The size required is of the order of microns. The interparticle distance must be sufficiently small to prevent the formation of a catastrophic crack. Craze formation is not increased by the interaction of neighbouring particles until the interparticle distance between centres is less than $0.9R$, where R is the particle radius [10].

Type II, the pseudo-ductile polymer matrices, so called because they are brittle under certain test conditions, fail by shear yielding. They have a high crack initiation energy but a low crack propagation energy. This gives them a high unnotched impact strength but a low notched impact strength, and they manifest a brittle-to-tough transition temperature, T_{BT}. Examples include polycarbonate, polyamides and polyethyleneterephthalate.

In these materials craze failure is completely suppressed in thin, notched samples (3 mm thick). In sharply notched samples or samples of large thickness, this is no longer true as plane strain crazing takes place before sufficient plastic flow can occur to relieve the dilatational stresses in the interior of the sample. This notch sensitivity is reduced by the dispersed elastomeric phase, which acts as a stress concentrator around each particle, which restores plastic flow in the absence of the large craze field. The minimum effective particle size is reduced to a few tenths of a micron or less, since the rubber domains promote shear banding rather than crack stopping. Stress overlap between particles is not important in these matrices because of the shear band initiation by the stress concentrations around the rubber particles [10].

It is possible to identify a class of materials with properties intermediate between type I and type II [1]. This class includes less brittle materials from the former, such as polymethylmethacrylate (PMMA), and less ductile materials from the latter, such as polyacetal (polyoxymethylene (POM)) and polyvinylchloride (PVC).

Such materials have comparable craze and yield stresses and a bimodal size distribution of rubber particles is the most efficient way of providing crack blunting and shear band initiation [11–15].

In continuation from work already reported in Chapter 1, it has been postulated by Wu [1] that the spectrum of plastic matrix behaviour between brittle and pseudo-ductile may depend on two molecular parameters characteristic of the polymer chains in the matrix. These are the entanglement density, ν_e, already defined in Chapter 1, and the characteristic ratio of the chain, C_∞, where

$$C_\infty = \lim_{n \to \infty} (R_0^2/nl^2) \tag{2.1}$$

in which R_0^2 is the mean square end-to-end distance of an unperturbed chain, n is the number of statistical skeletal units, and l^2 is the mean square end-to-end distance of a random flight chain. Thus C_∞ is a measure of the flexibility, rigidity and tortuosity of an unperturbed real chain. From Ref. 1 ν_e and C_∞ are related by

$$\nu_e = \frac{\rho_a}{3M_v C_\infty^2} \tag{2.2}$$

where ρ_a is the amorphous mass density and M_v is the average molecular weight per statistical unit.

As crazing is initiated by chain scission, the probability of crazing will be related to the entanglement density. The crazing stress σ_z is related to the entanglement density ν_e by [1]

$$\sigma_z \propto \nu_e^{1/2}. \tag{2.3}$$

A graph of σ_z against ν_e is shown in Fig. 2.1 where the solid line is the best-fit line with a slope of 0.5 [1]. The equation of this line is given by

$$\log \sigma_z = (1.83 \pm 0.03) + 0.5 \log \nu_e \tag{2.4}$$

where σ_z is in MN m^{-2} and ν_e is in mmol cm^{-3}. In the case of the yield stress σ_y, it has been shown that it is proportional to $\Delta T = T_g - T$ and δ^2, where T is the test temperature and δ^2 is the cohesive energy density [16]. A normalized yield stress $\bar{\sigma}_y$ is defined as

$$\bar{\sigma}_y = \frac{\sigma_y}{\delta^2 (T_g - T)}. \tag{2.5}$$

The denominator allows for the effect of interchain actions on the yield stress.

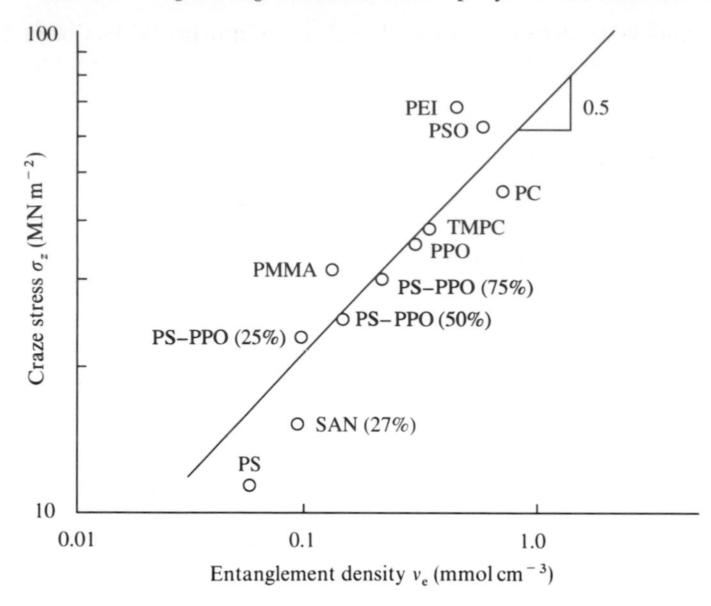

Fig. 2.1 Double-logarithmic plot of craze stress vs entanglement density [1].

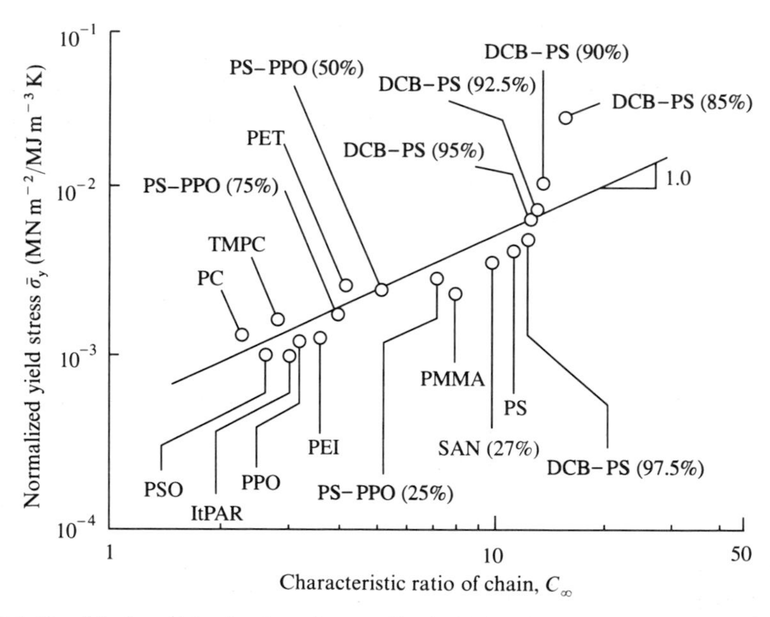

Fig. 2.2 Double-logarithmic plot of normalized yield stress vs characteristic ratio of a chain [1].

The normalized yield stress $\bar{\sigma}_y$ should be a function of C_∞, as shown in Fig. 2.2 and in

$$\log \bar{\sigma}_y = (-3.36 \pm 0.04) + \log C_\infty \tag{2.6}$$

where $\bar{\sigma}_y$ is in $\mathrm{MN\,m^{-2}(MJ\,m^3)^{-1}\,K^{-1}}$. Thus C_∞ controls the yield behaviour [1].

The competition between crazing and yielding can be expressed by the ratio $\sigma_z/\bar{\sigma}_y$. This gives the criterion for crazing–yielding:

$$\frac{\sigma_z}{\bar{\sigma}_y} \propto \left(\frac{\rho_a}{3M_v}\right)^{1/2} C_\infty^{-2}. \tag{2.7}$$

The factor $(\rho_a/3M_v)^{1/2}$ is fairly constant in many polymers except for those where ρ_a and/or M_v is large.

For low v_e or large C_∞, such that $\sigma_z/\bar{\sigma}_y$ is small, the matrix fractures by crazing (i.e. $v_e \leqslant 0.15\,\mathrm{mmol\,cm^{-3}}$ and $C_\infty \geqslant 7.5$); for all matrices in which v_e is high or C_∞ is low ($v_e \geqslant 0.15\,\mathrm{mmol\,cm^{-3}}$ and $C_\infty \leqslant 7.5$) the matrices yield.

This theory also applies to crosslinked PS, when the combined entanglement and crosslink density is used as the criterion [17]. Fuller details are given by Wu [1]. The work developed by Wu and coworkers provides a simple basis for judging the toughness of matrices, but the work has been tested only on polyamides. It will be interesting to see how the theory stands up to similar deep investigations in other engineering polymers.

So far the terms brittle and pseudo-ductile have been used without assigning a measure to them or indeed a type of measurement. In fracture mechanics G_c, the fracture toughness in brittle fracture, is the parameter measured. This involves the use of sharp notches. G_c is defined as the critical strain energy release rate per unit area of crack advance required to propagate a pre-existing sharp crack. Usually G_c is taken as the value of G, the strain energy release rate, at the onset of unstable rapid crack propagation. Cracks will, however, propagate slowly at values below G_c.

G is related to the stress intensity factor, K_I, for the crack from linear elastic fracture mechanics by

$$G = \frac{K_I^2}{E^*} \tag{2.8}$$

where $E^* = E$ for plane stress, and $E^* = E/(1 - v^2)$ for plane strain, in which E is Young's modulus and v is Poisson's ratio.

For a tensile stress, σ, normal to and far from the crack plane

$$K_I = Y\sigma a_0^{1/2} \tag{2.9}$$

where a_0 is the crack length and Y is a factor depending on the crack and specimen geometry. Experimental values of G_c are generally greater than $1000\,\mathrm{J\,m^{-2}}$, whereas the theoretical value for PS is $0.08\,\mathrm{J\,m^{-2}}$, which is twice the surface energy, allowing for the energy of two faces of the crack. The reason

Table 2.1 Standards governing notched Izod and Charpy tests

	BS	*ISO*	*ASTM*	*DIN*	*AFNOR*
Charpy	2782	R-179	D-256B	53453	NFT51-035
Izod	2782	R-180	D-256A	–	–

for this discrepancy is that the calculation is based on a perfectly brittle material, which does not take into account the energy of plastic deformation at the crack tip. This energy must be added to twice the surface energy and in practice greatly exceeds the surface energy.

In ductile polymers, the plastic zones around the crack tip are not small compared with the dimensions of the crack tip and those of the sample. Therefore, the crack tip stress field is no longer characterized by the stress intensity factor, K_I.

A much more widely used measure of impact strength by manufacturers and suppliers is the notched impact strength, either Izod or Charpy, depending on the notch configuration. The sizes of notch and the test configurations and methods must conform to specified standards such as those shown in Table 2.1. The notches used are much more blunt than those for fracture mechanics measurements, and the impact strength is defined as the energy required to initiate and propagate a crack from a blunt notch in a bar of square cross-section under impact conditions. As a result of the different notch configurations, there can be no direct comparison between Izod and Charpy values.

In order to give the reader a feeling for toughness, all that will be added here is that notched Izod impact strength values greater than $5\,\mathrm{J\,m^{-1}}$ and $530\,\mathrm{J\,m^{-1}}$ are designated tough and super tough respectively.

2.2 MISCIBILITY AND DISPERSION OF THE RUBBER PHASE

The rubber phase intended for rubber toughening must be dispersed as small particles in the plastic matrix. The particle size and size distribution of the dispersed particles will depend on the miscibility of the two phases and on the way in which they are mixed. If the miscibility is good the particles of the rubber will be too small to promote toughening and may even be distributed on a molecular scale. If the two phases are immiscible, which is more likely, the rubber may be dispersed as macroscopic particles too large to give toughening.

Miscibility between two species A and B may be examined using the Hildebrand solubility parameter, δ. There are several drawbacks to this approach, but nevertheless it can be used as a guide.

When molecules of species A are mixed with those of species B, a homogeneous mixture will form if the force of attraction between a molecule of A and a molecule of B, F_{AB}, is greater than or equal to F_{AA} or F_{BB}, the forces of attraction between like molecules. If F_{AB} is less than F_{AA} or F_{BB} a two-phase blend will occur, and the two materials will be immiscible.

The forces of attraction between the molecules are an important parameter, and one way of estimating these forces is to measure the energy required to separate the molecules by thermal means. The energy required to separate like molecules is obtained from the latent heat of vaporization, L. This overestimates the value required because it includes the energy expended in evaporation RT, where R is the universal gas constant and T is the temperature in kelvins. The required energy of vaporization is given by $L - RT$. If this is related to the molar volume, the cohesive energy density, E_c, is related to the latent heat of vaporization by

$$E_c(\mathrm{J\,m^{-3}}) = \frac{L - RT}{M/\rho} \tag{2.10}$$

where M is the molar mass and ρ is the density.

The Hildebrand solubility parameter, δ, is related to E_c by

$$\delta\,((\mathrm{J\,m^{-3}})^{1/2}) = E_c^{1/2} = \left(\frac{L - RT}{M/\rho}\right)^{1/2}. \tag{2.11}$$

Values of δ for various polymers, solvents and plasticizers are given in Refs. 18 and 19. Some typical values are given in Table 2.2. The SI units $(\mathrm{MJ\,m^{-3}})^{1/2}$ are used in the newer texts, but CGS units are still quoted $((\mathrm{cal\,cm^{-3}})^{1/2})$. To obtain the SI unit multiply the CGS unit by 2.04. The values quoted for δ are most accurate for values less than $19.4\,(\mathrm{MJ\,m^{-3}})^{1/2}$, when the polymers are amorphous and non-polar.

Table 2.2 Values of Hildebrand solubility parameter, δ, for several polymers

Polymer	$\delta\,((MJ\,m^{-3})^{1/2})$	*Polymer*	$\delta\,((MJ\,m^{-3})^{1/2})$
Polydimethylsiloxane	14.9	Polyvinylchloride	19.4
Ethylene–propylene rubber	16.1	Polycarbonate	19.4
Polyethylene	16.3	Polybutylene-terephthalate	21.5
Polypropylene	16.3	Polysulphone	21.6
Polyisoprene	16.5	Polyethylene-terephthalate	21.8
Polybutadiene	17.1	Polyetheretherketone	22.5
Styrene–butadiene	17.1	Polyethersulphone	25.1
Polymethylphenyl-siloxane	18.3	Polyamide 66	27.8
Polystyrene	18.7	Polyacrylonitrile	28.7

The shortcomings of using δ to establish miscibility are due to the assumption that the behaviour involved in F_{AA} and F_{BB} is repeated in F_{AB}. This may not be true and restricts blind usage to amorphous non-polar polymers. There may be complications when the following are involved: polar polymers, semicrystalline polymers, polar and semicrystalline polymers and crosslinked polymers.

To return to the simple case of the mixture of two amorphous polymers involving no specific interactions, such as hydrogen bonding, let it be assumed that F_{AB} will be the geometrical mean of F_{AA} and F_{BB}. If it is also supposed that $F_{AA} > F_{BB}$, then $F_{AA} > F_{AB} > F_{BB}$, and compatibility between the two species will occur when

$$F_{AA} \approx F_{AB} \approx F_{BB}.$$

Thus the two polymers must have similar δ values. In practice, it has been shown that if the δ values differ by less than 0.3 $(\mathrm{MJ\,m^{-3}})^{1/2}$ mixing will occur and a homogeneous blend will result. Phase separation will occur above 0.4 $(\mathrm{MJ\,m^{-3}})^{1/2}$ but there will be good interfacial adhesion for differences between δ values of less than 0.8 $(\mathrm{MJ\,m^{-3}})^{1/2}$ [20]. A glance at Table 2.2 will reveal that an elastomer such as polydimethylsiloxane (PDMS), a possible rubber toughener for high temperature engineering thermoplastics, will not mix well with much else. Methods of compatibilizing PDMS will be essential to the success of rubber toughening such matrices. This is discussed in Chapter 5.

The Hildebrand solubility parameter may be regarded as a composite of three component solubility parameters [21, 22]: δ_d due to dispersion forces, δ_p due to polar forces and δ_b due to hydrogen bonding:

$$\delta^2 = \delta_d^2 + \delta_p^2 + \delta_b^2. \tag{2.12}$$

Values of the three-dimensional solubility parameters are given for solvents only in Ref. 19 and further details of this subject are given in Refs. 13, 21, 22 and 23.

The necessity of obtaining two-phase blends with good interfacial adhesion may lead to seeking specific interactions, such as the above. A problem with this is that if the interaction between the species is too strong, a homogeneous mixture may occur or the particle size may be too small, particularly when toughening matrices that fail by crazing.

2.3 EFFECT OF THE DISPERSED RUBBER PHASE

2.3.1 Concentration effects

Increasing the concentration of the rubber phase decreases the blend modulus and tensile strength irrespective of whether the matrix is brittle or pseudo-ductile.

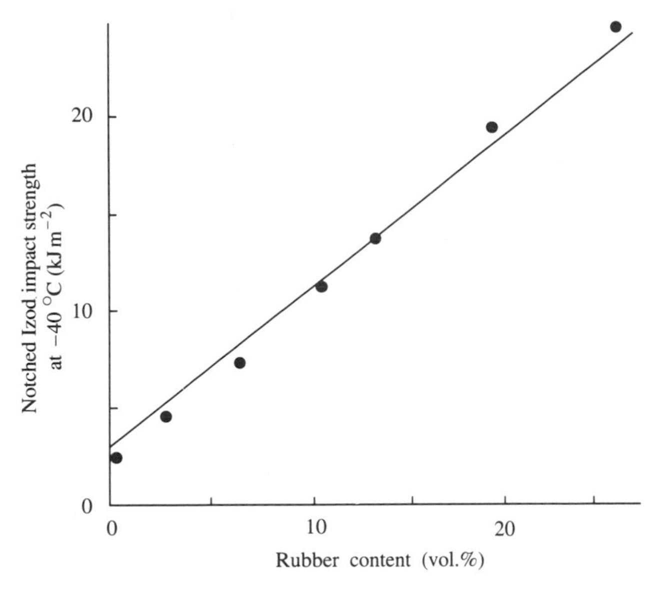

Fig. 2.3 Notched Izod impact strength at $-40\,^{\circ}\mathrm{C}$ vs volume fraction of EPDM in PA6. Particle size and interfacial adhesion are constant [24].

As most of the research has been carried out on polyamide blends, this is covered in Chapter 7. The main findings are that up to 30% rubber the T_g values of the two phases are unaffected by rubber concentration in PA6–EPDM blends, but the brittle–tough transition temperature T_{BT} is reduced as the concentration of rubber is increased [9, 24–26]. Figure 2.3 shows the variation of notched Izod impact strength with rubber content [24]. The relationship is linear up to 30% EPDM in PA6, but higher rubber loadings show a marked decrease in impact strength [26, 27]. In these results particle size and interfacial adhesion are constant.

2.3.2 Particle size and the single structural parameter

In brittle matrices, in which crazing is the predominant fracture mechanism, as v_e increases the intrinsic ductility increases, and the apparent interior structure of the craze appears to be finer [28]. Fibrillar crazes form in the very brittle PS with fibrils of about 10 nm diameter, whereas the less brittle PMMA forms homogeneous crazes devoid of internal structures coarser than a few nanometres. An intermediately brittle matrix such as SAN forms mixed homogeneous fibrillar crazes with some fibrils embedded in homogeneous crazes [28].

At a fixed rubber concentration and fixed interfacial adhesion, a graph of toughness vs rubber particle size gives a maximum toughness for an optimum particle size [16, 29], as shown in Fig. 2.4. The optimum particle sizes for

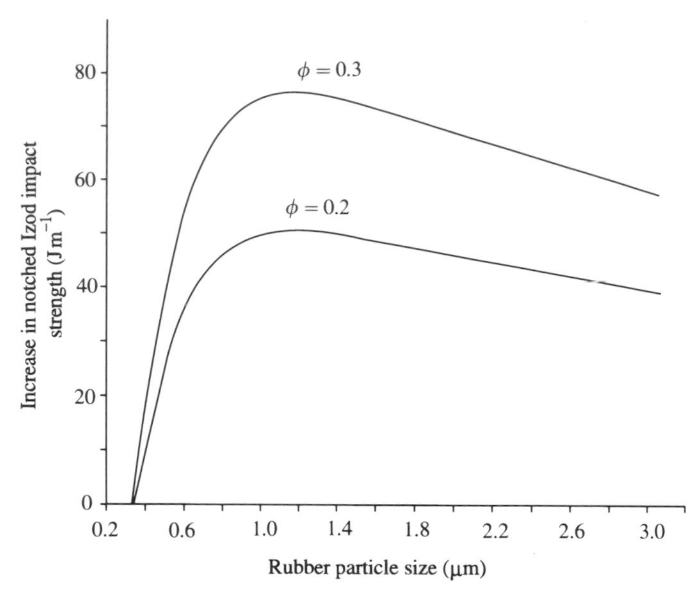

Fig. 2.4 Increased toughness vs rubber particle size [29].

Table 2.3 Brittle matrices and their rubber blends [1]

Matrix polymer	*PS*	*SAN*	*PMMA*
Neat matrix polymer			
ν_e (mmol cm^{-3})	0.056	0.093	0.127
Predominant fracture mechanism	Crazing	Crazing	Crazing
Apparent internal structure of craze	Fibrillar	Mixed homogeneous and fibrillar	Homogeneous
Typical notched Izod impact strength (J m^{-1} (ft-lb in^{-1}))	21 (0.4)	16 (0.3)	16 (0.3)
Polymer–rubber blend			
Fracture mechanism	Crazing	Crazing and yielding	Yielding and crazing
Optimum rubber diameter (μm)	2.5	0.75 ± 0.15	0.25 ± 0.05
Bimodal size synergism	Marginal	Pronounced	Marginal
Typical notched Izod impact strength (20–25 vol.% rubber) (J m^{-1} (ft-lb in^{-1}))	130 (2.5)	780 (15)	80 (1.5)

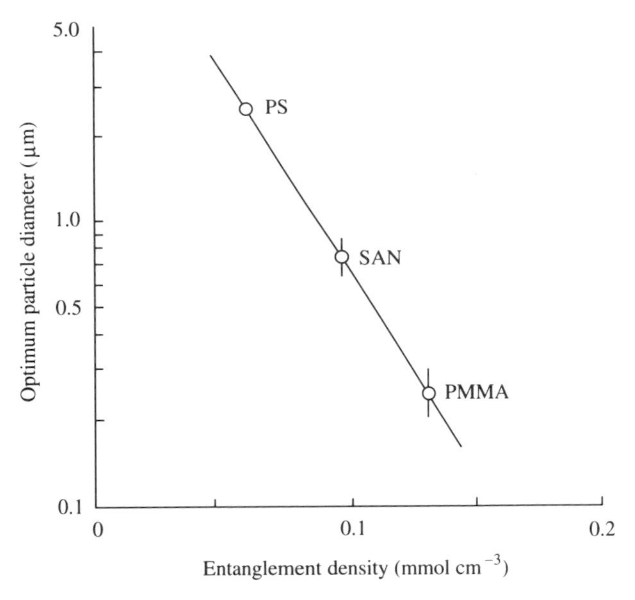

Fig. 2.5 The optimum rubber particle diameter, d, vs entanglement density, ν_e, of brittle matrices based on notched Izod impact strength [1].

toughening several brittle matrices are given in Table 2.3 [1], with the Izod impact strength of the matrix and values for the toughened blend.

The optimum rubber particle size depends on the entanglement density ν_e of the matrix, and this is shown in Fig. 2.5 [1]. The optimum particle size decreases as ν_e increases and the matrix becomes more ductile.

Bimodal-sized particles are particularly successful in the toughening of SAN [11, 12], possibly because of the balance of crazing and yielding that occurs in the fracture of this material [14, 15]. There is only a marginal improvement in toughness for PS [25] and PMMA [14, 15]. For this reason it may be surmised that matrices with $\nu_e = 0.09\,\text{mmol}\,\text{cm}^{-3}$ may be suited to toughening with a bimodal particle size distribution [1]. It may also be noted from Table 2.3 that SAN is far more conducive to toughening than PS or PMMA [1]. This suggests that the responsiveness of a polymer to toughening depends on its intrinsic brittleness, and hence ν_e, and its intrinsic ductility, and hence C_∞. Perhaps if all other factors such as particle shape, rubber concentration, T_g of the rubber and interfacial adhesion were optimized any brittle polymer matrix could be super toughened (notched Izod impact strength $> 530\,\text{J}\,\text{m}^{-1}$). At present it seems that the more brittle the matrix is, the more difficult it is to toughen.

In pseudo-ductile matrices, when the toughness is plotted against particle size for a given rubber concentration and interfacial adhesion, a sharp brittle–ductile transition occurs, as shown in Fig. 2.6 and in Chapter 7. When the

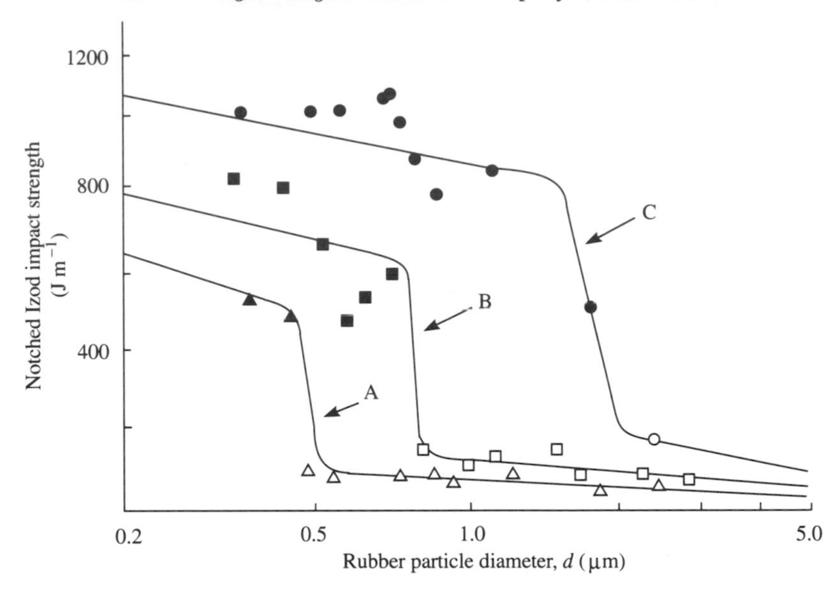

Fig. 2.6 Notched Izod impact strength vs rubber number-average particle diameter, d, at constant interfacial adhesion at rubber concentrations (by weight) of 10% (curve A), 15% (curve B) and 25% (curve C): ▲, ■, ●, tough fracture; △, □, ○, brittle fracture [9].

average particle size is smaller than the critical diameter, d_c, the blend is tough; if it is larger the blend is brittle. The critical particle size depends on the rubber concentration, being smaller for the lower concentrations, and on the type of rubber. It was hoped that the average particle size and rubber concentration could be linked in a single structural parameter on which the brittle–tough transition solely depends. Wu [9] examined three models to achieve this: (a) an interfacial area model, (b) a particle concentration model and (c) an interparticle distance or percolation model.

(a) Interfacial area model

It is assumed in this model that the brittle–tough transition occurs when the rubber–matrix interfacial area per unit volume of the blend is at a critical value. The critical particle diameter, d_c, is given by

$$d_c = \frac{6\phi_r}{A_c} \tag{2.13}$$

where ϕ_r is the rubber volume fraction and A_c the critical interfacial area per unit volume. By definition, A_c is independent of ϕ_r and d_c [9]; Fig. 2.7 shows experimental values and theoretical values from this model for d_c vs ϕ_r for $A_c = 1.508\ \mu m^{-1}$. The linear relationship predicted is not borne out by the experimental values.

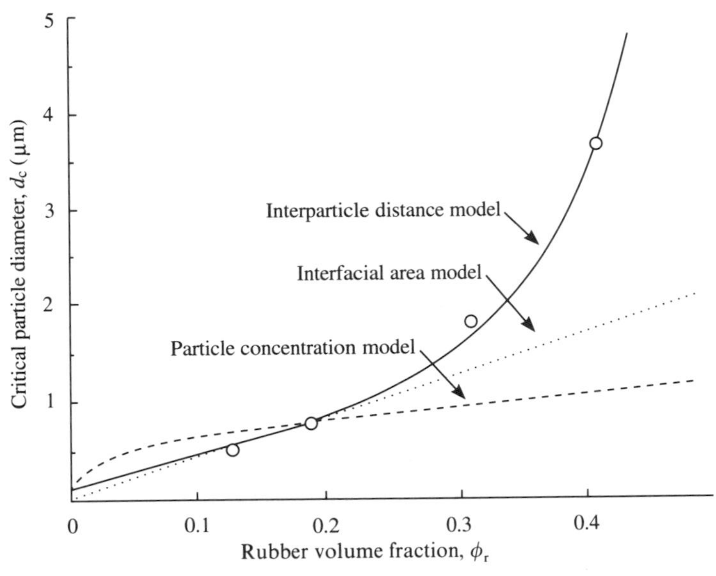

Fig. 2.7 Critical particle diameter, d_c, for toughening in notched Izod impact vs rubber volume fraction ϕ_r ($A_c = 1.508\ \mu m^{-1}$, $N_c = 0.831\ \mu m^{-3}$ and $I_D = 0.304\ \mu m$).

(b) Particle concentration model

In this model it is assumed that the brittle–ductile transition occurs when the particle concentration is at a critical value, N_c. The critical particle size is given by

$$d_c = \left(\frac{6\phi_r}{\pi N_c}\right)^{1/3}. \tag{2.14}$$

Again an examination of Fig. 2.7 shows that the experimental values do not support the theory [9].

(c) Interparticle distance or percolation model

In this case it is assumed that the brittle–tough transition occurs when the surface-to-surface interparticle distance between two nearest neighbours, I_D, is at a critical value, as shown in Fig. 2.8 [9]. d_c is given by

$$d_c = I_D\left[\left(\frac{\pi}{6\phi_r}\right)^{1/3} - 1\right]. \tag{2.15}$$

The interparticle distance, I_D, later called the ligament thickness [30–32], is independent of d_c and ϕ_r. This equation is discussed with particular reference to polyamides in Chapter 7.

Figure 2.7 shows a good agreement between theoretical and experimental

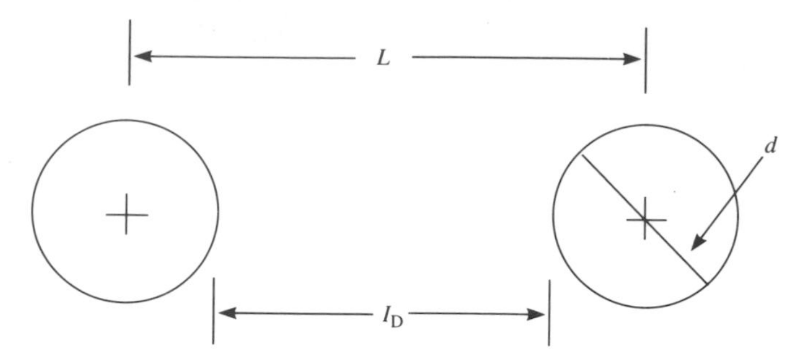

Fig. 2.8 Critical matrix ligament thickness, I_D, for particles of diameter d and distance L apart.

values for $I_D = 0.304\,\mu m$. By plotting notched Izod impact strength vs interparticle distance, a sharp tough–brittle transition occurs at $I_D = 0.304\,\mu m$, that is independent of d_c and ϕ_r. This is shown in Chapter 7 for polyamides (Fig. 7.14).

Thus a single parameter can be identified dependent on the pseudo-ductile polymer matrix. A polymer blend will be tough if the surface-to-surface interparticle distance is less than the critical value.

As the ductility of polymer matrices depends on C_∞, it is to be expected that I_D should correlate with C_∞.

The highest intrinsic ductility limit corresponds to $C_\infty = 2$ [1], which correlates with chains of freely rotating joints with tetrahedral skeletal bonds. The highly ductile PC has a value of $C_\infty = 2.4$, which accounts for its super tough behaviour without reinforcement for thicknesses of less than 3.2 mm with a notch radius larger than 0.2 mm [33].

The value of I_D depends on rate, temperature, method of loading, internal stress, etc. [9, 34]. The origin of I_D has been interpreted in terms of percolation (connectivity) of thin matrix ligaments such that I_D is greater than or equal to the interparticle distance [31, 32] or of a stress field overlap [9, 10, 34, 35]. The former interpretation assumes that at the critical ligament thickness the yielding process propagating through the ligaments changes from plane strain to plane stress. This is analysed more fully by Margolina [36].

The percolation theory predicts that uniform-sized rubber particles are more effective in toughening pseudo-ductile matrices; asymmetric particles, such as ribbons and networks, are more effective than spherical ones; flocculation to form isolated particle clusters harms toughening, while flocculation to form an interconnected rubber particle network is beneficial [31, 32, 37], which agrees with observation [37].

As a result of the differential shrinkage between the rubber and the matrix on cooling from the melt, internal stresses are generated in the matrix ligaments. Hoopwise compressive stresses and radial tensile stresses (twice the compressive stress) are induced in the matrix ligaments because the rubber shrinks more and has a lower T_g than the matrix. These internal

Table 2.4 Toughness of intermediate polymers and blends [1]

Matrix polymer	*PMMA*	*POM*	*PVC*
Neat matrix polymer			
v_e (mmol cm^{-3})	0.13	0.49	0.25
C_∞	8.2 ± 0.4	7.5	7.6 ± 1.0
Notched Izod impact strength (J m^{-1} (ft-lb in^{-1}))	16 (0.3)	110 (2)	40 (0.8)
Polymer–rubber blend			
Notched Izod impact strength with discrete rubber phase (J m^{-1} (ft-lb in^{-1}))	80 (1.5)	200 (4)	200 (4) 1300 * (24 *)
Notched Izod impact strength with intermeshed rubber phase (J m^{-1} (ft-lb in^{-1}))	–	1000 (19)	1300 (24)

*With core–shell particles of butadiene–styrene rubber core and grafted shell of polymethylmethacrylate.

stresses lower the local yield stress of the matrix and affect I_D [38–42]. Different rubbers may give rise to different amounts of internal stress, and so may influence I_D.

With matrices of intermediate ductility, such as PMMA ($v_e = 0.13$ mmol cm^{-3} and $C_\infty = 8.2 \pm 0.4$), POM ($v_e = 0.49$ mmol cm^{-3} and $C_\infty = 7.5$) and PVC ($v_e = 0.25$ mmol cm^{-3} and $C_\infty = 7.6 \pm 1.0$), certain special rubber phase morphologies are needed to obtain super tough blends, as seen in Table 2.4 [1].

POM reaches super toughness (910 J m^{-1}) when the rubber particles form an intermeshed structure, whereas discrete rubber particles give a notched Izod impact strength of 200 J m^{-1} [43, 44]. A similar situation occurs with PVC matrices [45–47]. In fact, PVC matrices have been successfully super toughened with discretely dispersed core–shell particles of butadiene–styrene rubber core with a grafted shell of PMMA [48]. In this case the PMMA shell is miscible with the PVC matrix. Therefore the two dissimilar chains can interdiffuse to form an interphase layer that promotes the energy dissipation of the matrix. The control of the chain structure, interchain interaction and craze–yield behaviour of the interphase region is important. These kinds of toughening–compatibilizing agents are discussed fully in Chapter 5.

2.3.3 Particle size and distribution of the rubber

When highly dispersed, the rubbery phase can act as an effective stress concentrator and enhances both crazing and shear yielding in the matrix. Ultimately the optimization of the size distribution of the dispersed phase depends on identifying the preferred deformation mechanism of the matrix polymer. A crazing mechanism is better suited to a higher particle size than a shear yielding mechanism.

Suitable processing conditions constitute one route to the desired size distribution. Although the mechanical effects on the morphology of the blend during processing are complex, by considering a modified Taylor's relationship one can describe the dispersion process:

$$d = (\sigma_i/\dot{\gamma}\eta_m) f(\eta_d/\eta_m) \tag{2.16}$$

where η_m and η_d are the viscosities of the matrix and of the dispersed phase respectively, σ_i is the interfacial tension, d the diameter of the dispersed phase and $\dot{\gamma}$ the shear rate. The shear rate is inversely proportional to the particle diameter. This was confirmed in blends of polybutyleneterephthalate (PBT) and ethylene–propylene rubber (EPR) with rubber particle size and size distribution decreasing with shear rate [49]. However, an optimum shear rate beyond which there was no further morphological variation was observed, although this was partly assigned to coalescence during moulding.

A further consequence of the Taylor expression is that by decreasing the interfacial tension (by increasing the interfacial adhesion) a resulting decrease in the size of the dispersed phase is expected. Furthermore, it has been shown [50] that coalescence of the dispersed phase during blending can only be prevented if the interfacial tension between the two phases is sufficiently low. Borggreve and Gaymans [51] found that the size distribution of EPDM rubber in PA6 improved greatly when utilizing maleic anhydride as a coupling agent between the two phases.

2.3.4 Effect of temperature

Temperature has a significant effect on impact behaviour in rubber toughened polymers. At very low temperatures the rubber phase is below its T_g and is effectively a rigid filler with little or no effect on the host polymer. The rubber toughened polymer behaves as a brittle glass. This was demonstrated by Bucknall [52] for a series of ABS polymers (Fig. 2.9). At temperatures below $-75\,^\circ$C the rubbery phase is below its T_g and the ability of the dispersed rubber phase to act as sites of stress concentration is poor. The impact strengths of the ABS materials are similar to the unmodified styrene–acrylonitrile copolymer. Above $-75\,^\circ$C, the T_g of polybutadiene, the impact strength of the ABS polymers begins to rise, accompanied by stress whitening around the notch, the extent of which increases with rubber content. At about $-10\,^\circ$C, the ABS containing 20% rubber shows a further sharp rise in impact strength together with stress whitening of the entire fracture surface. Similar increases occur at $-5\,^\circ$C and at $20\,^\circ$C for ABS materials containing 14% and 10% rubber respectively.

Bucknall [52] has suggested that these secondary transitions occur at higher temperatures and rubber contents because the energy required for

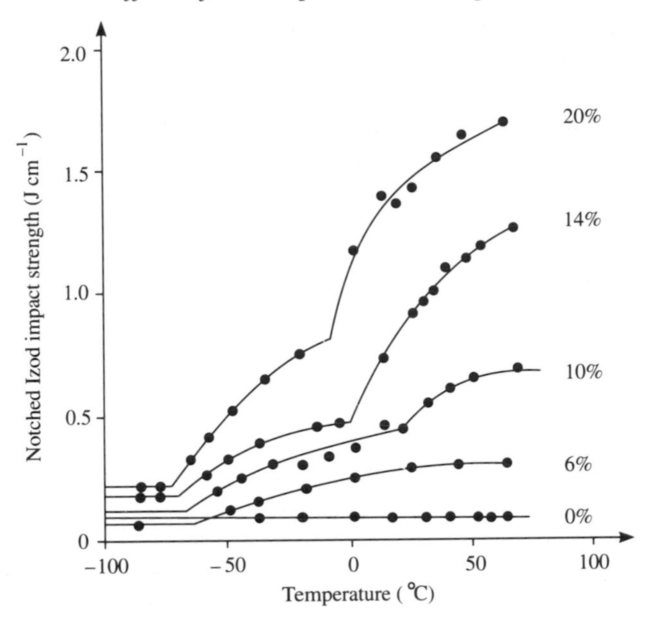

Fig. 2.9 Notched Izod impact strengths over a range of temperatures for styrene–acrylonitrile copolymer (0% rubber) and for a series of ABS polymers containing nominally 6–20% PB [52].

crack propagation is greater than the energy stored elastically in the specimen when the crack is initiated. Therefore, additional energy is taken from the pendulum during the propagation stage. At lower temperatures, the crack propagation energy is smaller and there is sufficient elastic energy stored to complete the fracture of the specimens.

Similar impact strength relationships have been observed in other rubber toughened polymers. For example, in toughened PVC [53], the transition from moderate to high temperature behaviour is accompanied by a very significant increase in impact strength.

2.3.5 Effect of the rubber

Very little work has been carried out on the influence of the rubber on toughening for a stated particle size and interfacial adhesion. This is partly due to other parameters.

One aspect that will influence the toughening is the stresses caused on cooling the blend from the melt. These arise from the differential thermal contraction between the matrix and the rubber, and they will be dependent on the rubber used.

With respect to rubber modulus, the effect of the rubber is difficult to evaluate as it depends on the mechanism by which toughening is believed to occur, be it as a stress concentrator to initiate crazes or shear bands [9, 10] or to alleviate local hydrostatic stresses by cavitation [54, 55]. Again, most of the work has been carried out on polyamides.

Du Pont patents [56, 57] suggest that the rubber modulus must be less than or equal to that of the polyamide so as to be suitable as a stress concentrator. The modulus difference will cause the concentration of stress around the particles, leading to a nucleation of crazes or shear bands. Additional toughening occurs if the stress fields overlap.

Oxborough and Bowden [58] calculated the stress concentration factors around low modulus spheres and found that the maximum stress concentration at the particle–matrix interface occurred when $G_R/G_M = 0.1$, where G_R and G_M are the shear moduli of the rubber and matrix respectively. For $G_R/G_M < 0.1$, there will be no improvement in toughening.

Bucknall [59] stated that crosslinking the rubber is desirable, as, during impact, the rubber undergoes high strains. Crosslinking would allow the rubber to reach high strains by fibrillation, and the fibrils would have high strength.

In all these cases a fast fracture of the rubber will cause a fast crack propagation and prevent extensive plastic deformation. Donald and Kramer [54] and Yee [55] state that, for shear yielding matrices, the rubber functions by providing voids, enabling a relief of local hydrostatic pressure. The relief of the triaxial stress would promote plane stress conditions even in relatively thick specimens.

Bucknall, Heather and Lazzeri [60] and Ramsteiner and Heckmann [61] showed that particles initiate voiding. Initially, shear deformation occurs until yield of the matrix. In Bucknall, Heather and Lazzeri's experiment yielding occurred at 4% strain, giving an applied stress greater than $30\,\mathrm{MN\,m^{-2}}$ and a hydrostatic tension of $10\,\mathrm{MN\,m^{-2}}$. Gent and Lindley [62] calculated a cavitation stress, P_H, of $5E_R/6$, where E_R is the rubber modulus. Gent and Cho [63] found a P_H/E_R value of between 0.38 and 0.7. Therefore, in Bucknall's work this gave the onset of cavitation during yield when E_R was between 3 and $15\,\mathrm{MN\,m^{-2}}$, which was consistent.

Borggreve, Gaymans and Schuijer's [64] work shows that the action of the rubber is not merely as a stress concentrator, but neither does this work show that cavitation is the sole mechanism, in agreement with Ref. 61. Certainly, cavitation is assisted by the different Poisson's ratios of the rubber and the matrix [65].

2.3.6 Interfacial adhesion

Results from some of the early work were inconsistent because of the variation of the interfacial adhesion [18]. Wu [9] prepared blends with two levels of surface adhesion ($G_a = 8100$ and $140\,\mathrm{J\,m^{-2}}$). During the melt extrusion, the reactive PR rubber molecules were grafted onto the polyamide matrix.

Electron microscopic measurements and solvent extraction showed that the interfacial zone thickness was 500 Å. The interfacial tension, γ_{12}, between the PR rubber and polyamide was calculated from [66]

$$\gamma_{12} = 55\,\lambda^{-0.86} \tag{2.17}$$

where γ_{12} is in dyn cm^{-1} and λ (Å) is the interfacial thickness.

The interfacial tension between the reactive PR rubber and polyamide was calculated to be 0.25 dyn cm^{-1}, at 300–325 °C. The interfacial tension at 325 °C for the unreactive NR rubber and PA was 8.8 dyn cm^{-1}, and with a temperature coefficient $d\gamma_{12}/dT = -0.02$ dyn cm^{-1} °C^{-1}. The interfacial thickness was 8 Å.

There is no satisfactory method of measuring interfacial adhesion *in situ* under test, and so Wu [9] used a peel test to evaluate the PA–rubber adhesion. The adhesion fracture energy G_a is independent of geometry but dependent on rate [10]. In the peel test, G_a is given by

$$G_a = (\chi - \cos\theta)P \tag{2.18}$$

where χ is the extension ratio of the flexible part (the rubber), θ is the peel angle and P the peel force per unit width. As G_a is rate dependent it should be evaluated at the impact speed, but this was not feasible.

A 25 mm wide and 0.5 mm thick rubber strip was backed with a 0.1 mm thick polyester–cotton fabric and bonded to a 32 mm thick polyamide strip at 180 °C, 2 MN m^{-2} pressure for 12 min. The bonds were peeled at $\theta = 180°$ at a rate of 0.85 mm s^{-1}. The fabric backing restricted the extension of the rubber such that $\chi = 1$. The peel force was found to be independent of the rubber thickness in the ranges tested.

From his work, Wu [9] found that the minimum surface adhesion necessary for toughening was about 10^3 J m^{-2}, which is typical for van der Waals adhesion. The interfacial adhesion does not influence yield stress and Young's modulus [24].

Wu [8] found that the interfacial energy for adhesion was the same as the tearing strength of the rubber (10^3 J m^{-2}), which meant that during fracture the rubber failed by cavitation rather than by debonding. The latter occurred only in the unreactive rubber–PA blends.

Walker [67] felt that the PBT–PDMS blends with good interfacial adhesion had better notched Izod impact strength. The adhesion was judged from a knowledge of the block copolymer interfacial agent and from electron microscopy which revealed fewer debonded rubber particles and more strain imparted to the resident rubber particles.

2.4 TOUGHENING MECHANISMS

A number of quite different mechanisms for toughening have been proposed but all rely on a dispersion of rubber particles within the glassy matrix. These have included energy absorption by rubber particles [68, 69], debonding at the rubber–matrix interface [70], matrix crazing [52, 59, 71], shear yielding [72–74] or a combination of shear yielding and crazing [8, 52, 75].

2.4.1 Energy absorption by rubber particles

Merz, Claver and Baer [68] first proposed the idea of rubber particles absorbing energy in order to toughen polymers. They observed that in HIPS an increase in volume and stress whitening accompanied elongation of the material and concluded that these phenomena were associated with the formation of many microcracks. It was suggested that the fibrils of styrene–butadiene copolymer bridged across the fracture surface of a developing crack and in so doing prevented the crack growing to a catastrophic size. This resulted in more energy being absorbed than an equivalent volume of the polystyrene matrix. The amount of energy absorbed in impact was attributed to the sum of the energy to fracture the glassy matrix and the work to break the rubber particles.

More recently Kunz-Douglass, Beaumont and Ashby [69, 76] have proposed a similar mechanism for rubber modified epoxies in which the elastic energy stored in the rubber particles during stretching is dissipated irreversibly when the particles rupture.

However, the main disadvantage of these proposed theories is that they are concerned primarily with the rubber rather than with the matrix. It has been calculated [77] that the total amount of energy associated with the deformation of the rubbery phase accounts for no more than a small fraction of the observed enhanced impact energies. Consequently, this mechanism plays only a minor role in the toughening of multiphase polymers. Further toughening theories concentrated on the deformation mechanisms associated with the matrix, which are enhanced by the presence of the rubber phase.

2.4.2 Matrix crazing

Rubber particles have been shown [52, 78] both to initiate and to control craze growth. Under an applied tensile stress, crazes are initiated at points of maximum principal strain, typically near the equator of rubber particles (maximum concentration of triaxial stresses), and propagate outwards normal to the maximum applied stress, although interactions between the particles' stress fields can introduce deviations. Craze growth is terminated when a further rubber particle is encountered, preventing the growth of very large crazes. The result is a large number of small crazes in contrast to a small number of large crazes formed in the same polymer in the absence of rubber particles. The dense crazing that occurs throughout a comparatively large volume of the multiphase material accounts for the high energy absorption observed in tensile and impact tests and the extensive stress whitening which accompanies deformation and failure. Microscopy studies, both optical and electron, while confirming that crazes frequently initiate from the rubber particles, have aided in elucidating further information concerning the constituents of a craze and the detailed mechanisms of craze initiation, growth and breakdown around rubber particles.

Although a craze appears to be similar to a crack owing to its lower refractive index than its surroundings, it actually contains fibrils of polymer drawn across, normal to the craze surfaces, in an interconnecting void network [79–81]. The void network is established at the craze tip. The craze tip advances by a finger-like growth produced by the meniscus instability mechanism [79]. The fibrils are formed at the polymer webs between void fingers and contain highly oriented polymer with the chain axis parallel to the fibril axis [79, 82]. As the craze tip advances the craze thickens by drawing in more polymer from the craze surfaces. It is the presence of the relatively strong craze fibrils that makes the craze load bearing and consequently differing from a crack. Cracks can form, however, by the breakdown of craze fibrils to form voids. These voids expand slowly by the rupture of surrounding craze fibrils until the void becomes a crack of critical size that can propagate catastrophically.

It has been consistently demonstrated experimentally in polymer matrices which fail by crazing that the optimum rubber particle size is in the range 1–5 μm [83–86]. Donald and Kramer [86] examined the craze initiation mechanism in HIPS and elucidated that ease of initiation is related to particle size and that crazes are rarely nucleated from particle sizes less than about 1 μm. This has been observed by other workers [84].

Donald and Kramer [86] proposed two criteria necessary for craze initiation from a rubber particle:

1. the initial elastic stress concentration at the rubber particle must exceed the stress concentration at a static craze tip;
2. the distance over which this critical stress acts must extend at least three fibril spacings from the particle into the glassy matrix.

The stress field around the particle is independent of particle size but the spatial extent of the stress enhancement scales with the particle diameter. The second criterion therefore explains the inability of small rubber particles to initiate crazes.

Further work studying HIPS by Donald and Kramer [87] considered the effect of the rubber particles' internal composition and its effect on craze breakdown and subsequent crack initiation. HIPS is usually prepared by polymerization of a styrene monomer containing about 5–10% of dissolved rubber. As the polystyrene forms, phase separation of the rubber occurs as a dispersion in the polystyrene solution. However, the dispersed rubber particles are swollen with styrene monomer, which continues to polymerize and forms a dispersion of polystyrene within the dispersed rubber particles.

Ultimately this system has a three-phase morphology. It was discovered that the larger particles only contained occluded polystyrene in the particle while the smaller ones tended to be rubber alone. This non-uniformity of particles appears to reduce the chances of premature craze breakdown. When crazes form around the solid rubber particles, either by initiating or by intersecting them, significant lateral contraction accompanied elongation of the particle in the applied stress direction. As the contraction proceeded, debonding at the particle–craze interface resulted and a void was thus formed. Under increasing load the void grew, resulting in premature craze breakdown and subsequent crack initiation and propagation. In the non-uniform particles, occluded polystyrene accommodates the displacements due to crazing by local fibrillation of the rubber surrounding each subinclusion, without the formation of large voids. It was therefore concluded that the optimum morphology of the rubber particles in HIPS should consist of the following:

- particles of a size just greater than the critical diameter (therefore maximum number of particles to initiate crazes);
- particles should contain a large number of small polystyrene occlusions each surrounded by a thin layer of rubber (therefore limiting the number of inherent flaws during crazing).

It has been proposed [78, 88] that rigid particulate fillers can be used to increase the toughness of brittle polymers by initiating multiple crazing. Under an applied stress rigid particles do induce tensile stress concentrations in the matrix but become debonded from the matrix readily as they are unable to deform to any significant degree. Since there is limited adhesion between the rigid particulate fillers and the matrix they are not particularly effective craze or crack terminators, resulting in poorer toughening performance when compared with well-bonded rubber particles.

2.4.3 Shear yielding

Shear yielding in the matrix phase also plays a major role in the mechanism of rubber toughening in polymer blends. Shear yielding, as localized shear bands or more generalized and diffuse regions of shear yielding, usually occurs in addition to elastic deformation. Not only does this phenomenon act as an energy absorbing process but the shear bands also present a barrier to the propagation of crazes and hence crack growth [52], therefore delaying failure of the material.

Newman and Strella [73] first proposed that shear yielding in the matrix was responsible for rubber toughening. Mechanical property and optical property studies, on ABS materials, showed that in tensile tests necking, drawing and orientation hardening occurred together with localized plastic deformation of the matrix around the rubber particles, indicative of shear yielding. It was proposed that the function of the rubber particles was to produce enough triaxial tension in the matrix so as to increase the local free volume and consequently to initiate extensive shear yielding and drawing of the matrix. However, this proposal did not account for stress whitening, characteristic of rubber toughening, and the fact that triaxial tension promotes crazing and brittle fracture rather than shear yielding.

Further studies have elucidated a better understanding of the detailed micromechanisms of systems which deform preferentially by shear yielding. The major toughening mechanisms are thought to be cavitation of rubber particles and shear yielding of the matrix, but whether cavitation occurs first followed by shear yielding or shear yielding initially occurs followed by cavitation is still not truly defined.

However, it is now generally accepted that the shear yielding mechanism constitutes cavitation of the rubber particles followed by extensive shear yielding throughout the matrix [89–92]. The cavitation of the rubber particles explains the observed stress whitening as light scattering occurs which is enhanced by the holes enlarging. Cavitation is followed by the onset of shear yielding, because on cavitation of the rubber particles the buildup of hydrostatic tension is locally relieved, lowering the yield stress. After cavitation the constrained conditions, triaxial stresses, disappear and the matrix behaves as if it were under plane stress conditions. Shear yielding deformations occur more readily under a biaxial stress state rather than the craze-favouring triaxial state. The voids created by the cavitated rubber particles act further as stress concentrators [90].

Although cavitation of the rubber particles does involve energy absorption, the enhanced shear yielding of the matrix is the major energy absorbing mechanism. However, cavitation of the rubber particles is a prerequisite for enhanced toughness where shear yielding is the principal mechanism.

2.4.4 Crazing and shear yielding

Crazing and shear yielding may occur simultaneously in many rubber toughened plastics. The dominant mechanism is the one by which the unmodified matrix would typically fail. However, the contribution of each mechanism to the toughening of the system depends on a number of variables such as the rubber particle size and dispersion, the concentration of the rubber particles and the rate and temperature of the test. The contribution of each mechanism to the toughening process can be assessed to some extent by using tensile dilatometry. It is assumed that deformations such as voiding and crazing are dilatational processes, which manifest themselves by an increase in volume strain. Unfortunately, if both voiding and crazing occur simultaneously, it is impossible to separate their contributions to volume strain. However, when shear yielding occurs, a decrease in the volume strain rate occurs since shear yielding is a non-dilatational or constant volume process.

An early example of the occurrence of crazing and shear yielding simultaneously was recognized by Bucknall, Clayton and Keast [75]. The observation that shear bands formed at 45° to the stress axis appeared to intersect crazes and to run between rubber particles led to the conclusion that shear bands, as well as being generated from rubber particles, were effective craze stoppers.

2.4.5 Cavitation and rumples

The occurrence of cavitation in the presence of shear yielding has been observed in rubber toughened amorphous and semicrystalline polymers (Refs. 90, 91 and 93 for amorphous polymers; Refs. 59 and 61 for semicrystalline PA66; Ref. 94, PBT; Refs. 24 and 95, PA6; Refs. 96–100, epoxies), whereas it is absent in the untoughened matrices.

Shear yielding is not accompanied by expansion of the matrix, but in rubber toughened blends the presence of the rubber gives rise to volume increase if the strain rate is sufficiently high [90]. This expansion is caused by the cavitation of the rubber particles [91]. The rubber particles dissipate the bulk strain energy by cavitation, leading to a reduction in local hydrostatic stress and a reduction in the yield stress of the blend. Thus shear band formation is enhanced by the voids in the matrix caused by the cavitated rubber particles [101, 102].

In the presence of a sharp crack, a triaxial stress exists ahead of the crack tip. This gives rise to rapid cavitation and growth of the resulting voids [90, 91]. A zone of voids and shear bands is formed ahead of the crack tip on the opening up of the crack faces. This voided zone blunts the crack tip and further tension causes an even larger plastic zone to form. The increasing size of this large plastic zone acts as the principal toughening mechanism (Refs. 90 and 91, epoxies). The rubber particles toughen by acting as stress concentrators, enhancing shear yielding, and then cavitate, dissipating energy and giving rise to more shear yielding. As mentioned in section 2.3.5, the cavitation stress depends on the modulus of the rubber ($P_H = 5E_R/6$) [62] and results of Bucknall, Heather and Lazzeri [60] confirm this in polyamides.

The occurrence of a rumpled surface accompanying cavitation was noted in Refs. 65 (PA66), 103 (unmodified PA in the presence of water) and 104 (EPDM modified PA66). The rumples lie parallel to the notch and give rise to tufts of highly drawn material.

The following model was proposed by Hahn, Hertzberg and Manson [65]. The considerable ductility of the toughened PA66 matrix gives rise to considerable drawing ahead of the crack tip before unstable fracture sets in. Consider Fig. 2.10. As material is drawn along the y axis in the y–z plane, there is a likelihood of the material weakening in a direction perpendicular to this plane. The highly drawn material is susceptible to delamination in the y direction owing to the normal stress σ_x. This causes secondary fissures or

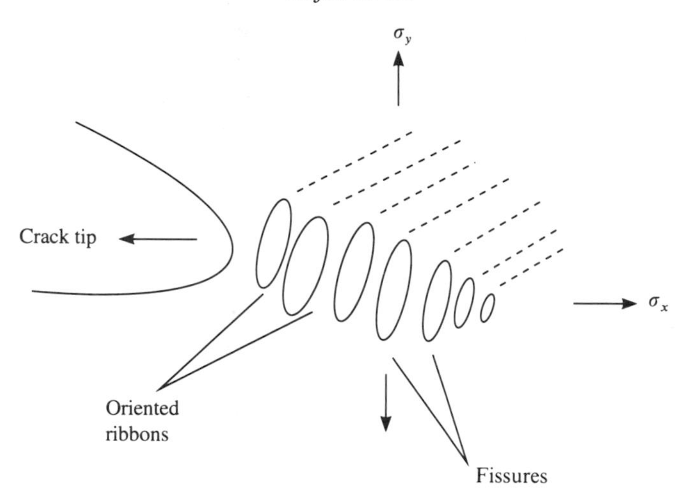

Fig. 2.10 Schematic of the model for rumpled fracture surface formation. (Reproduced from M. T. Hahn, R. W. Hertzberg and J. A. Manson, *Journal of Materials Science*, 1986.)

voids in the y–z plane, parallel to the rumple bands. The secondary fissures surround the oriented material and they link up, causing an advance of the crack tip in the x direction. This occurs for highly ductile blends.

Speroni *et al.* [105] attribute the nature of the rumples to the effect of an octahedral shear mechanism proposed by Yee and Pearson [90] and extensive cavitation. The existence of rumples is clearly linked by them to cavitation.

REFERENCES

1. Wu, S. (1990) *Polym. Eng. Sci.*, **30**(13), 753–61.
2. Kambour, R. P., Faulkner, D., Kampf, E. E., Miller, S., Niznik, E. E. and Schultz, A. R. (1976) *Adv. Chem. Ser.*, **26**, 312.
3. Miller, N. J. (1976) *J. Mater. Sci.*, **11**, 363.
4. Fraser, R. A. W. and Ward, I. M. (1977) *J. Mater. Sci.*, **12**, 459.
5. Narisawa, I., Ishikawa, M. and Ogata, H. (1980) *J. Mater. Sci.*, **15**, 2059.
6. Narisawa, I., Ishikawa, M. and Ogata, H. (1983) *J. Mater. Sci.*, **18**, 2826.
7. Wellinghoff, S. T. and Baer, E. (1978) *J. Appl. Polym. Sci.*, **22**, 2025.
8. Wu, S. (1983) *J. Polym. Sci., Polym. Phys. Ed.*, **21**, 699.
9. Wu, S. (1985) *Polymer*, **26**, 1855.
10. Hobbs, S. Y., Bopp, R. C. and Watkins, V. H. (1983) *Polym. Eng. Sci.*, **23**(7), 381.
11. Matsuo, M. (1969) *Polym. Eng. Sci.*, **9**, 206.
12. Morbitzer, L., Kranz, D., Humme, G. and Ott, K. H. (1976) *J. Appl. Polym. Sci.*, **20**, 2691.
13. Riew, C. K., Rowe, E. H. and Siebert, A. R. (1976) *ACS Adv. Chem. Ser.*, **154**, 326.
14. Fowler, M. E., Keskkula, H. and Paul, D. R. (1987) *Polymer*, **28**, 1730.
15. Fowler, M. E., Keskkula, H. and Paul, D. R. (1988) *J. Appl. Polym. Sci.*, **35**, 1563.
16. Kambour, R. P. (1983) *Polym. Commun.*, **24**, 292.

17. Henkee, C. S. and Kramer, E. J. (1984) *J. Polym. Sci. B, Polym. Phys.*, **22**, 721.
18. Brydson, J. A. (1975) *Plastics Materials*, Newnes Butterworth, London.
19. Brandrup, J. and Immergut, E. H. (1975) *Polymer Handbook*, Vol. 9, Wiley, New York, pp. 337–59.
20. Stehling, F. C., Huff, T., Speed, C. S. and Wissler, G. (1981) *J. Appl. Polym. Sci.*, **26**, 2693.
21. Crowley, J. D., Teague, G. S. and Lowe, J. W., Jr (1966) *J. Paint Technol.*, **38**, (496), 269.
22. Crowley, J. D., Teague, G. S. and Lowe, J. W., Jr (1967) *J. Paint Technol.*, **39**, (504), 19.
23. Hansen, C. M. (1967) *J. Paint Technol.*, **39**, (505), 511.
24. Borggreve, R. J. M., Gaymans, R. J., Schuijer, J. and Ingen Housz, A. J. (1987) *Polymer*, **28**, 1489–96.
25. Sederel, L. C., Mooney, J. and Weese, R. H. (1987) Polymer Blends and Alloys Conf., Strasbourg, June 1987.
26. Neuray, D. and Ott, K. H. (1981) *Angew. Makromol. Chem.*, **98**, 213.
27. Hobbs, S. Y., Dekkers, M. E. J. and Watkins, V. H. (1989) *J. Mater. Sci.*, **24**, 2025.
28. Michler, G. H. (1989) *Colloid Polym. Sci.*, **267**, 377.
29. Cigma, G., Lomellini, P. and Merlotti, M. (1989) *J. Appl. Polym. Sci.*, **37**, 1527.
30. Hobbs, S. Y. (1986) *Polym. Eng. Sci.*, **26**, 74.
31. Margolina, A. and Wu, S. (1988) *Polymer*, **29**, 2170.
32. Wu, S. and Margolina, A. (1990) *Polymer*, **31**, 972.
33. Flexman, E. A. (1979) *Polym. Eng. Sci.*, **19**, 564.
34. Borggreve, R. J. M., Gaymans, R. J. and Luttmer, A. R. (1988) *Makromol. Chem., Macromol. Symp.*, **16**, 195.
35. Sjoerdsma, S. D. (1989) *Polym. Commun.*, **30**, 106.
36. Margolina, A. (1990) *Polym. Commun.*, **31**, 95.
37. Wu, S. (1988) *J. Appl. Polym. Sci.*, **35**, 549.
38. Newman, S. (1973) *Polym. Plast. Technol. Eng.*, **2**(1), 67.
39. Beck, R. H., Grath, S., Newman, S. and Rusch, K. C. (1968) *Polym. Lett.*, **6**, 707.
40. Morbitzer, L., Ott, K. H., Schuster, H. and Kranz, D. (1972) *Angew. Makromol. Chem.*, **27**, 57.
41. Booj, H. C. (1977) *Br. Polym. J.*, **9**(3), 47.
42. Low, I. M. (1990) *J. Appl. Polym. Sci.*, **39**, 759.
43. Flexman, E. A., Hwang, D. D. and Snyder, H. L. (1988) *Am. Chem. Soc. Div. Polym. Chem. Polym. Prepr.*, **29**(2), 189.
44. Wadhwa, L. H., Dolce, T. J. and La Nieve, H. L. (1985) SPE RETEC, Columbus, Ohio, December 12, 1985.
45. Siegmann, A. and Hiltner, A. (1984) *Polym. Eng. Sci.*, **24**, 869.
46. Siegmann, A., English, L. K., Baer, E. and Hiltner, A. (1984) *Polym. Eng. Sci.*, **24**, 877.
47. Menges, G., Berndtsen, N. and Oppermann, J. (1979) *Plast. Rubber Proc.* **4**, 156.
48. Petrich, R. P. (1973) *Polym. Eng. Sci.*, **13**, 248.
49. Cecere, A., Greco, R., Ragosta, G., Scarinzi, G. and Taglialatela, A. (1990) *Polymer*, **31**, 1239.
50. Elmendorp, J. J. (1986) PhD Thesis, University of Technology, Delft.
51. Borggreve, R. J. M. and Gaymans, R. J. (1989) *Polymer*, **30**, 63.
52. Bucknall, C. B. (1977) *Toughened Plastics*, Applied Science, London.
53. Bucknall, C. B. and Street, D. G. (1967) *SCI Monogr.*, **26**, 272.
54. Donald, A. M. and Kramer, E. J. (1982) *J. Mater. Sci.*, **17**, 1765.
55. Yee, A. F. (1985) Proc. Conf. on Toughening of Plastics, PRI, London.
56. Du Pont, Br. Patent 998,439, July 14, 1965.

57. Du Pont (Epstein, B. N.), US Patent 4,174,358, November 13, 1979.
58. Oxborough, R. J. and Bowden, P. B. (1974) *Philos. Mag.*, **30**, 171.
59. Bucknall, C. B. (1979) In *Polymer Blends*, (eds D. R. Paul and S. Newman), Academic Press, New York, p. 91.
60. Bucknall, C. B., Heather, P. S. and Lazzeri, A. (1989) *J. Mater. Sci.*, **16**, 2255.
61. Ramsteiner, F. and Heckmann, W. (1985) *Polym. Commun.*, **26**, 199.
62. Gent, A. N. and Lindley, P. B. (1969) *Proc. R. Soc. London, Ser. A*, **249**, 2520.
63. Gent, A. N. and Cho, K. (1988) *J. Mater. Sci.*, **23**, 141.
64. Borggreve, R. J. M., Gaymans, R. J. and Schuijer, J. (1989) *Polymer*, **30**, 71.
65. Hahn, M. T., Hertzberg, R. W. and Manson, J. A. (1986) *J. Mater. Sci.*, **21**, 31, 39–45.
66. Wu, S. (1982) *Polymer Interface and Adhesion*, Dekker, New York.
67. Walker, I. (1991) PhD Thesis, Sheffield City Polytechnic.
68. Merz, E. H., Claver, G. C. and Baer, M. (1956) *J. Polym. Sci.*, **22**, 325.
69. Kunz-Douglass, S., Beaumont, P. W. R. and Ashby, M. F. (1980) *J. Mater. Sci.*, **15**, 1109.
70. Sultan, J. N. and McGarry, F. J. (1973) *Polym. Eng. Sci.*, **13**, 29.
71. Bucknall, C. B. and Smith, R. R. (1965) *Polymer*, **6**, 437.
72. Newman, S. (1978) In *Polymer Blends*, Vol. 2 (eds D. R. Paul and S. Newman), Academic Press, New York, pp. 63–89.
73. Newman, S. and Strella, S. (1965) *J. Appl. Polym. Sci.*, **9**, 2297.
74. Petrich, R. P. (1977) *Polym. Eng. Sci.*, **12**, 757.
75. Bucknall, C. B., Clayton, D. and Keast, W. E. (1972) *J. Mater. Sci.*, **7**, 1443.
76. Kunz-Douglass, S., Beaumont, P. W. R. and Ashby, M. F. (1981) *J. Mater. Sci.*, **16**, 3141.
77. Bucknall, C. B. (1978) *Adv. Polym. Sci.*, **27**, 121.
78. Donald, A. M. and Kramer, E. J. (1981) *Philos. Mag. A, Ser. 8*, **43**, 857.
79. Beahan, P., Bevis, M. and Hull, D. (1974) *J. Mater. Sci.*, **8**, 162.
80. Kramer, E. J. (1982) In *Polymer Compatibility and Incompatibility Principles and Practices* (ed. K. Solc), Symp. Series, Vol. 22, Harwood Academic Publishers, New York, p. 251.
81. Brown, H. R. (1979) *J. Polym. Sci., Polym. Phys. Ed.*, **17**, 1431.
82. Moore, J. D. (1971) *Polymer*, **12**, 478.
83. Silberberg, J. and Man, C. D. (1978) *J. Appl. Polym. Sci.*, **22**, 599.
84. Morton, M., Cizmecioglu, M. and Lhila, R. (1984) *Adv. Chem. Ser.*, **206**, 221.
85. Turley, S. G. and Keskkula, H. (1980) *Polymer*, **21**, 466.
86. Donald, A. M. and Kramer, E. J. (1982) *J. Appl. Polym. Sci.*, **27**, 3729.
87. Donald, A. M. and Kramer, E. J. (1982) *J. Mater. Sci.*, **17**, 2351.
88. Kinloch, A. J. and Young, R. J. (1983) *Fracture Behaviour of Polymers*, Applied Science, London.
89. Borggreve, R. J. M., Gaymans, R. J. and Eichenwald, H. M. (1989) *Polymer*, **30**, 79.
90. Yee, A. F. and Pearson, R. A. (1986) *J. Mater. Sci.*, **21**, 2462.
91. Pearson, R. A. and Yee, A. F. (1986) *J. Mater. Sci.*, **21**, 2475.
92. Parker, D. S., Sue, H. J., Huang, J. and Yee, A. F. (1990) *Polymer*, **31**, 2267.
93. Breuer, H., Haaf, F. and Stabenour, J. (1977) *J. Macromol. Sci. Phys. B*, **14**, 387.
94. Polato, F. (1985) *J. Mater. Sci.*, **20**, 1455.
95. Borggreve, R. J. M. (1988) PhD Thesis, University of Twente.
96. Bascom, W. D. and Hunston, D. L. (1978) Proc. Int. Conf. on Toughening of Plastics, PRI, London, July 1978, p. 22.
97. Bascom, W. D., Cottington, R. L., Jones, R. L. and Peiper, P. (1975) *J. Appl. Polym. Sci.*, **19**, 2545.

98. Bascom, W. D. and Cottington, R. L. (1976) *J. Adhes.*, **7**, 333.
99. Bascom, W. D., Cottington, R. L. and Timmins, C. O. (1977) *Appl. Polym. Symp.*, **32**, 165.
100. Bascom, W. D., Ting, R. Y., Moulton, R. J., Riew, C. K. and Siebert, A. R. (1981) *J. Mater. Sci.*, **16**, 2657.
101. Tamamoto, H. (1978) *Int. J. Fract.*, **14**, 347.
102. Tvergaard, V. (1981) *Int. J. Fract.*, **17**, 389.
103. Bretz, P. E. (1980) PhD Thesis, Lehigh University.
104. Flexman, E. A., Jr (1978) Proc. Int. Conf. on Toughening of Plastics, PRI, London, July 1978, p. 14.
105. Speroni, F., Castoldi, E., Fabbri, P. and Casiraghi, T. (1989) *J. Mater. Sci.*, **24**, 2165.

3

Fracture and toughening in fibre reinforced polymer composites

G. C. McGrath

3.1 INTRODUCTION

In order to understand the behaviour of polymeric materials for use as matrices in fibre reinforced composites and structural adhesives it is necessary to determine the fracture mechanisms of such polymers not only in the bulk form but also in the form anticipated for their expected use. Even if the resin displays excellent properties in bulk form, they may not be translated to laminated composites. This has been demonstrated by Bersch [1] who identified 24 polymers which show higher strains to failure than current epoxies. However, only five of those resins under investigation provided a higher strain to failure when contained in a composite system, as examined by the residual compression strength after impact technique. Further evidence of the inability to read across information into the composite system was provided by other shortcomings in properties, especially the elastic modulus, which rendered many of the polymers unsuitable for use as matrices in a composite system. Other attempts have been made to improve the composite toughness by improving the toughness of the polymer systems. These also have had disappointing results, in that a large increase in polymer toughness has not necessarily been found to give a proportionate increase in composite toughness. Scott and Phillips [2] found that a tenfold increase in resin toughness increased composite toughness by a factor of 2. Similar results were reported in different systems by Bascom and coworkers [3, 4], Vanderkley [5] and Bradley and Cohen [6, 7]. Thus, thorough investigations are necessary if polymers are to be used successfully for composite systems.

3.2 MATRICES FOR COMPOSITE SYSTEMS

Composite systems are often supplied as pre-impregnated tapes or prepregs which are either ready-to-mould materials in sheet form or ready-to-wind materials in roving form, which may be cloth, mat, unidirectional fibre, or paper impregnated with resin and stored for use. The resin is partially cured to a B stage and supplied to the fabricator, who lays up the finished shape and completes the cure with heat and pressure. The two distinct types of prepregs available are

- commercial prepregs, where the roving is coated with a hot melt or solvent system to produce a specific product to meet specific customer requirements, and
- wet prepreg, where the basic resin is installed without solvents or preservatives but has limited room temperature shelf life.

Toughness and fracture resistance are important requirements in most load bearing applications of materials. Resistance to impact damage is particularly important in advanced composites used for aerospace applications and other uses, including automotive purposes. The difficulty is not simply to enhance the toughness, but to achieve this without excessive reduction in other attractive properties, i.e. the balance of properties has to be improved, and this includes processability, cost and availability. Toughness can be enhanced by a variety of techniques, including modification of the matrix, fibre coatings and interleaves between the plies. The goal of such techniques is to combine the stiffness and processability of thermosetting matrices with the fracture resistance of elastomers or thermoplastics.

The most significant development in the evolution of continuous fibre composites has been the introduction of thermoplastic matrices which are creep resistant, tough and have a high deflection under load temperature (Table 3.1). One of the earliest improvements in this area was polyethersulphone (PES) but, being amorphous, PES is subject to environmental attack under adverse conditions. Polyetheretherketone (PEEK) was introduced later and its semicrystalline nature proved advantageous against environmental attack. The use of PEEK with carbon, aramid and glass fibres has produced various composites. Aromatic Polymer Composite (APC-2) is an outstanding example, but other products are available including a fabric-like material, Filmix. Many other thermoplastic matrices are now available, including polyetherimide (PEI), polyarylether, polyamideimide (PAI), polyphenylenesulphide (PPS), polyetherketone (PEK) and polyphenyleneoxide (PPO). There is clearly scope for further developments in this area, including thermoplastics formed by *in situ* polymerization.

Concurrent developments to increase the fracture toughness of thermoset matrices by the addition of rubber-like particles to the matrix have taken place. An essential requirement for toughness enhancement in this manner is that the additive should be soluble in the uncured resin. A phase separation during the curing cycle is preferred as phase separated blends are generally tougher than homogeneous blends. However, when elastomer blends are used, caution is necessary because, if any additive remains after the cure cycle,

Table 3.1 Thermoplastic matrix composite material properties

		Flexural modulus (MPa)		*Heat deflection temperature under flexural load of 1.81 MPa (°C)*		*Notched Izod impact strength ($J\,m^{-1}$)*	
Base resin	*Glass content (%)*	*23 °C*	*177 °C*	*Resin*	*Plus reinforcement*	*Resin*	*Plus reinforcement*
PES	30	8.70	7.60	201	216	77	82.5
PEI	30	9.53	7.74	199	211	55	104.5
PPS	40	13.12	5.18	135	254	30	66
PEEK	30	10.36	3.80	155	300	82.5	127
PEK	30	10.57	4.49	165	310	82.5	94
PAI	30	11.74		278	282	148.5	82.5

the glass transition temperature of the resin is lowered, with deleterious effects on the ultimate performance of the composite.

Epoxy resins blended with carboxyl-terminated butadiene acrylonitrile (CTBN) are among the most widely studied. The nitrile group raises the glass transition temperature, but makes the rubber insoluble in the resin, and the terminal carboxyl groups react with the epoxy groups (Chapter 6). The solubility decreases as the molecular weight of the blend constituents increases, and low molecular weight rubbers are therefore easier to process. The necessary phase separation occurs during the cure cycle of the epoxy for two reasons:

- the molecular weight increases, and
- the formation of polar groups causes the epoxy to become less miscible with CTBN rubber.

The degree of phase separation is determined by the thermodynamic driving force supporting the dissociation, and also the limits represented by the gelation curve and the glass transition of the partly cured resin. Similar factors affect phase separation in resins containing dissolved thermoplastics, some of which also have reactive end groups.

As previously discussed, correlations between the fracture energies of matrices and corresponding composites are not simple, and the capacity of a resin to develop a substantial plastic zone at the crack tip is significantly restricted in the composite.

An alternative technique to produce toughening enhancement of composite systems is to utilize a coating of either thermoplastic or thermoset resin on the fibres. This can be achieved by electrocoating carbon fibres or, more readily for all fibres, from a solution treatment. Good adhesion between the coating and the fibre is essential for an effective toughening mechanism. The solution coating technique produces a thin film of epoxy size to the fibre, whereas electrocoating creates chemical bonds between the carbon and the CTBN. The creation of bonds prevents the rubber from dissolving in the epoxy resin during and after the pre-impregnation process, contributing to the toughness of the cured laminate. The application of elastomeric coatings to fibres enhances the interlaminar fracture energy of the composite system, and significantly improves the resistance to impact damage. However, the interlaminar shear strength is reduced.

A further approach for producing tougher composites is to introduce thin layers of a more ductile matrix between the plies of highly crosslinked matrix prepreg. These may be continuous, or discrete. The use of discrete additions promotes the establishment of direct chemical bonds between the individual plies of the composite. This technique increases the impact resistance, as measured by residual compressive strength after impact. Some reduction in the compressive strength of the undamaged laminate may occur, although in

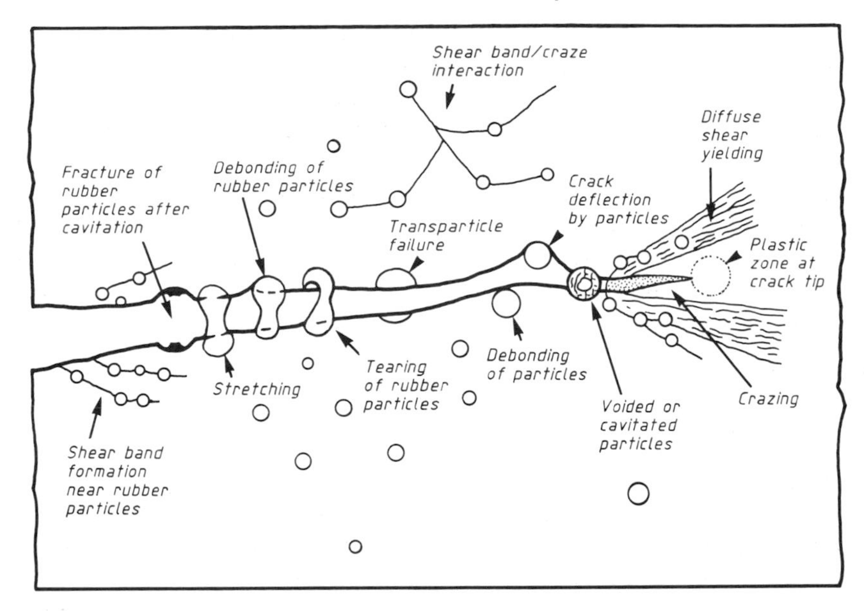

Fig. 3.1 Crack toughening mechanisms in rubber filled modified polymers.

some cases this may be more than fully offset by the greater compressive strength when used in adverse environmental conditions.

The toughening mechanisms produced in the polymer are presented in Fig. 3.1, and need to be operative in the composite for successful development of fracture resistance in the laminate.

Toughening mechanisms in the laminate are characterized by two distinctly different processes, which are categorized by the propagation of the crack path. At the crack tip, processes in operation are crazing, matrix microcracking, plastic deformation, fibre or particle debonding, and phase transformation. As the crack propagates another set of toughening mechanisms becomes operative. These can be summarized by fibre or particle bridging, matrix bridging, fibre pull-out and precipitate elongation. Of course, these mechanisms do not operate in isolation and two or more processes may combine to produce a synergistic toughening effect. The interaction between processes can increase the effective area for toughening mechanisms to operate on, and therefore enhance toughness. The toughness enhancement achieved by a specific mechanism is produced by the energy change in a volume element as a crack propagates through the composite.

3.3 MICROMECHANICAL ANALYSIS

To advance the understanding of the translation of toughening mechanisms from polymer to composite, it is necessary to develop quantitative expres-

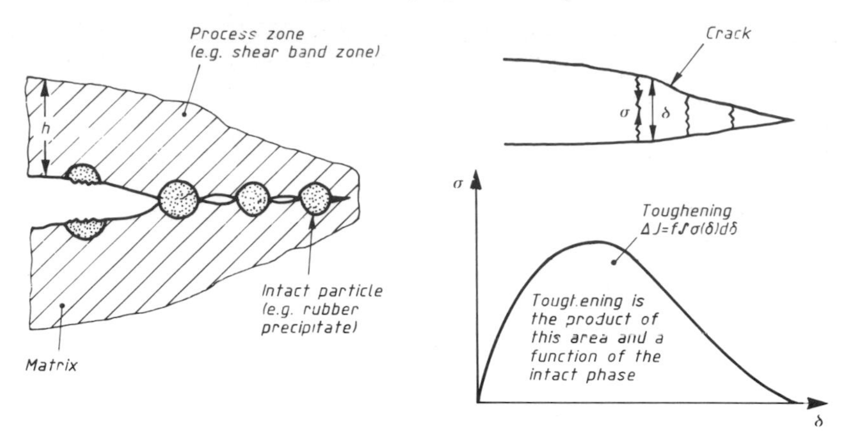

Fig. 3.2 The interaction between mechanisms can increase the process zone size and bridging zone size to promote toughening.

sions for toughness with regard to the microstructure. This is represented schematically in Fig. 3.2, where the area under the curve represents a portion of the toughening effect, the remainder being provided by consideration of the intact phase. Thus, the microstructure must be controlled to maximize the area in order that a tougher composite results. Any toughness enhancement by a specific mechanism is related to the energy change in a volume element as a crack progresses through the composite. As previously discussed, two categories of mechanism are in operation: one at the crack tip, and the other in the crack path which is dependent on the microstructural integrity in the crack path.

The perception of the crucial influence of the fibre–matrix interface on composite toughness has initiated a series of models to explain the phenomena. These models characterize initial fibre debonding at the crack tip, followed by fibre fracture in the crack path, which produces a synergistic toughening by fibre bridging and fibre pull-out.

Unfortunately, the inherent heterogeneities in composite materials, especially short fibre materials, mean that the application of a classical fracture mechanics approach as applied to isotropic metals is not always relevant. Kanninen, Rybicki and Brinson [8] report that composite fracture research can be categorized broadly into two divisions:

- a continuous analysis for homogeneous anisotropic linear elastic material containing an internal flaw of known length;
- a semiempirical analysis of the micromechanical details of the crack tip region in a unidirectional composite (this is more applicable to composite materials).

Wang *et al.* [9] utilized the latter approach to analyse, by fracture mechanics techniques based on conservation laws of solid mechanics, the notch tip stress intensities in the composite. This provides information on fracture characterization of the composite material and is useful in the design and analysis of the composite.

During tensile fracture, composites can fail in one of two ways:

- fibre breakage;
- fibre pull-out.

Folkes [10] highlights that, for optimized performance, maximum fibre breakage is necessary. To prevent fibre pull-out, the fibre must be sufficiently long for the frictional energy of pull-out to exceed the energy of fibre breakage.

The length at which these two energies are equal is called the critical fibre length, l_c, and Blumentritt, Vu and Cooper [11] give

$$l_c = \frac{\sigma_{uf} r}{\tau} \tag{3.1}$$

where l_c is the critical fibre length, σ_{uf} is the ultimate strength of the fibre, r is the radius of the fibre and τ is the shear strength of the composite. When this critical fibre length is exceeded, then the major fracture mechanism should be the result of fibre fracture. In practice, fibre pull-out still exists at lengths three to four times the critical length [12] owing to anomalies in the bonding of the fibre to the matrix. A detailed insight into the mechanisms of fibre pull-out is given by Folkes [10], Cooper [13] and Piggot [14].

Although short fibres tend to be deleterious to the strength of the composite, it has been shown that they improve the toughness of the composite [11, 15]. This is due to the work required firstly to debond the fibre from the matrix, the frictional work required to pull the fibres from the matrix, and the crack blunting mechanism produced as a result of the resin-rich areas in the composite.

3.3.1 Debonding

As the load on the composite is increased, matrix and fibre at the crack tip attempt to deform differentially and a relatively large local stress begins to build up in the fibre. This stress generates the local Poisson contraction which, aggravated by lateral contraction of tensile stress (normal to the interface) ahead of the crack tip may initiate fibre–matrix decohesion, or debonding as presented in Fig. 3.3(c).

The interfacial shear stress resulting from the fibre–matrix modulus mismatch will then cause extension of the debond along the fibre in both directions away from the crack plane. This will permit further opening of the matrix crack beyond the fibre, and the process will be repeated at the next fibre. An upper limit to the energy of debonding is given by the total elastic

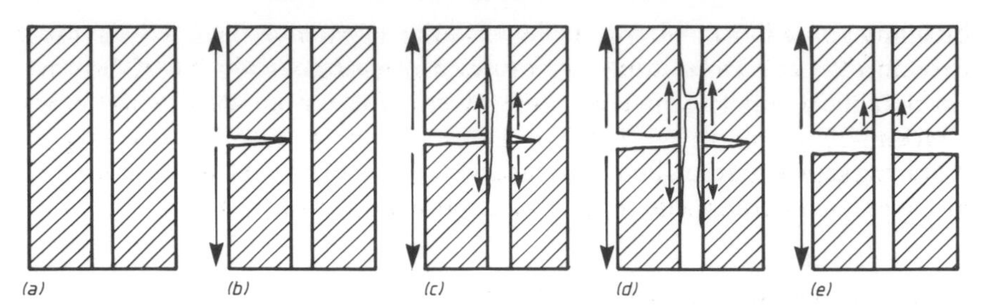

Fig. 3.3 Schematic representation of stages in crack growth in a fibre composite.

energy that will subsequently be stored in the fibre at breaking load, i.e. $\sigma_f^2/2E_f$ per fibre per unit volume, or with N fibres bridging the crack

$$W_{\text{debond}} = \frac{N\pi r^2 y \sigma_f}{2E_f} \tag{3.2}$$

where W_{debond} is the energy of debonding, r is the radius of the fibre, σ_f is the

Fig. 3.4 Side view of a polished DCB specimen. A PEEK-rich zone has formed at the confluence of three particles.

breaking stress of the fibre, E_f is the modulus of the fibre and y is the mean debonded length. Debonding may produce large-scale deviation of the crack tip parallel with the fibres, resulting in an effective blunting of the crack. Cracking may then be produced on another plane remote from the original crack plane, with a resultant increase in the complexity of the fracture face and increase in composite toughness as presented in Fig. 3.4.

3.3.2 Frictional work following debonding

After debonding, the fibre and matrix move relative to each other as crack opening continues and work must be done against frictional resistance during the process. The magnitude of this work is difficult to assess because the extent of the frictional force is not accurately known. It can be estimated, assuming that the interfacial frictional force, λ, acts over a distance equal to the fibre extension, as

$$W_{\text{friction}} = N\lambda\pi r y^2 \varepsilon_f \tag{3.3}$$

where ε_f is the fibre failure strain. This contributes substantially to the toughness of the composite.

3.3.3 Fibre pull-out

After debonding, a continuous fibre is loaded to failure over a gauge length, and it may break at any point as shown in Fig. 3.3(d), with a strength statistically characteristic of that gauge length. The broken ends then retract and resume their original diameter, and will be held by the matrix. In order to prevent further opening of the crack, which will ultimately separate the two parts of the material, these broken ends must be pulled out of the matrix (Fig. 3.3(e)). Further frictional work is required to achieve this, and the resulting fracture surface will often have a brush-like appearance as shown in Fig. 3.5, characteristic of many fractured composites.

Approximate estimates for the work of pull-out, $W_{\text{pull-out}}$, are given by

$$W_{\text{pull-out}} = \frac{N\lambda r l_c^2}{6}. \tag{3.4}$$

The distance over which the fibre end is pulled out is approximately $l_c/4$. It can be shown that in aligned short fibre reinforced composites, the work of pull-out (and consequently the work of fracture) is a maximum when the reinforcing fibres are of exactly the critical length [13].

If the fibres are at an angle to the crack face, brittle fibres will fail prematurely without pull-out, but plastically deformable ones will undergo an extra work of shearing as they pull out, thus contributing extra toughness. These two effects of fibre length and fibre orientation are seen in Fig. 3.6.

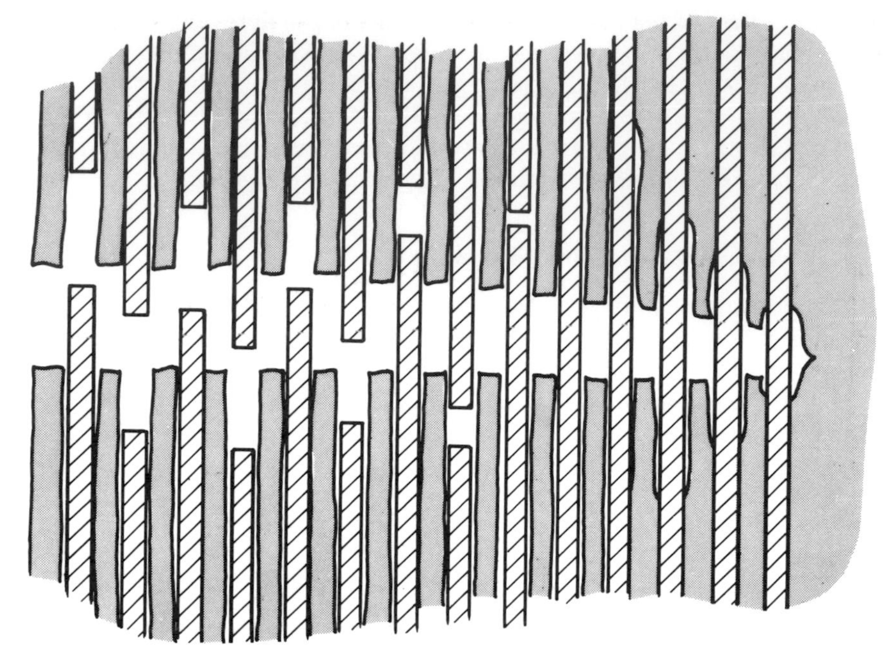

Fig. 3.5 Schematic of a crack tip damage zone.

During delamination a composite can fail in three ways:

- fracture through the matrix;
- fracture through the fibre–matrix interface;
- fracture of the fibres.

The energy for fracture of the fibres is much higher than for other fracture types, and composites which have good fibre–matrix adhesion are more likely to fail through the matrix. This was demonstrated by Barlow, Ward and Windle [16] who presented an optimized matrix where polyetherether-ketone (PEEK) particles and waves were adhering to the fibres and failure was through the matrix. The non-optimized matrix, however, revealed a clean fibre surface, indicating substantial fibre–matrix debonding and failure along the fibre–matrix interface.

The investigations of Chai [17] report that both types of failures may occur in the same fracture, mainly because of different fibre orientations, but also partially as a result of a variation in thickness of the matrix layer between the fibres. In multidirectional composites the direction of wave markings, characteristic of brittle fracture, is across the fibre, regardless of the delamination direction. The region for the brittle fracture mode, Chai proposes, is the interference of fibres on the ductile matrix; this has also been reported by Bascom *et al.* [3] for woven fibre reinforcements, which also tend to have a greater fracture toughness than unidirectional equivalents.

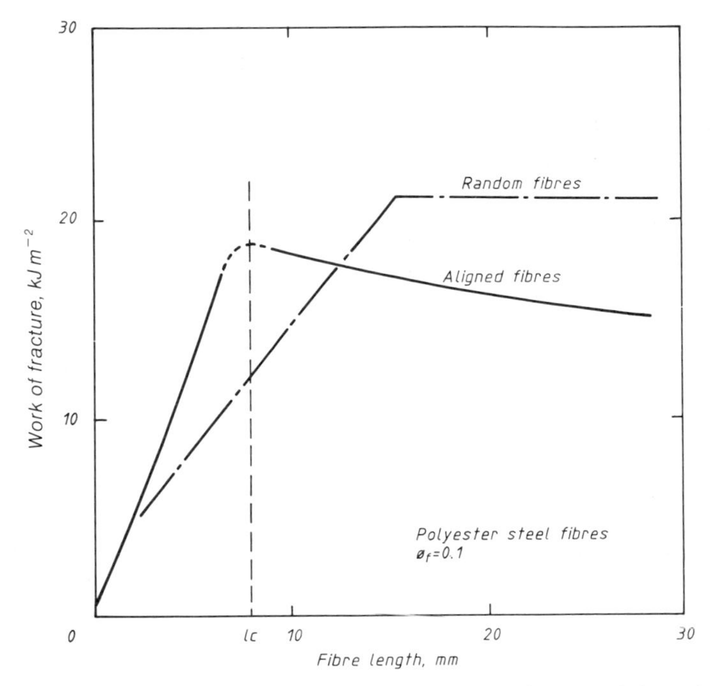

Fig. 3.6 Work of fracture of random and aligned composites consisting of chopped steel fibres ($\phi_f = 0.1$) in a polyester matrix. (Redrawn from B. Harris, *Engineering Composite Materials*; published by the Institute of Metals, 1986.)

The proposed explanation, by Greenhalgh and McGrath [18], for these increases in fracture toughness is twofold. Firstly, regions which have an increased volume fraction of matrix act as crack arresters and therefore deflect the crack, so increasing the fracture toughness because the crack path is more complicated. Secondly, to maintain the macroscopic volume fraction of fibres, it is necessary for regions of increased volume fraction of fibres to exist. The propagation of the crack through these regions requires more energy, because these are regions of plane stress and therefore need more energy to fracture. These deductions are only applicable to tough matrices because with a brittle matrix a deleterious effect will be produced.

The fibre bridging effect often found during delamination of short fibre composites was demonstrated by Wang, Suemasu and Zahlan [19]. The bridge took the form of fibres, or fibre bundles, bridging the gap between the two halves, tending to retard crack growth and so increasing fracture toughness. Fracture toughness was also increased through the fibre pull-out mechanism. Exaggerated fibre bridging was produced by the formation of subcritical cracks ahead of the main crack front and so a considerable increase in energy absorption arose. It was suggested that the subcritical crack formations might be related to multiaxial stresses induced by the main crack and local microscopic weak spots in the composite.

Thus, the fracture mechanism for unidirectional composites stressed in the fibre direction is substantially different to the mechanism for delamination between fibre layers.

The fracture toughening mechanisms in operation for the composite will depend on the nature of the reinforcement and its orientation. The three major types of reinforcement for composite materials are fibres, whiskers and platelets, the theory and application of which are discussed in detail by Rayson, McGrath and Collyer [20].

3.4 DAMAGE IN COMPOSITE MATERIALS

The continuing development of composite materials has produced a demand for a theoretical understanding of their mechanical properties. Both continuous and short fibres are currently available as reinforcements, in a comprehensive variety of matrices, to produce a diverse range of composite materials for widespread use. However, acceptance for use is often only achieved after a demanding and expensive evaluation programme. Progress in the understanding of the micromechanics of composite failure is thus advantageous from two specific aspects:

- to facilitate the development of enhanced material properties from an appreciation of existing materials;
- to provide a basis for the development of more efficient test procedures.

The type of damage in the composite material is critical to the consideration of the failure mechanism. Composite materials are generally of two distinct types, containing reinforcements of either a continuous or a discontinuous nature. A composite with continuous reinforcing fibres is generally used in the form of a laminate where the individual layers are arranged at an angle to the load.

The main consideration here is orientation of the layers of fibres within a structure. The orientation of each layer of a laminated structure is easily defined with respect to a control axis. To identify the construction of such a laminate it is necessary to indicate the ply orientation with respect to the control axis and the number of such laminates. Thus 0_4 indicates that four plies are laid up with their axes parallel to the control axis. By contrast, 90_4 indicates that the four plies have axes perpendicular to the control axis.

A set of parentheses indicates a repeated sequence, the numerical suffix being the number of times that sequence is repeated. A further suffix, s, indicates that the lay-up is symmetrical about the central axis, e.g. $(0, 90)_{2s}$ expands to

$$0, 90, 0, 90, \quad 90, 0, 90, 0.$$

Symmetry about a central axis is essential when constructing laminates from the anisotropic prepreg, otherwise buckling may result from thermally generated residual stresses.

In the case of a symmetric lay-up with an odd number of plies, the central ply which is not repeated is indicated by a bar, e.g. $(0, 90, \bar{0})_s$ expands to

$$0, 90, 0, 90, 0.$$

The quasi-isotropic laminate is popular, having approximately uniform properties in all directions in the plane of the sheet, e.g. $(+45, 90, -45, 0)_{Ns}$ or $(0, 90, +45, -45)_{Ns}$ where N is any number. For convenience a woven cloth is identified as 0/90 or ± 45.

The stresses in each ply during loading can be determined from laminated plate theory [21–23] which can be modified to include thermal stresses, and stresses arising from moisture uptake. Short fibre reinforced composite materials provide an additional complexity because both the fibre orientation and length may vary throughout the material. This is particularly true in injection mouldings, and predictions of material properties are necessarily more arduous.

3.4.1 Damage classification

When a composite laminate is considered, various damage mechanisms occur under load which produce a redistribution of the load and introduce a non-linearity into the stress–strain response. Cyclic loading produces a similar progression of damage, which can be defined in three categories:

- matrix cracking,
- delamination, and
- fibre breakage.

This is represented diagrammatically in Fig. 3.7.

The first significant damage is debonding of the fibres lying at an angle to the loading direction, followed by matrix cracking in a similar orientation. The density of these cracks intensifies as the load increases and appears to stabilize at a unique value for a given laminate, which has been defined as the characteristic damage state by Reifsnider and coworkers [24–26]. The mismatch of Poisson's ratio between adjacent plies produces matrix cracking in the 0° plies perpendicular to the fibre direction.

An important mode of failure in composite systems is delamination, resulting from interlaminar stresses which develop in a boundary layer region along the free edges of bonded dissimilar materials. Even though bonded dissimilar materials have been used for many years, delamination failures were not a serious problem prior to the development of advanced composites, primarily because the mismatch in material properties was small. With the advent of advanced composites this is no longer true. The variation in properties from ply to ply produces non-uniform internal stresses in laminates. This non-uniform distribution of internal stresses must be equilibrated by interlaminar stresses when free edges are present, and analysis shows that

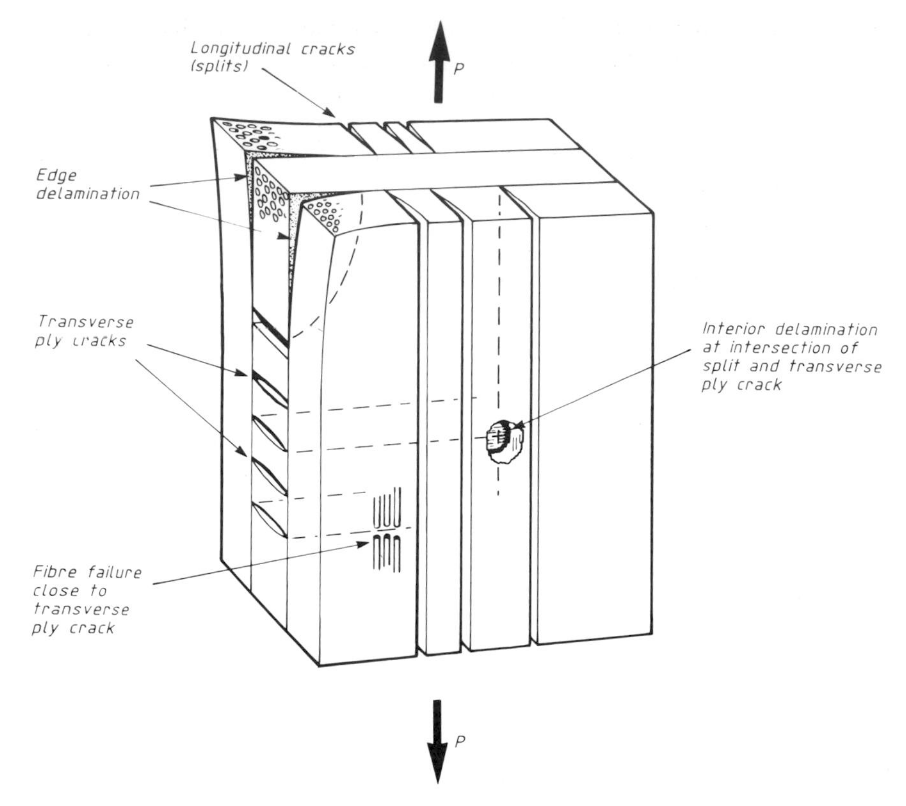

Fig. 3.7 Schematic of damage mechanisms in a cross-ply laminate.

these interlaminar stresses are concentrated in a boundary region along a free edge. It is the stress concentrations associated with these interlaminar stresses that may cause delamination failures at loads which otherwise would be well below the load capacity of the composite structure. Edge effects are present at all free edges. Thus, the failure characteristics of a laminate with a hole or other type of notch can be very dependent on interlaminar stresses in the region of the notch. The notched strength of a laminate does not exhibit the same degree of anisotropy as the strength of the unnotched laminate, because the edge effects around the notch vary continuously, whereas those along the straight side of the coupon are constant. Delamination failures can be more critical for compressively loaded laminates than for those loaded in tension. If delamination occurs, the effective moment of inertia of the cross-section is reduced with a corresponding reduction in the buckling load. However, if delamination occurs in a laminate under tensile load it is possible that the laminate will continue to carry the load.

Fibre fracture under static loading, prior to ultimate failure of the composite, is a function of the statistical distribution of flaws along the fibre length.

However, a stress concentration due to a matrix crack in an adjacent ply can cause fibre fracture, and this can be observed under cyclic loading. When fibre fracture occurs, maximum stress occurs in the adjacent fibres and thus the stress transfer becomes advantageous in a composite system containing a uniform distribution of fibres. The broken fibres are prone to debonding because of the high shear stresses developed at the fibre–matrix interface in the vicinity of the fracture.

Short fibre reinforced composites exhibit similar forms of damage: matrix cracking, fibre fracture and debonding. The interaction of fibre fracture and

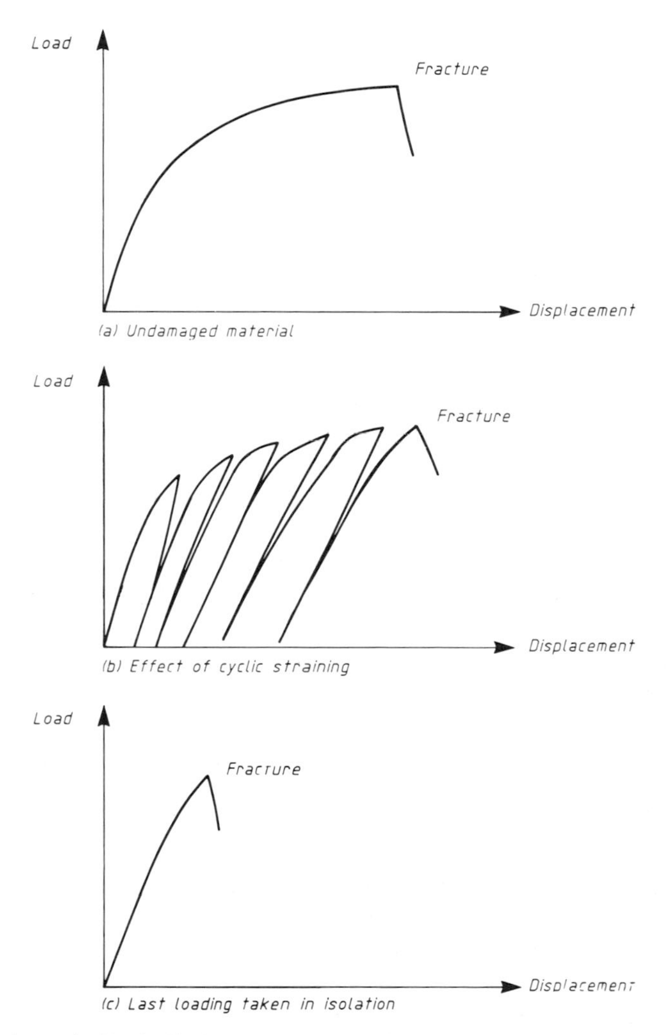

Fig. 3.8 Schematic load–displacement curves for ductility exhaustion effect: (a) undamaged material; (b) effect of cyclic straining; (c) last loading taken in isolation.

matrix debonding is much more important in these materials because of the increased volume fraction of fibre ends in the short fibre reinforced composite.

Damage which develops in a composite under load is a significant concern for subsequent structural integrity considerations, because of the critical changes in mechanical response so induced. The potential changes in mechanical response can be categorized broadly into three distinct areas:

- modulus reduction,
- hysteresis losses, and
- residual strength reduction.

The reduction in residual strength of the composite system during fatigue loading is common and the consequence is the proposal of a wear-out model of fatigue failure (Fig. 3.8). The residual strength of a composite is assumed to decrease gradually as damage accumulates, until the residual strength of the composite is lower than or equal to the maximum cyclic stress, whereupon failure occurs.

3.4.2 Damage accumulation in composite materials

Damage accumulation is a complex operation, but it is possible to identify certain general features which aid the prediction of damage mechanisms. Damage accumulation has three basic constituents: local, regional and global. The local damage state initiates in discrete regions which are non-interactive at first. As these zones develop and expand, the degree of interaction increases, first gradually and then eventually becomes severe enough to cause regional damage development, depending on the lay-up of the composite and the applied load. Further accumulation of the damage involves instability and loss of structural integrity. This is global damage, and is statistically governed. Currently, damage modelling describes only the local damage accumulation, and subsequent developments are not yet fully understood with sufficient clarity to perform realistic modelling.

The complexity of the composite response was demonstrated by Boniface and Bader [27]. The carbon fibre composite system under investigation had a lay-up of $(0_2/90_2/+45/-45)_s$ and was the subject of both fatigue loading and static failure. During the static failure process, the first observable indication of failure was at 0.4% strain in the form of matrix debonding around the fibres near to the 90/45 interface. At 0.8% strain transverse ply cracks developed, and at 1.0% strain the onset of delamination was observed at the 90/45 interface where the transverse ply crack intersected the $+45°$ ply. Matrix cracks in the $+45°$ ply developed above 1.0% strain, providing initiation sites for delamination in the $+45/-45$ interface.

Fatigue loading at 0.7% strain produced very similar damage accumulation, with fibre debonding close to the 90/45 interface after about 100 cycles.

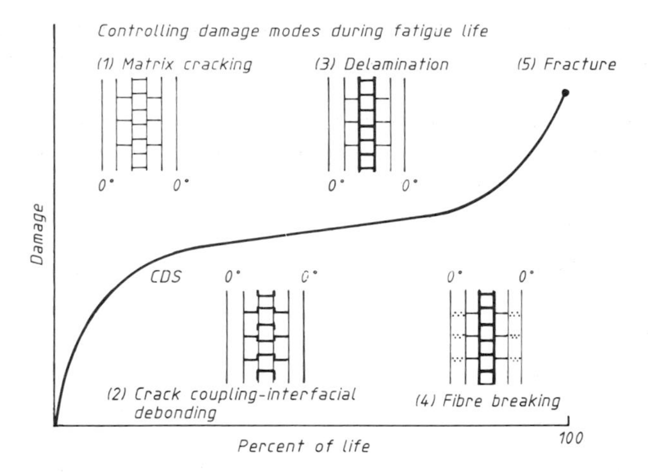

Fig. 3.9 Development of damage in composite laminates, where CDS represents the characteristic damage state. (Redrawn from K. L. Reifsnider *et al.*, *Mechanics of Composite Materials, Recent Advances*, 1983.)

Transverse ply cracks appeared along the length of the coupon after about 1000 cycles, initiating cracks in the adjacent $+45^\circ$ ply. Delaminations, associated with the matrix cracking, extended along the 90/45 and $+45/-45$ interfaces.

Damage accumulation in laminates is a complex series of interactions, but has been schematically represented by the work of Reifsnider *et al.* [28] (Fig. 3.9).

3.5 FAILURE OF LAMINATES

The micromechanical approach to modelling the failure behaviour attempts to determine the overall response of a body containing damage from the local stress or strain fields at a damage site.

3.5.1 Unidirectional systems

In order to calculate the tensile strength and modulus of a composite system it is possible to use the 'rule of mixtures'. This ignores any interaction between the constituents of the composite and is normally sufficiently accurate for engineering practice. When applied to modelling criteria, however, it is wrong to disregard interactions between the reinforcements. It is necessary, initially, to assume that a continuous fibre reinforced composite system has two basic constituents and, if the two constituents have the same strain to failure, then the strength of the unidirectional composite is the weighted sum of the strength of the fibres, σ_f, and the strength of the matrix, σ_m:

$$\sigma_c = \sigma_f \phi_f + \sigma_m (1 - \phi_f) \tag{3.5}$$

where ϕ_f is the fibre volume fraction. When considering cross-ply laminates, further complications arise, but similar considerations are possible with reference to the lay-up and ratio of orientations. However, the strains to failure are rarely equal and the strength of the composite is dependent on the load being supported by the intact constituent, which is a function of the volume fraction of the intact constituent. The two possible conditions for treatment are where the strain to failure of the matrix is less than the strain to failure of the fibre, and vice versa.

Where the strain to failure of the matrix is less than the strain to failure of the fibres and a high volume fraction of fibres is also present, as with many engineering composites, the fibres can sustain the load when the matrix fails and the final fracture depends on the fibre strength:

$$\sigma_c = \sigma_f \phi_f. \tag{3.6}$$

If the matrix remains bonded to the fibres then extensive fracture of the matrix will occur under increasing load. This is not often observed with fibre reinforced polymers but is a feature of brittle matrix reinforced systems, e.g. reinforced concrete and glass.

The more common situation for polymeric composites is when the strain to failure of the matrix is greater than the strain to failure of the fibres, and when the high volume fraction of fibres situation exists the matrix cannot sustain the applied load after fibre fracture. The composite failure stress is given by

$$\sigma_c = \sigma_f \phi_f + \sigma'_m (1 - \phi_f) \tag{3.7}$$

where σ'_m is the tensile strength in the matrix at failure.

These elementary equations neglect the strength variation associated with the reinforcement, and the individual fibres often fail at 50% of the actual composite strength. Two models by Rosen [29] and Zweben [30] have attempted to explain this phenomenon.

Rosen put forward the cumulative weakening model, in which individual fibre fracture produces a stress redistribution over the remaining unbroken fibres as discussed earlier. The final failure occurs when the remaining ligament, progressively weakened by the accumulating fractures, can no longer support the applied load. This predicts a higher composite strength than would be predicted using the mean fibre strength in the above expressions.

The model suggested by Zweben is more realistic because the stress redistribution is throughout the fibres adjacent to the fracture. This is the fibre break propagation model, and although prediction of the number of fibre fractures is possible, it is difficult to use the model to predict final failure. By comparing the theoretical predictions with experimental data, a failure criterion is possible, with the critical load represented by the occurrence of two fibre fractures in a specific group, but validation is difficult.

Unidirectional composite materials exhibit excellent fatigue resistance when stressed parallel to the fibre axes. In these materials, static strengths up to 3 GPa can be achieved, and although application is limited because of the poor mechanical properties in other directions, some use is made of composites with a high percentage of fibres in one direction, e.g. in helicopter blades. The fatigue behaviour of unidirectional composite materials might be expected to depend exclusively on the fibres, and as the fibres are not usually sensitive to fatigue loading, good fatigue behaviour should result. However, experimental evidence suggests that strain in the matrix determines the fatigue loading response [31, 32]. Therefore, mean strain representation of the cycles to failure is more useful information when considering composite systems. The statistical distribution of fibre strengths, discussed earlier, results in premature failure of individual fibres, promoting high local stresses in the matrix and at the fibre–matrix interface, leading to cumulative fatigue damage. Manufacturing defects including voids, resin-rich zones and misaligned fibres may also promote failure.

Recent developments have sought to increase the performance of the matrix and the fibres. Improvement of the fibre properties has very little effect on the fatigue life of the composite. However, development of tougher matrices for composite systems has usually resulted in poorer fatigue behaviour. Consequently fatigue behaviour in the next generation of composite systems may be somewhat more important than previously. The reduction in fatigue expectation appears to be associated with the poorer fatigue performance of the tougher matrices, but other failure processes may be involved.

The prediction of the mechanical properties of composite systems is the subject of much current research, which extends to the analysis of impact damage mechanisms. However, a further complication is introduced by the anisotropic nature of composites; the transverse tensile strength and the in-plane shear strength need to be addressed in conjunction with cross-ply laminates.

3.5.2 Cross-ply systems

Composite materials are usually employed in a laminated form, with plies arranged so that fibres are oriented in the principal load directions. Therefore, typically a large percentage of the plies will have fibres in the 0° direction to support tensile and compressive loads, a smaller percentage at ±45° to support torsional loads and perhaps even fewer at 90° to impart transverse rigidity to the structure and to reduce lateral contraction on axial loading. Increasing the non-axial fibre ratio in the laminate reduces the tensile modulus and strength.

Comprehensive investigations have been carried out on cross-ply laminates using micromechanical analysis techniques [32–36]. A major factor of such considerations is the initiation of cracks due to mechanical loading, and the associated complications from the superimposed residual thermal strains and Poisson mismatch strains. The thermal strains in carbon fibre reinforced composite systems can be very large and are superimposed on the applied mechanical strain:

$$\varepsilon_{22} = \varepsilon_m + \varepsilon_t \tag{3.8}$$

where ε_{22} is the transverse failure strain, ε_m is the applied mechanical strain and ε_t is the thermal strain in the 90° ply in the 0° direction.

Thermal stresses and strains are therefore important considerations, which originate from the mismatch of the coefficients of thermal expansion in the 0° and 90° directions and are greater in carbon fibre reinforced composite systems than in glass fibre reinforced composite systems. Their origin is readily demonstrated by the consideration of three discrete plies with the outer plies of constant length; if the laminate is free from stress at an elevated temperature, as the laminate cools the inner ply tries to contract, but is restricted by the outer plies. Consequently residual tensile stress is generated in the centre ply. Although the coefficient of thermal expansion for composite systems is not zero, the value in the 0° direction is considerably lower than that in the 90° direction and residual tensile stress results in the 90° ply parallel to the 0° direction [34]:

$$\varepsilon_t = \frac{\varepsilon_{11} b(\alpha_{11} - \alpha_{22})\Delta T}{E_{22} d + E_{11} b} \tag{3.9}$$

where E_{11} is Young's modulus parallel to the fibre, E_{22} is Young's modulus perpendicular to the fibre, α_{11} is the coefficient of thermal expansion parallel to the fibre, α_{22} is the coefficient of thermal expansion perpendicular to the fibre, ΔT is the temperature change from the stress-free temperature, b is the thickness of the 0° ply and $2d$ is the thickness of the 90° ply.

Therefore, $(\alpha_{11} - \alpha_{22})\Delta T$ must be known to calculate the thermal strains for a cross-ply laminate; Bailey, Curtis and Parvizi [34] abridged this difficulty by use of practical measurements on a (0, 90) laminate. On cooling, a curvature develops, as previously discussed, because of coupling, and

$$\Delta T(\alpha_{11} - \alpha_{22}) = \frac{p}{12R}\left(\frac{E_{11}}{E_{22}} + \frac{E_{22}}{E_{11}} + 14\right) \tag{3.10}$$

where R is the radius of curvature and p is the ply thickness.

Thus, thermal strains are fabrication considerations, while the Poisson mismatch strains arise from the application of mechanical loads. When a load is applied to a cross-ply laminate, the 0° plies try to extend parallel to the direction of the applied load, and contract perpendicularly to this. The contraction is restricted by the 90° plies and tensile stresses result in the 0°

ply, perpendicular to the fibre, which tends to split the 0° plies. These strains are larger in glass fibre reinforced composite systems than in carbon fibre reinforcement systems.

Crack initiation begins with debonding at or near the fibre–resin interface, followed by the combination of debonds to produce a transverse ply crack and then by an increased crack density until the system has a regular array of cracks, with a progressive reduction of modulus.

The strain at which cracks appear decreases as the individual transverse ply thickness increases and it is thus advantageous to reduce the ply thickness as far as practically possible. This phenomenon can be explained by fracture mechanics or energetics approaches and, from a practical stance, crack constraint is vital since matrix cracking can be suspended or prevented totally by the use of very thin plies to construct the laminate.

Plies with off-axis fibres are more easily damaged in fatigue because of their increasing dependence on matrix properties and cross-ply composite laminates are therefore less fatigue resistant (Fig. 3.10). Indeed, transverse loaded plies can develop transverse cracks in the first tensile fatigue cycle. However, because they do not support major axial loads, this has little effect on axial modulus or strength of the composite. The transverse ply crack density increases with subsequent cycles and, once the cracks are initiated, they propagate across the laminate width, with fibre fracture in the 0° ply adjacent to the matrix cracking. The residual strength of the laminate falls with cycling, and delamination at the 0/90 interface can occur as a result of free edge stresses and stresses associated with transverse ply cracking.

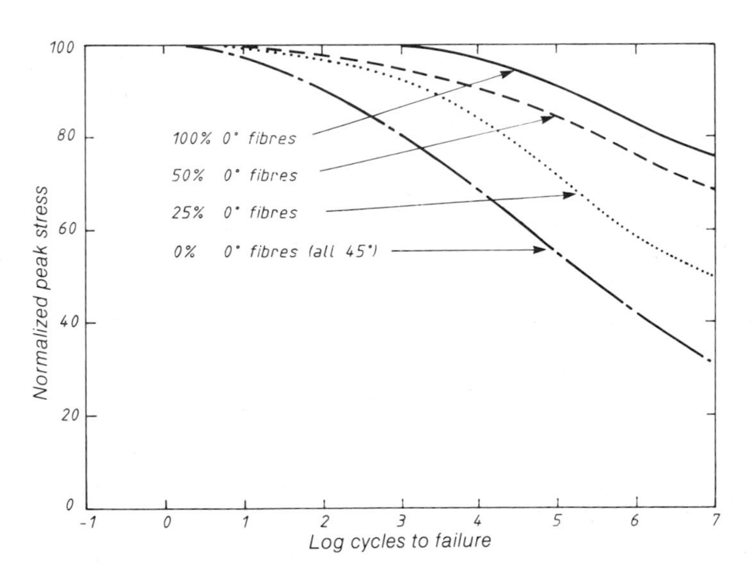

Fig. 3.10 Normalized $S-N$ curves for CFRP laminates with varying percentages of 0° fibres.

A further consideration for angled ply laminate, with fibres typically at ±45°, is the development of intraply damage which causes reductions in strength and stiffness. Stress concentrations at the end of such cracks can result in delamination initiating between the plies, usually causing the decoupling of the 0° principal load bearing layers, which may produce a general loss of structural integrity. Alternatively the cracks may propagate into adjacent load bearing layers and seriously weaken the material. Ultimate tensile fatigue failure is still determined by the 0° plies, and thus the tensile S–N curves are still relatively shallow.

The progressive failure of glass reinforced composite systems can be identified also. Initially, there is a rapid reduction in modulus, associated with increasing transverse ply crack density. Fibre fracture may occur in the 0° plies. The increase in crack density occurs much more slowly after this initial phase, and the modulus reduction is linear. Simultaneously, edge delamination at the 0/90 interface and fibre fracture in the 0° ply may be happening. Small whitened regions may be observed, which contain large numbers of broken fibres. The whitening is probably the product of matrix damage or fibre–matrix debonding initiating at the fibre fracture sites. Finally, a rapid modulus reduction is associated with the continued growth of a whitened area to form a macrocrack, which propagates prior to ultimate failure.

Therefore, crack initiation and propagation is of paramount importance to the successful application of composite materials and techniques are necessary to assess the delamination resistance of these materials.

3.6 FRACTURE TOUGHNESS

The use of fracture mechanics to analyse the failure of composite materials has been treated with scepticism, because of the incorrect deduction that it is unsuited to dealing with the complicated fracture process found in composite systems. The subject is still in its infancy as applied to composite materials, but nevertheless has found acceptance as a method for characterizing the toughness of laminates in terms of their weakest link, delamination. Fracture mechanics has been used to define toughness in laminates both in mode I and mode II.

The subject of fracture toughness has developed from the basic concept of Griffith [37] that the strength of a brittle material is governed by small defects that act as stress concentrators.

If an artificial crack is introduced, a measurement of the strength should allow the fracture surface energy to be calculated, but the experimentally derived values for many materials are orders of magnitude higher than would be expected from calculations based on atomic or molecular constants. Cracks growing in normally brittle plastics seem to be associated with fracture surface energies of about $1\,\mathrm{kJ\,m^{-2}}$ whereas estimates based on assumptions of chain scission would be in the region of $1\,\mathrm{J\,m^{-2}}$. This discrepancy, coupled with the presence of red and green coloured regions on the fractured surface of PMMA, led Berry [38] to postulate that there is appreciable plastic flow at the tip of a crack growing in a polymer in its glassy state. This promotes the formation of a plastic zone at the crack tip which toughens the material in three ways:

- it dissipates the high local stresses to some degree;
- it absorbs more energy than would be required for growth of a brittle crack of the same length;
- it dissipates energy, because energy absorbed in the plastic zone is not recoverable as it would be from elastic deformation, and therefore this energy is not available to support further crack growth.

3.6.1 Mode I characterization

In respect of damage tolerance it is resistance to delamination which is of major importance and this has been the basis of research founded on tests such as the double cantilever beam (DCB) [39], where the energy required to propagate a crack between two plies can be measured directly.

The investigations of thermoplastic composite were originated by Hartness [40] and Carlile and Leach [41] on carbon fibre reinforced PEEK, and produced results for the interlaminar strain energy release rate which were an order of magnitude greater than for epoxy systems. These results have since been substantiated by Donaldson [42] and others. The resistance to delamination appears to be associated with the ductile character of PEEK. Carlile and Leach indicate that the high resistance to crack propagation is one reason for the good damage tolerance as exemplified by the retention of compressive strength after impact.

As discussed previously, the matrix may be tough, but this toughness is not always realized in the composite; indeed short fibre additions to a tough matrix may result in a moulding which will shatter on impact [10]. The most pronounced problems are the geometry of the resin phase and the constraints imposed upon the resin by the fibres.

Fracture mechanics suggests that the thin layers will be tougher than the thick ones, which prompted a hypothesis for a criterion for toughness in composites [43]. This is based on comparing the plastic zone size of the resin phase with the characteristic thickness of the resin layers. However, the criterion is beset with difficulties as these thin layers are subject to thermal stresses induced during fabrication, causing warp in an unbalanced lay-up.

The DCB test is basically the controlled propagation of a pre-positioned crack by means of mode I stressing. Data so collected can be used to assess G_{IC}, the critical strain energy release rate; there have been numerous data reduction methods put forward which are reviewed here.

(a) The area method

This method is based on the definition of G_{IC} proposed by Irwin [44]:

$$G_{IC} = -\frac{1}{b}\frac{du}{da} \tag{3.11}$$

where u is the total strain energy stored in the test specimen, b is the crack width and a is the crack length. du is the change in strain energy due to crack extension from a to $a + da$, the area between loading and unloading curves (dA) on a load–displacement plot from the DCB test:

$$-du = dA + \tfrac{1}{2}(P\,d\delta - \delta\,dP) \tag{3.12}$$

where P is the applied load and δ is the resultant deflection. Combining equations (3.11) and (3.12) gives

$$G_{IC} = \frac{1}{b}\frac{dA}{da} = \frac{1}{2b}\left(\frac{P\,d\delta}{da} - \frac{\delta\,dP}{da}\right) \tag{3.13}$$

This same approach is also applicable to the case of finite crack extension and provides the basis for a straightforward data reduction scheme.

A typical DCB plot is presented in Fig. 3.11(a): at P_1, the crack starts extending, and simultaneously the load falls off to P_2, whereupon the beam is unloaded, the crack having grown from a to $a + \Delta a$. The loss of strain energy due to this crack extension is the area, ΔA, between the loading and unloading curves. For cases in which the load–deflection curve during crack propagation can be approximated by a straight line one can determine the critical strain energy release rate, G_{IC}, from the relationship

$$G_{IC} = \frac{1}{b}\frac{\Delta A}{\Delta a} = \sum_{i=1}^{N} \frac{1}{2b\,\Delta a}(P_1\delta_{2i} - P_2\delta_{1i}). \tag{3.14}$$

An average value of G_{IC} can be determined by measuring P_1, P_2, δ_1 and δ_2 for

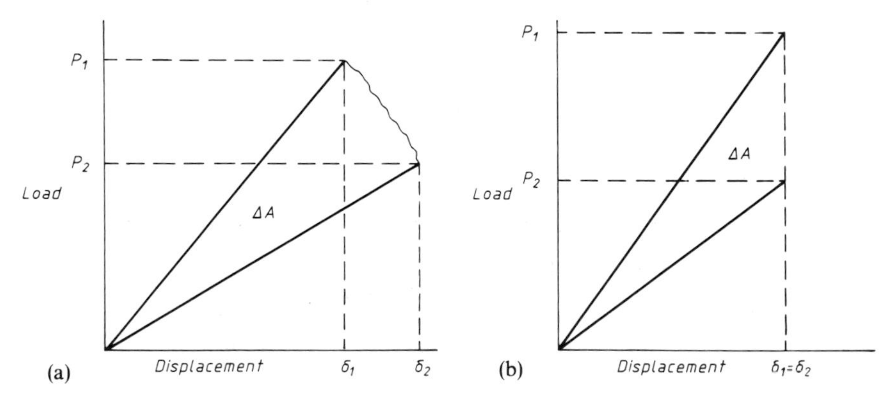

Fig. 3.11 Diagrammatic representation of the calculation for G_{IC}.

a series of N crack extensions of length Δa. Thus

$$G_{IC} = \frac{1}{2bN\Delta a}(P_1 - P_2). \tag{3.15}$$

Therefore, in the case of Fig. 3.11(b) when the crack propagates by cleavage, the load decreases instantaneously, with no increase in deflection:

$$G_{IC} = \frac{1}{2b\Delta a}\delta(P_1 - P_2). \tag{3.16}$$

When the loading and/or unloading curves are non-linear then the G_{IC} values can be determined by measurement of the area between the two curves.

(b) Compliance method based on linear beam theory

As discussed by Wang, Suemasu and Zahlan [19] this method only applies if the material is linear elastic and the beam displacement is small.

The theory assumes that each cracked half behaves as a conventional beam as illustrated in Fig. 3.12 and therefore

$$\delta = BPa^3 \tag{3.17}$$

where B is a constant defined by

$$B = \frac{64}{E_x^b bh^3} \tag{3.18}$$

where E_x^b is the effective bending modulus in the axial direction of a cantilever beam of thickness $h/2$.

Combination of equations (3.13) and (3.17) gives

$$G_{IC} = \frac{3BP^2a^2}{2b} = \frac{3P\delta}{2ba}. \tag{3.19}$$

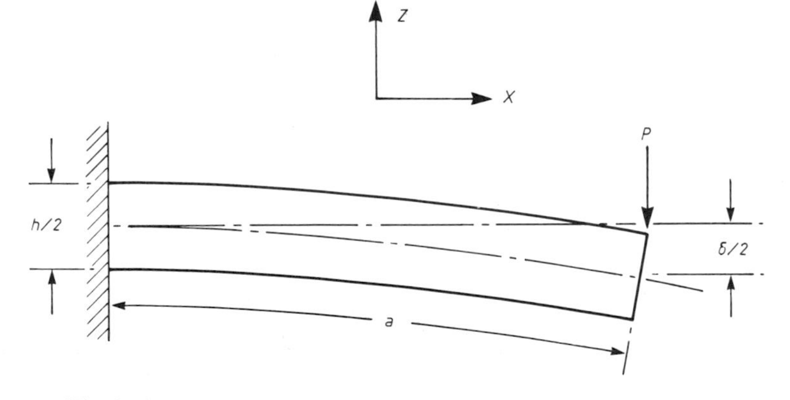

Fig. 3.12 Cantilever beam model for analysis of DCB specimens.

Therefore, for stable crack growth, as is usually found in displacement controlled tests on ductile materials, G_{IC} is given by

$$G_{IC} = \frac{3BP_c^2 a^2}{2b} \tag{3.20}$$

or

$$G_{IC} = \frac{3P\delta_c}{2ba} \tag{3.21}$$

where P_c is the critical applied load and δ_c is the critical displacement.

It is also possible to incorporate shear deformation into the calculation. This is of particular interest when considering orthotropic materials such as unidirectional reinforced composites. Applying the beam theory of Timoshenko [45] equation (3.17) becomes

$$\delta = BPa^3(1 + S) \tag{3.22}$$

where S is the expression for shear deflection

$$S = \frac{3E_x^b h^2}{32G_{xz}a^2} \tag{3.23}$$

where G_{xz} is the interlaminar shear modulus relative to the x–z plane. Substitution into equation (3.13) gives

$$G_{IC} = \frac{3BP^2a^2}{2b}\left(1 + \frac{S}{3}\right). \tag{3.24}$$

This requires a knowledge of E_x^b and G_{xz}. However, a precise value of these laminate properties is difficult to obtain; the ratio E_x^b/G_{xz} can be estimated allowing estimation of S. If this is the case a more conventional form of G_{IC} can be obtained by combination of equations (3.22) and (3.24):

$$G_{IC} = \frac{3P_c\delta_c}{2ba}\frac{1 + S/3}{1 + S}. \tag{3.25}$$

An estimated value of S allows use of a data reduction scheme as described in conjunction with equation (3.21).

Large deflections complicate the analysis and equations (3.20) and (3.21) are no longer valid. Devitt, Schapery and Bradley [46] proposed a non-linear beam analysis method.

(c) Generalized empirical analysis

Once again this is based on compliance, and a generalized form of equation (3.17):

$$\delta = RPa^n \tag{3.26}$$

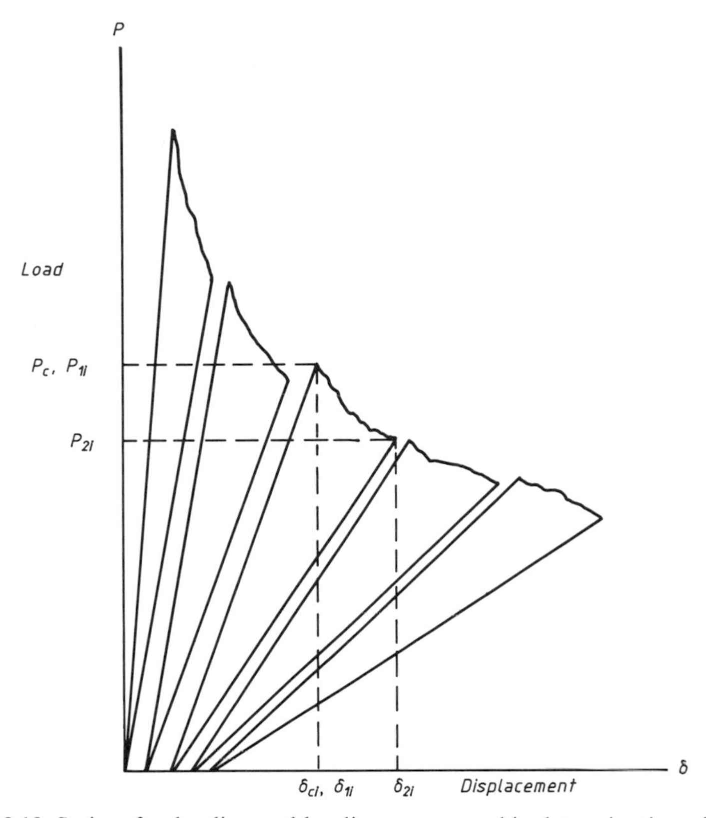

Fig. 3.13 Series of unloading and loading curves used in determination of G_{IC}.

where R and n are constants determined experimentally from the relationship

$$\log(P/\delta) = -\log R - n \log a. \tag{3.27}$$

A least-squares fit to equation (3.26) for a series of loading and unloading curves, as shown in Fig. 3.13 allows R to be determined. Obviously, if $n = 3$ beam theory is recovered.

Substitution into equation (3.13) of (3.26) produces

$$G_{IC} = \frac{nRP_c^2 a^{n-1}}{2b} \tag{3.28}$$

or

$$G_{IC} = \frac{nP_c\delta_c}{2ba}. \tag{3.29}$$

This method was successfully applied by Chai [17] to DCB results on graphite epoxy laminates with different lay-ups. Values of n ranged from 2.74

to 2.97. Using the same method, Whitney, Browning and Hoogsteder [39] obtained a value for n of 3.74 for a different unidirectional epoxy system. This large value was probably due to damage at the crack tip.

Wang, Suemasu and Zahlan [19] compared these methods and put forward an alternative reduction method based on a non-linear analysis. This method is applicable to non-linear load–displacement curves and large displacements. The geometrically non-linear analysis was based on the Ritz method [47]. The derivation is complex and its inclusion is not relevant here. The result is

$$G_{IC} = -1\left(\frac{\partial u}{\partial a}\right)_{\delta} \tag{3.30}$$

$$\delta = \text{constants}$$

Wang believed non-linear conditions largely invalidated the area and linear compliance methods, and speculated that the differences in G_{IC} values for the same method may be due to difficulty in measuring crack length and viscoelastic effects. The problems of viscoelasticity are discussed by Devitt, Schapery and Bradley [46].

(d) Double cantilever beam tests on composite materials

There have been a number of studies on the mode I delamination fracture toughness of continuous fibre composites. Carlile and Leach [41], Whitney, Browning and Hoogsteder [39], Devitt, Schapery and Bradley [46] and Chai [17] were in agreement that a DCB is suitable for studying the interlaminar fracture toughness of continuous fibre reinforced composite systems. Wang, Suemasu and Zahlan [19] and Bascom *et al.* [3] also employed DCB methods to study random short fibre and woven composites respectively.

Only the approach of Bascom *et al.* was markedly different, in that they used a width tapered beam, maintaining a constant crack length-to-width ratio. The general expression for G_{IC} is

$$G_{IC} = \frac{P^2}{2b}\frac{dc}{da} \tag{3.31}$$

where for ideal elastic beams

$$\frac{dc}{da} = \frac{24a^2}{E_B h^3 b} \tag{3.32}$$

where E_B is the bending modulus of the composite and h is the beam height. If the specimen is width tapered for a constant ratio of a/b, G_{IC} can be determined from P_c and E_B alone, being independent of crack length, because

$$G_{IC} = \frac{12P^2a^2}{E_B h^3 b}. \tag{3.33}$$

This avoids the problem of accurate crack length measurement. However, to offset this advantage, a knowledge of E_B is required. Also, more material is needed and fabrication problems are encountered with fibre reinforced composites as directionality is critical.

Barlow and coworkers [16, 48] presented an interesting variation in the approach to DCB theory. A razor blade was forced into the composite normal to the fibre axis, thus creating a split and hence a cantilever beam. G_{IC} is given by

$$G_{IC} = \frac{3E_B d^3 h_b^2}{8(a')^4} \tag{3.34}$$

where h_b is the blade thickness, d is the section thickness, a' is the effective crack length and

$$E_B = E\frac{a^2}{a^2 + 3d^2}. \tag{3.35}$$

The measured crack length, a, is related to the effective crack length, a', by a graphically determined factor:

$$a' = a + \Delta a \tag{3.36}$$

where Δa is material dependent and may be positive or negative.

This work is particularly important because it highlights the need for an optimized matrix for maximum fracture toughness. Chai [17], for example, found that in the quasi-isotropic material he tested, the fibre orientation made considerable differences to G_{IC}.

Wang, Suemasu and Zahlan [19] found that the different beam thicknesses played a considerable part in the value of G_{IC}, ranging from 875 N m^{-1} for 5 mm beams to 2000 N m^{-1} for 14 mm beams. He proposed this difference to be a result of variation in lateral constraints and the likelihood of different properties due to longer curing times for thicker beams. Even when measuring G_{IC} for a constant beam thickness, calculated values differed considerably, e.g. for a beam thickness of 11 mm, the mean G_{IC} value was 1700 N m^{-1}, the standard deviation being 672 N m^{-1}. Devitt, Schapery and Bradley [46] also reported a variation in results (a range of G_{IC} from 500 to 1000 N m^{-1}). This demonstrates the inherent problems of characterizing composites, as properties may vary from one batch to another. As expected, the fracture toughness of thermoplastic matrix composites is higher than that of corresponding thermoset matrix composites.

Two types of fracture are evident in the load–displacement plots presented by these workers: firstly, a 'saw tooth' type where cracks propagate in uncontrolled steps representing brittle failure; secondly, a steady crack growth where the load falls steadily with crack extension representing ductile failure.

3.6.2 The fracture mechanics of delamination tests

Thus, a wealth of dedicated research has been devoted to the subject of delamination and its assessment by fracture mechanics. The experience and methods of analysis used have produced some very precise tests, and these will form the basis of design methodology. The analysis based on beam theory, as described, is generally very accurate because many laminates can be tested as elastic, slender beams as required by beam theory. However, several complications may occur because of the nature of the material under test. The tougher composite systems, when tested in this manner, often undergo large displacements during crack growth which cause alterations in geometry, especially the shortening of the bending arms, and hence errors in the calculations. This results in non-linear load–deflection curves, and correction factors are required. These can be determined from large deflection beam theory [49, 50].

A further factor is that the shear modulus of laminates is much lower relative to the axial modulus than in isotropic materials. In the latter case the ratio E/G is approximately 2.6, while for laminates it may be up to 50. This produces a much greater relative contribution of shear to the displacements and a modification of the correction factors. This is especially true for this type of analysis and is addressed by Williams [51]. Further complicating effects are observed with reference to the geometry, rate and temperature and these criteria are addressed by Hashemi, Kinloch and Williams [52].

The application of conventional elastic beam analysis can provide accurate solutions for the calculation of G_{I} and G_{II} components of test geometries. However, a complicated interaction of the various modes exists at the crack tip when growth occurs and there is a great deal of research required to clarify such mechanisms.

3.7 CONCLUDING REMARKS

Recent developments have led a rapid development in the applications of composite materials and the ability to characterize their behaviour. Sophisticated techniques for stress analysis now exist; micro- and macrochemical analysis provides understanding of the relationship between fibre, matrix and interface. However, this understanding remains incomplete and it is not yet possible to predict the properties of the composite system from the properties of the constituents. The intricate failure mechanisms are one of the major restraints to completing this understanding and a committed, interdisciplinary approach is vital for completion of the picture.

The interaction of non-destructive testing, failure analysis, micromechanics, fracture mechanics, damage development, delamination and fractographic analysis will provide an analytical background for the continuing expansion of composite technology.

REFERENCES

1. Bersch, C. F. (1982) What we have done. Proc. Critical Review: Techniques for Characterisation of Composite Materials, Army Materials and Mechanics Research Centre, MA. *AMMRC MS Report 82-3*, pp. 487–9.
2. Scott, J. M. and Phillips, D. C. (1975) Carbon fibre composites with rubber toughened matrices. *Journal of Materials Science*, **10**, 551–62.
3. Bascom, W. D., Bitner, J. L., Moulton, R. J. and Siebart, A. R. (1980) The interlaminar fracture of organic-matrix, woven reinforcement composites. *Composites*, **1**, 9–18.
4. Bascom, W. D., Ting, R. Y., Moulton, R. J., Riew, C. K. and Siebart, A. R. (1981) The fracture of an epoxy composite containing elastomeric modifiers. *Journal of Materials Science*, **16**, 2657–64.
5. Vanderkley, P. S. (1981) Mark I–mode II delamination fracture toughness of a uni-directional graphite/epoxy composite. Masters Thesis, Texas A & M University, College Station, TX, December 1981.
6. Cohen, R. N. (1982) Effect of resin toughness on fracture behaviour of graphite/epoxy composites. Masters Thesis, Texas A & M University, College Station, TX, December 1982.
7. Bradley, W. L. and Cohen, R. N. (1983) Matrix deformation and fracture in graphite reinforced epoxies. ASTM Symposium – Delamination and Debonding of Materials, Pittsburgh, PA, November 8–10, 1983.
8. Kanninen, M. F., Rybicki, E. F. and Brinson, H. F. (1977) A critical look at current applications of fracture mechanics to the failure of fibre reinforced composites. *Composites*, **1**, 17–22.
9. Wang, S. S., Chin, E. S., Yu, T. P. and Goetz, D. P. (1983) Fracture of random short fibre SMC composite. *Journal of Composite Materials*, **17**, 299–315.
10. Folkes, M. J. (1982) *Short Fibre Reinforced Plastics*, Research Studies Press, Wiley, New York.
11. Blumentritt, B. F., Vu, B. T. and Cooper, S. L. (1975) Fracture in orientated short fibre reinforced thermoplastics. *Composites*, **5**, 105–14.
12. Harris, B. (1986) *Engineering Composite Materials*, Institute of Metals.
13. Cooper, G. A. (1970) The fracture toughness of composites reinforced with weakened fibres. *Journal of Materials Science*, **5**, 645–54.
14. Piggot, M. R. (1970) Theoretical estimation of fracture toughness of fibrous composites. *Journal of Materials Science*, **5**, 669–75.
15. McGrath, G. C. (1988) Structure and properties of carbon fibre reinforced aromatic thermoplastics. PhD Thesis, Sheffield City Polytechnic.
16. Barlow, C. Y., Ward, M. V. and Windle, A. H. (1985) The influence of microstructure on the toughness of carbon fibre/plastic composites. Proceedings 6th International Conference on Deformation Yield and Fracture of Polymers, April 1985, pp. 14.1–14.4.
17. Chai, H. (1984) The characterisation of mode I delamination failure in non woven, multi-directional laminates. *Composites*, **15**, 277–90.
18. Greenhalgh, E. S. and McGrath, G. C. (1991) Fracture analysis of thermoplastic welds. 1st International Conference on Deformation and Fracture of Composites, Manchester, March 25–27, 1991.
19. Wang, S. S., Suemasu, H. and Zahlan, N. M. (1984) Interlaminar fracture of random short fibre SMC composite. *Journal of Composite Materials*, **18**, 574–94.
20. Rayson, H. W., McGrath, G. C. and Collyer, A. A. (1986) Fibres, whiskers and flakes for composite applications, in *Mechanical Properties of Reinforced Thermoplastics* (eds D. W. Clegg and A. A. Collyer), Applied Science, London, Chapter 2.

21. Jones, R. M. (1975) *Mechanics of Composite Materials*, McGraw-Hill, New York.
22. Ashton, J. E., Halpin, J. E. and Petit, P. H. (1969) *Primer on Composite Materials: Analysis*, Technomic, Stamford, CT.
23. Greaves, L. J. (1987) Stiffness matrices of a carbon fibre cloth laminate. *RAE Technical Report 87047*.
24. Reifsnider, K. L. and Talung, A. (1980) Analysis of fatigue damage in composite laminates. *International Journal of Fatigue*, **2**, 3–11.
25. Reifsnider, K. L. and Highsmith, A. (1981) Advances in fracture research. ICF5, Cannes, Vol. 1.
26. Reifsnider, K. L. and Jamison, R. (1982) Fracture of fatigue-loaded composite laminates. *International Journal of Fatigue*, **4**, 187–97.
27. Boniface, L. and Bader, M. G. (1986) The micromechanics of damage initiation and development under static and fatigue loading of CFRP XAS/914 and E-glass laminates. *D/ERI/9/4/2064/066 XR/MAT*.
28. Reifsnider, K. L., Henneke, E. G., Stinchcomb, W. and Duke, J. C. (1983) Damage mechanics and NDE of composite laminates. *Mechanics of Composite Materials, Recent Advances*, **19**, 399–420.
29. Rosen, B. W. (1964) Tensile failures of fibrous composites. *AIAA Journal*, **2**, 1985–91.
30. Zweben, C. (1968) Tensile failures of fibrous composites. *AIAA Journal*, **6**, 2325–31.
31. Curtis, P. T. (1986) A comparison of the mechanical properties of improved carbon fibre composite materials. *RAE TR 86021*.
32. Talreja, R. (1981) Fatigue of composite materials: damage mechanisms and fatigue life diagrams. *Proceedings of the Royal Society of London, Series A*, **378**(1775), 461–75.
33. Bailey, J. E. and Parvizi, A. (1981) On fibre debonding effects and the mechanism of transverse-ply failure in cross ply laminates of glass fibre/thermoset composites. *Journal of Materials Science*, **16**, 649–59.
34. Bailey, J. E., Curtis, P. T. and Parvizi, A. (1979) *Proceedings of the Royal Society of London, Series A*, **366**, 599–623.
35. Hashin, Z. (1985) Analysis of cracked laminates: a variational approach. *Mechanics of Materials*, **4**, 121–36.
36. Ogin, S. L. and Smith, P. A. (1987) A model for matrix cracking in cross-ply laminates. *ESA Journal*, **11**, 45–60.
37. Griffith, A. A. (1921) The phenomena of rupture and flow in solids. *Philosophical Transactions of the Royal Society of London, Series A*, **221**, 163–98.
38. Berry, J. P. (1961) Fracture processes in polymeric materials: I. The surface energy of polymethylmethacrylate. *Journal of Polymer Science*, **50**, 107–15.
39. Whitney, J. M., Browning, C. E. and Hoogsteder, W. (1982) A double cantilever beam test for characterising mode one delamination of composite materials. *Journal of Reinforced Plastics and Composites*, **1**, 297–313.
40. Hartness, J. T. (1982) Polyetheretherketone matrix composites. 14th National SAMPE Technical Conference, October 12–14, 1982, pp. 26–37.
41. Carlile, D. R. and Leach, D. C. (1983) Damage and notch sensitivity of graphite/PEEK composites. 15th National SAMPE Technical Conference, October 1983, pp. 82–93.
42. Donaldson, S. L. (1985) Fracture toughness testing of graphite/epoxy and graphite/PEEK composites. *Composites*, **16**(2), 103–12.
43. Blundell, D. J. and Osborn, B. N. (1983) The morphology of poly(aryl-ether-ether-ketone). *Polymer*, **24**, 953–58.
44. Irwin, G. R. (1957) Analysis of stresses and strains near the end of a crack traversing a plate. *Journal of Applied Mechanics*, **24**, 361–4.

45. Timoshenko, S. D. (1955) *Strength of Materials*, Part 1, 3rd edn, Van Nostrand, New York.
46. Devitt, D. F., Schapery, R. A. and Bradley, W. L. (1980) A method for determining the mode I delamination fracture toughness of elastic and viscoelastic composite materials. *Journal of Composite Materials*, **14**, 270–85.
47. Crandall, S. H. (1956) *Engineering Analysis*, McGraw-Hill, New York.
48. Barlow, C. Y. and Windle, A. H. (1985) Razor blade test for composites toughness. *Journal of Materials Science Letters*, **4**, 233–4.
49. Williams, J. G. (1987) Large displacements and end block effects in the DCB interlaminar test in modes I and II. *Journal of Composite Materials*, **21**, 330–48.
50. Williams, J. G. (1987) Large displacement effects in the DCB test for interlaminar fracture in modes I and II. Proc. ICC MVI/ECCM, Vol. 3, Applied Science, Barking, pp. 233–42.
51. Williams, J. G. (1989) End corrections for orthotropic DCB specimens. *Composite Science and Technology*, **35**, 367–79.
52. Hashemi, S., Kinloch, A. J. and Williams, J. G. (1990) The effects of geometry, rate and temperature on the mode I, mode II and mixed mode I/II interlaminar fracture of carbon-fibre/polyetheretherketone composites. *Journal of Composite Materials*, **24**, 918–56.

4

Methods of measurement and interpretation of results

A. Savadori

4.1 INTRODUCTION

Rubber toughened engineering polymers are materials designed for conditions requiring good impact resistance at room and low temperature. Many years ago it was discovered that the fracture resistance of polystyrene (PS) was increased by adding some amounts of rubber. This discovery became of paramount interest in tailoring the microstructure and the related property profile [1] and led to the commercialization of the ever-broadening family of high impact thermoplastics.

Now there are many rubber toughened grades of commodities and engineering plastics. The inclusion of rubber in polymers improves the impact resistance but reduces the elastic modulus and the yield stress. Moreover, rubber toughening generally affects the total mechanical behaviour of the material.

A qualitative evaluation of rubber toughened polymers indicates that phase separation between the polymer and the rubber is a predominant requirement, and mechanical resistance increases if the rubber has the following properties:

- low elastic modulus in relation to the matrix, good adhesion to the matrix, adequate crosslinking;
- average particle size and distribution optimized;
- low glass transition temperature.

For example, recent work of Wu [2] has also pointed out, with reference to the Izod impact strength at room temperature of several toughened polymers, that the optimum phase morphology for toughening correlates with two chain parameters of the matrix, and for the same amount of rubber the extent to which a matrix can be toughened depends on them.

The simplest evaluation of mechanical performance involves two tests, one associated with stiffness (elastic modulus) and the other with toughness (for

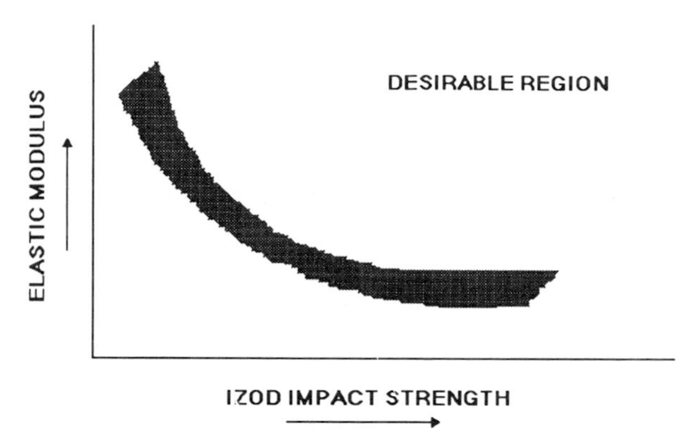

Fig. 4.1 Schematic diagram of relationship between elastic modulus and toughness (Izod impact strength). (Reproduced with permission from S. Turner, *Mechanical Testing of Plastics*; published by Longman, 1983.)

example, Izod impact strength). If these two characteristics are plotted against each other (Fig. 4.1) [3], the mechanical performance of a given material can be estimated. The right-hand region in the diagram represents the generally desirable place.

Of course, this rudimentary approach allows only the comparison of different materials. In fact Izod impact strength and elastic modulus depend on many parameters and are not fully independent of each other. The fracture energy is indeed in some way influenced by the elastic behaviour of the material (more details are given in the appropriate sections).

An improvement is the evaluation of the ductile–brittle transition temperature (DBTT), which has long been accepted as the measure of ductility of steels by impact tests. Typical results for low strength structural steel are indicated in Fig. 4.2 [4]. The energy absorbed in fracture is increasing from a lower value typical of brittle fracture to a higher value typical of ductile fracture. The surface fracture analysis is used together with energy measurements. Two 'transition temperatures' have been indicated in this figure:

- the NDT is the temperature at which the energy curve first begins to rise;
- the FATT is the temperature at which the fracture is 50% brittle and 50% ductile.

The rubber toughened polymers present the same behaviour and the position of the curve with reference to the T axis is an indication of the effectiveness of rubber toughening (better if shifted to the left). An extensive analysis of the DBTT of plastics can be found in the book *Failure of Plastics* [5], in particular with reference to the influence of free volume. The position of this curve with reference to the T axis is influenced by many factors:

- matrix and rubber parameters (intrinsic variables);

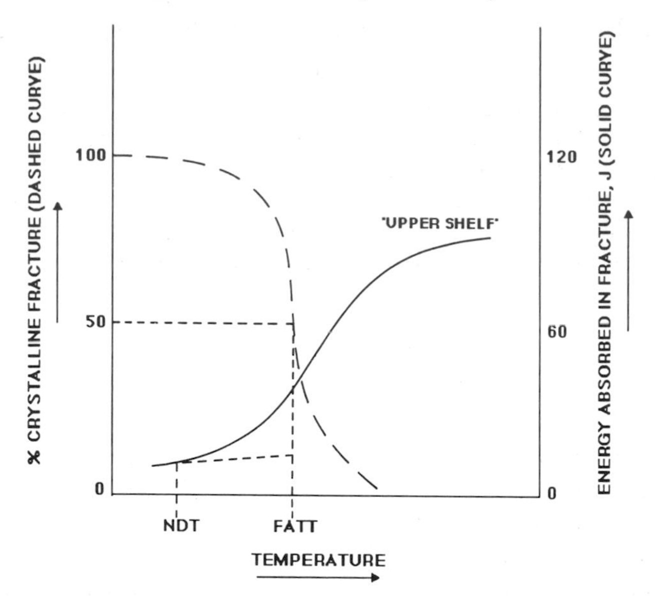

Fig. 4.2 Schematic impact transition curves (NDT, nil ductility temperature; FATT, fracture appearance transition temperature). (Reproduced from J. F. Knott, *Fundamentals of Fracture Mechanics*; published by Butterworth-Heinemann Ltd. ©.)

- processing conditions and aging;
- loading time;
- specimen geometry and loading mode.

The last group of factors requires a more appropriate approach if the mechanical potential of rubber toughened engineering polymers is to be evaluated for design purposes. This is the reason for the increasing attention to fracture mechanics and in general to the continuum mechanics theories in order to create a rational framework for the methods of measurements of toughness.

4.2 BASIC MECHANICAL PARAMETERS

4.2.1 Elastic moduli

The elastic modulus represents the intrinsic capacity of the material to resist an applied load and is defined as the ratio between the stress and strain. In an isotropic material there are three elastic constants, two of which can be considered as being independent of each other: E, the Young's or tensile modulus, G_M, the shear modulus, and v, Poisson's ratio. (A more rigorous analysis would require also the bulk modulus K, which describes volumetric strain: $K =$

$E/3(1-2\nu)$.) These constants are connected by the following relationship:

$$E = 2G_M(1+\nu). \tag{4.1}$$

Thus, the elastic behaviour of an isotropic material can be described by the relationships between stress (σ) and strain (e) and two independent elastic constants (E and G_M, or E and ν).

For viscoelastic materials, such as polymers, these two parameters are functions of temperature and time (frequency, loading rate). With reference to the elastic modulus in shear (G) and in tension (E) the most widely used methods for their evaluation are [6]

- dynamic–mechanical analysis,
- creep and relaxation tests, and
- stress–strain tests.

Poisson's ratio, which reflects material response to a change in shape and volume, is determined through stress–strain tests using a second extensometer for measuring the lateral contraction. For polymers the value of this parameter ranges from 0.35 to 0.42 [6].

The elastic modulus from creep and relaxation tests can also be determined from data from dynamic–mechanical tests but at very low strain rates. The elastic modulus in each case can be compared using the relationship based on linear viscoelasticity. However, for heterogeneous materials, such as rubber toughened polymers, whose phases generally exhibit different viscoelastic behaviours, the interconversion methods of viscoelastic functions cannot be easily applied [7].

Another useful technique is the rebound test proposed by Casiraghi [8]. It can be performed on the same specimen to be used for impact testing. The time between the ends of a typical rebound curve reported in Fig. 4.3 is the half-period to be used in the following formula for bending geometry:

$$E = M_e \frac{L^3}{48I} \frac{\pi^2}{t_r^2}. \tag{4.2}$$

In the case of heterogeneous blends the elastic response depends on blend morphology, on the degree of molecular mixing or interpenetration and on the size of the phase separated regions as well as on the molecular relaxation processes characteristic for the blend constituents.

There are three principal groups of models for predicting the composition dependence of the modulus [9]:

1. mechanical coupling models, which imply the mechanical properties of the constituents;
2. 'self-consistent' models, derived by analysis of the stress deformation behaviour around the inclusions;
3. models giving threshold values (limits or bounds) for the modulus.

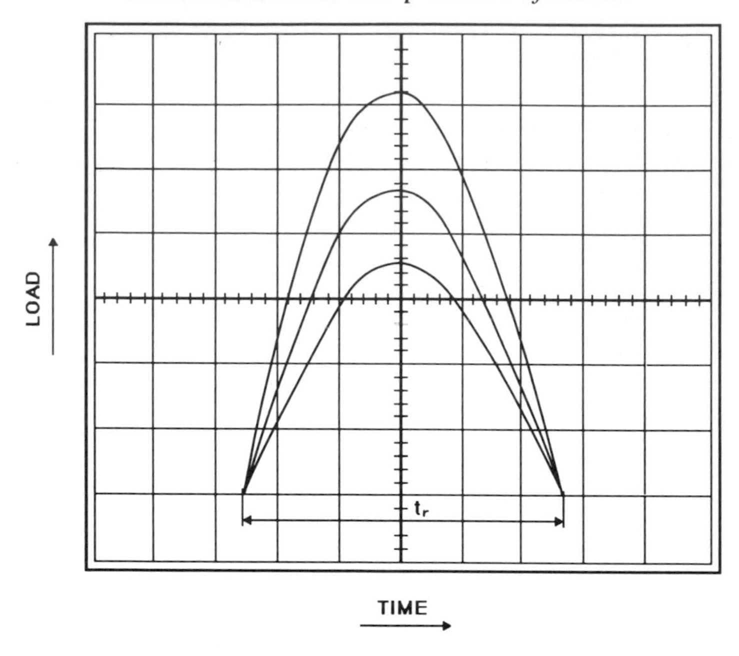

Fig. 4.3 Examples of typical rebound curves.

Hwang *et al.* [10] found in the case of rubber toughened epoxies that both Takayanagi's [11] and Hashin and Shtrikman's [12] models, belonging respectively to groups 1 and 3, show greater deviations from the experimental results than Kerner's model [13], represented by the following equation

$$\frac{E}{E_c} = \frac{E_d\psi_d/A + \psi_c/(15 - 15v_c)}{E_c\psi_d/A + \psi_c/(15 - 15v_c)} \tag{4.3}$$

where

$$A = (7 - 5v_c)E_c + (8 - 10v_c)E_d$$

E is the modulus of the composite, E_c the modulus of the continuous phase, E_d the modulus of the dispersed phase, v_c Poisson's ratio of the continuous phase (taken as 0.35), ψ_c the volume fraction of the continuous phase and ψ_d the volume fraction of the dispersed phase.

Hsu and Wu have shown very recently [14] that the tensile properties of plastic–rubber blends depend critically on the morphology and connectivity of the two phases. There is a transition (percolation transition) from low plastic concentration to high plastic concentration whose exact value depends on the morphology of the blend.

4.2.2 Yield and fracture stress

Depending on experimental conditions and material composition, the stress–strain curve of rubber toughened engineering polymers may show a maximum which can be identified as yield stress. This maximum as shown in Fig. 4.4 is strongly dependent on temperature and strain rate. Increasing the strain rate or decreasing the temperature normally results in the embrittlement of the polymer and the yield stress is then replaced by the brittle stress. The presence of a rubber phase and the matrix morphology can influence the shape of the stress–strain curve and this modifies the behaviour.

With reference to high impact PS (HIPS), Bucknall [15] proposed the following equation for the yield stress as a function of a rubber phase volume and temperature:

$$\sigma_y = (1 - 1.21\psi^{2/3})\frac{kT}{v}\left(\log\frac{\dot{e}}{A_y} + \frac{\Delta H}{kT}\right). \quad (4.4)$$

Data in tension and compression in HIPS from Oxborough and Bowden [16] confirm this trend in composition but the predicted values are lower than the experimentally determined results.

A classical scheme for this behaviour is that reported in Fig. 4.5 in which it is possible to identify a temperature range under which the yield strength (σ_y) is replaced by the brittle strength (σ_B) of the material. The position of this transition from yield strength to brittle strength is also sometimes called the DBTT.

Since the brittle strength increases with strain rate less than the yield strength, the brittle–ductile transition will be shifted to higher temperatures by increasing the speed or, as we shall see later, by notching the specimen to be impacted. The yield strength indicates a higher elongation which results in an increase in energy. The phenomenon of yielding in rubber toughened polymers is complex. In many cases the material is heterogeneous and therefore the

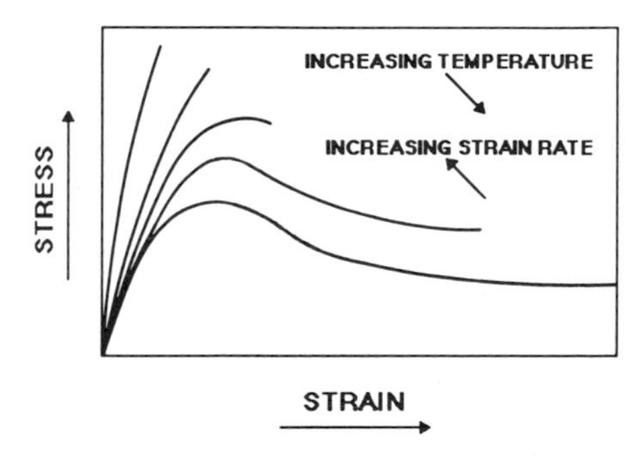

Fig. 4.4 Tensile stress–strain curves of a ductile plastic, showing schematically the effect of strain rate and temperature.

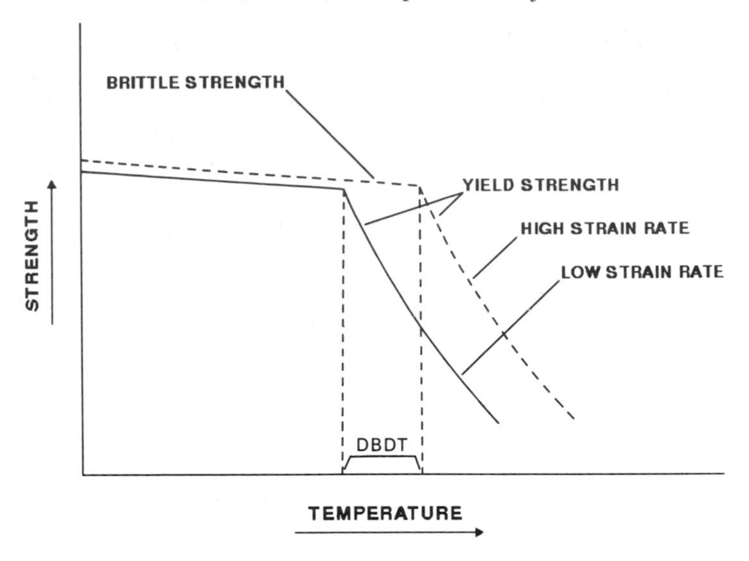

Fig. 4.5 Schematic diagram of effect of temperature and strain rate on brittle strength and yield strength.

stress on a macroscopic level is different from that on a microscopic level, i.e. in the vicinity of rubber particles.

At present, two methods are used to improve the toughening of certain plastics:

- understanding the micromechanism before yielding in simple stress situations and using the information ascertained to redesign the material with improved toughness;
- application of the theories of continuum mechanics which are currently used for steel and other construction materials.

For the first case, according to many researchers [17–21], the determination of volume strain provides a satisfactory quantitative basis for a more complete understanding of this behaviour. Contrary to the method used by Bucknall, which relied on creep tests, Heikens and coworkers [19, 20] have immersed specimens in water and measured displacement of the water in the capillary during tensile tests under constant strain rate. The latter method appears to be less reliable. Neverthless, it permits the evaluation of the volume change at higher strains before and after yielding. The mechanics of deformation can be studied from a plot of volume strain vs tensile deformation (e). The slope of the line indicates the mechanism of deformation. A slope of 1.0 or near 0 signifies that deformation is occurring by multiple crazing or by shear yielding respectively.

The work of Truss and Chadwick on ABS [18] has shown that the slope of the volume strain (ΔV) vs percentage elongation increases with increasing strain rate (Fig. 4.6). This suggests that the mechanism changes from one that is predominantly shear yielding to one which is crazing. The latter mechanism becomes more important under various experimental conditions. The method described above can be used to study more complex situations, such as stress gradients around notches. However, the application of this method to the study of multiaxial stress states is open to question.

Using continuum mechanics to study the behaviour of amorphous polymers it can be shown that the following relationships apply [22].

- The shear yield stress is related to the shear modulus by

$$\tau_y = C_G G_M \tag{4.5}$$

  where C_G ranges from 0.03 to 0.13.
- The shear yield stress is also dependent on the hydrostatic component of the stress tensor. The modified Von Mises criterion is

$$\tfrac{1}{\sqrt{6}}[(\sigma_1 - \sigma_2)^2 + (\sigma_2 - \sigma_3)^2 + (\sigma_3 - \sigma_1)^2]^{1/2} = \tau_y - \mu P. \tag{4.6}$$

- The yield stress in tension is less than that in compression and the difference ranges between 10% and 20%.
- In specific cases, e.g. at low temperature or high strain rate, the above yield criterion should be replaced by the Sternstein and Ongchin criterion [23] in the tensile stress region.

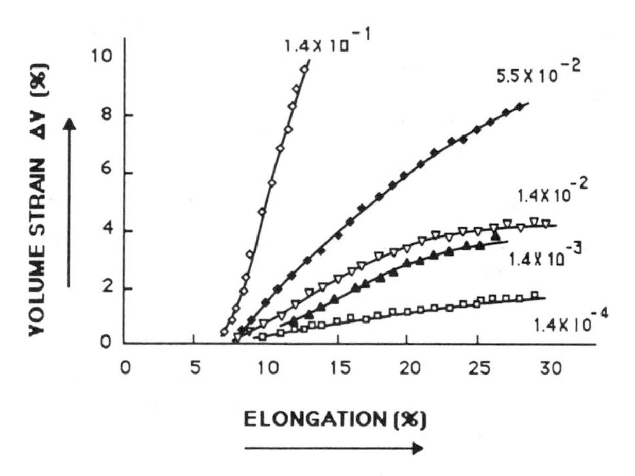

Fig. 4.6 Relationship between ΔV and percentage elongation, showing the mechanism of tensile deformation as a function of strain and strain rate. (Reproduced with permission from R. W. Truss and G. A. Chadwick, *Journal of Materials Science*; published by Chapman & Hall, 1976.)

Sultan and McGarry [24] found that the yield criterion of toughened rubber epoxies is

$$\tau_{oct} = \tfrac{1}{3}[(\sigma_1 - \sigma_2)^2 + (\sigma_2 - \sigma_3)^2 - (\sigma_3 - \sigma_1)^2]^{1/2}$$
$$= \tau_y - \mu P \qquad (4.7)$$

This relationship is consistent with equation (4.6). In this case two pressure coefficients were necessary to describe the failure locus: $\mu = 0.210$ in the first quadrant (tension–tension) and $\mu = 0.175$ in the rest. The value 0.175 remains constant on changing to a smaller rubber particle size.

4.3 A TRADITIONAL APPROACH TO STRENGTH EVALUATION

4.3.1 Static tests

These tests are important for determining the basic mechanical parameters for engineering design. The standard method for measuring the strength of polymers is the tensile test which is normally carried out at constant crosshead speed. Tests can be carried out in bending, torsion and compression. The parameters which can be obtained from such tests are yield stress, strength (brittle and yield), stress at breakage and the related strains. Tensile, compression and torsion tests are complementary from the point of view of the mechanical description. Nevertheless, one has to be cautious with respect to the heterogeneity of the stress field when comparing data for rubber toughened polymers. Among the static tests the creep failure test represents the strength evaluation of the material at constant load (it can also be called the static fatigue test). The parameters and properties that could be measured are strain vs time, time to rupture and type of fracture (brittle or ductile). When the test is made by monitoring the crack growth (natural or artificial) the approach used is based on the fracture mechanics concept. The creep behaviour in the linear range can be analysed using the theory of viscoelasticity but in the non-linear range other approaches ought to be used (e.g. micromechanics evaluation, semiempirical treatments). The Andrade law proposed by Kausch [25] can usually be used to describe the non-linear range.

4.3.2 Fatigue tests

The important aspects of the failure of materials under cyclic loading have been known for a long time. Under these conditions failure loads can be lower than those under monotonic load.

As reported by Bucknall [26], the amount of information available about fatigue properties of specific polymer grades is at present rather limited, in particular with respect to the rubber toughened materials. It seems, however, that the improvements in impact strength produced by particle toughening result in better resistance to other forms of fracture loadings including fatigue fracture.

Three principal mechanisms contribute to fatigue failure:

- thermal softening;
- excessive creep or flow;
- initiation and propagation of cracks.

Following the book of Hertzberg and Manson [27] fatigue tests can be classified into five different types according to the stress–strain control:

- periodic loading (between fixed stress limits, in tension or compression);
- periodic loading (between fixed strain limits, in tension or compression);
- reversed bending stress (flexing in one dimension);
- reversed bending stress in two dimensions (by rotary deflection of a cylindrical specimen);
- reversed shear stresses (obtained by torsional deformation).

Nevertheless, the majority of the fatigue tests reported in the literature have been carried out by controlling the stress amplitude during the test and the response is often presented in the form of S (average stress)–N (number of cycles) curves.

In the S–N curves of several polymers (e.g. PC) a peculiar fatigue lifetime inversion was observed. One possible explanation of this phenomenon is

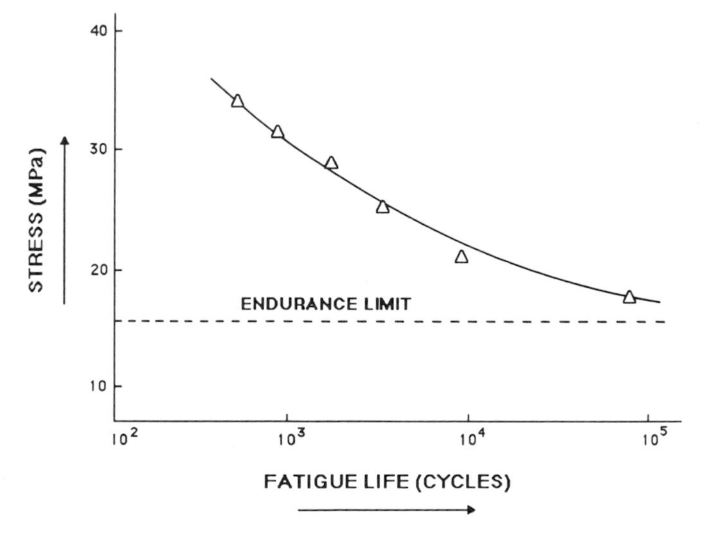

Fig. 4.7 Stress amplitude vs number of cycles to failure for ABS samples tested at 0.2 Hz: the endurance limit should be noted. (Reproduced from J. A. Sauer and C. C. Chen, *Advances in Polymer Science*; published by Springer, 1983.)

based on a ductile–brittle transition during fatigue [28]. Another argument is that the polymer (e.g. PC) undergoes shear deformation and crazing simultaneously at the crack tip. This process is reflected by the formation of a unique epsilon type plastic zone the characteristics of which were reviewed recently [29]. The curve for ABS [30] is reported as an example in Fig. 4.7. Below a certain stress level, the so-called endurance limit, no failure is expected.

Depending on the experimental conditions, failure of the samples can occur through thermal softening and/or plastic flow. The stress and frequency range where thermal failure is expected can be predicted from energy input, geometry and thermal and viscoelastic properties of the material. According to Ferry [31] the energy dissipation rate can be described by

$$\dot{U} = \pi f J''(f, T)\sigma^2 \tag{4.8}$$

where $\dot{U}$ is the energy dissipation rate per unit time, f the frequency, J'' the loss compliance and σ the peak stress. The Ferry equation can also be expressed in terms of temperature change per unit time:

$$\Delta\dot{T} = \pi f J''(f, T)\sigma^2/\rho C_p \tag{4.9}$$

where C_p is the specific heat and ρ the density.

4.3.3 Impact testing

Impact testing is a common laboratory practice for evaluating the impact strength of polymers and, by extension, their toughness.

At present the most well-known methods are

- Izod (ASTM D 256, ISO R180),
- Charpy (DIN 53453, ASTM D256, ISO R179),
- falling weight (FW) or driven dart (DIN 53443, ASTM D3029, D1709), and
- tensile impact (DIN 53448, ASTM D1822).

The impact resistance is evaluated in energy terms, i.e. the energy absorbed by the specimen during the impact process, and is given by the difference between the potential energies of the hammer or striker before and after impact. In the case of a falling weight statistical analysis has been used. The weight (probit method) or the height (staircase method) is changed and failure or survival can be examined in order to determine the energy for 10%, 50% and 90% of failure.

By changing the test temperature the evaluation can be extended for the determination of the DBTT in combination with the nature of the fractured surface, as is currently used with steel [4]. The probit and staircase procedures are time consuming, they require large numbers of specimens and at present have been substituted by instrumented tests. In the past the main reason for the use of regular impact tests was to characterize the materials (steel at that time) at very high deformation rates and also to study the effect of notches as underlined by Wullaert [32]. This may explain why Charpy or Izod tests are so widely used for characterizing materials sensitive to strain rate, and in particular for rubber toughened products. The FW and tensile impact tests are also performed at high strain rates, but in the first case the stress state is mainly biaxial and in the second case uniaxial instead of triaxial [32] as in Charpy and Izod tests. The latter tests represent in many cases more critical conditions. FW tests may result in DBTT values different from those of Charpy and Izod owing to some degree of anisotropy, or orientation in the specimens, striker geometry or speed used. The main advantages of FW and tensile impact tests are the detection of inherent anisotropy of parts and the measurement of the tensile properties at high strain rates respectively.

With respect to the application of traditional impact analysis to rubber toughened engineering plastics, a good example is found in the work of Wu on toughened nylon resins [33]. As reported in Fig. 4.8 a sharp brittle–ductile transition is found at a critical particle size at constant rubber volume fraction and adhesion between the rubber and matrix. There are many references in the literature which pertain to the effect of rubber content and particle size on the DBTT [34] in pseudo-ductile matrices. Nevertheless, Wu's interpretation relates only to the interparticle distance (matrix ligament thickness) whose critical value correlates with one of the chain parameters regarding the intrinsic ductility of the matrix. The related model is still under discussion

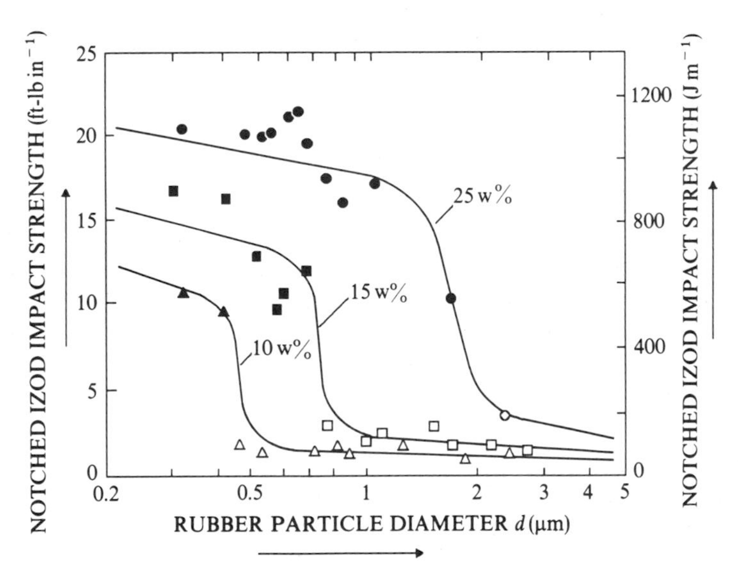

Fig. 4.8 Notched Izod impact strengths vs number-average diameter particles for PA66–rubber blends. (Reproduced from S. Wu, *Polymer*, 1985, by permission of the publishers, Butterworth-Heinemann Ltd. ©.)

[35, 36]. However, the conclusions are not general and only relate to the Izod impact geometry (see Chapters 2 and 7).

An important step towards the comprehension of impact test phenomenology was the introduction of force–displacement transducers in impact machines [37] and the application of fracture mechanics [38].

The instrumented impact test has permitted the measurement of the force during the impact process and the application of standard formulae for supported and clamped beams and plates [37]. Fracture mechanics has 'rationalized' the data from different testing methods.

The nature of the dynamic character of the impact test, however, has to be taken into consideration in order to interpret the data using the transducer. Williams [39] has examined this problem and concluded that at low loads (brittle behaviour) the dynamic problem is very significant with regard to the validity of the results. The calculation of the energy from force and time in instrumented impact testing can be made through equations connecting the impulse to the change of momentum according to the relationship

$$\int_0^t F\,\mathrm{d}\tau = M_e(V_0 - V_F). \tag{4.10}$$

Since the energy absorbed in the impact process is related to the variation of the kinetic energy, i.e.

$$\Delta U_0 = \tfrac{1}{2}M_e(V_0^2 - V_F^2) \tag{4.11}$$

we have

$$\Delta U_0 = V_0 \int_0^t F\,\mathrm{d}\tau \left(1 - \frac{V_0 \int_0^t F\,\mathrm{d}\tau}{4U_0}\right) \tag{4.12}$$

where F is the load, M_e is equivalent to the striker mass, V_0 is the striker speed before impact, V_F is the striker speed after impact, ΔU_0 is the energy absorbed by the impact process and U_0 is the striker energy.

In the case of FW tests this relationship has to be modified with respect to the influence of gravity on the final speed. With reference to the curve shown in Fig. 4.9, pertaining to Izod and Charpy tests, the following information can be obtained:

- load vs time;
- energy vs time;
- displacement vs time;
- load vs displacement;
- energy vs displacement.

With regard to displacement, one has to consider the possible compliance of the testing apparatus. Before testing, compliance has to be determined for

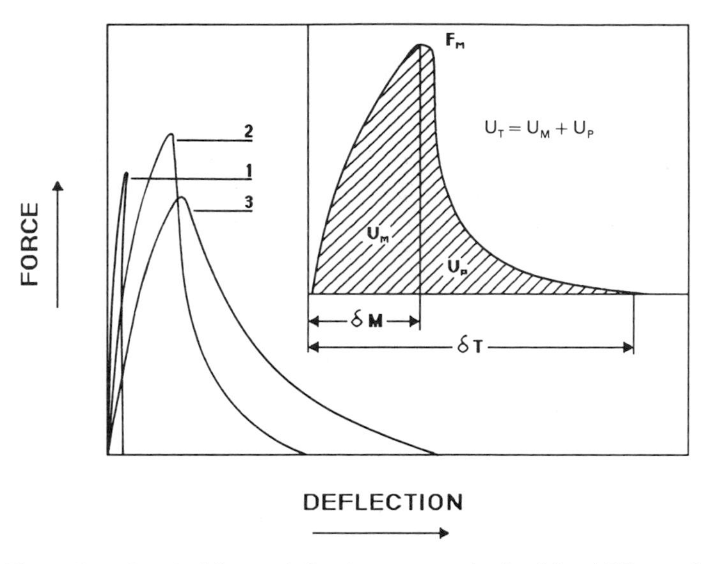

Fig. 4.9 Examples of typical force–deflection curves obtained for ABS samples: curve 1, brittle fracture; curve 2, ductile fracture; curve 3, intermediate behaviour. (Reproduced from W. Lubert, M. Rink and A. Pavan, *Journal of Applied Polymer Science*; published by Wiley Interscience, 1976.)

high stiffness specimens in order to avoid misinterpretation of the specimens' deflection during the impact.

Tests on polymeric materials can generate different curves depending on speed, temperature, clamping systems and the particular equipment used. With reference to Lubert, Rink and Pavan's work on ABS [40] the load–time curve, or better the force–displacement curve, in the case of instrumented Izod or Charpy tests can be characterized by the following: F_M, the maximum force developed, which in a tensile test could be converted to yield stress and/or tensile strength; δ_M or t_M, the displacement or time corresponding to F_M; U_M, the energy absorbed up to F_M; U_T, the total energy absorbed during the whole test; δ_T or t_T, the total displacement or time. The FW and tensile tests can have similar curves (Fig. 4.9). A more precise description of impact curves and the formula may be found in Ref. 37. The load–time curve from an FW test is difficult to interpret if the material is not brittle [41]. Studies by Braglia and Casiraghi [42] have shown also that a peculiar texture of craze is obtained under biaxial loading (FW) in comparison with triaxial loading (sharply notched tensile impact test). Simple plate bending theory could be used for evaluation of the strength of plates of brittle material [43, 44]. The impact strength for brittle material can be represented by

$$U = ABr_c^2\sigma_c^2 \qquad (4.13)$$

where

$$A = \frac{3(3+v)(1+v^2)}{8(1+v)E}\left[(1+v)\,(0.485\log\frac{r_c}{B}+0.52)+0.48\right]^{-2}, \quad (4.14)$$

$$\sigma_c = [(1+v)\,(0.485\log\frac{r_c}{B}+0.52)+0.48]\frac{F}{B^2}, \quad (4.15)$$

B is the plate thickness, σ_c is the critical value of the maximum stress and r_c is the clamping radius, and a linear relationship between FW impact strength and plate thickness can be defined.

It is important to emphasize the presence of Young's modulus E and Poisson's ratio v in these formulae and the influence of these material constants on impact performance. Rink *et al.* [45] described the changes of F_M, U_T, U_p and U_M with rubber content ψ and temperature T for ABS (Izod tests). In Figs. 4.10 and 4.11, constructed from the data of Rink *et al.*, there are maxima in energy and force owing to a transition in the failure mechanisms. The interpretation of this behaviour could be as follows: the force at the peak F_M goes through a maximum with rubber content because of an increasing amount of plastic deformation at the notch tip. This reduces the stress concentration and a higher load is therefore required for breaking the specimen. As the number of particles increases, the breakage of the specimen becomes more controlled by the yield stress, which decreases with rubber content. To support this view, comparative experiments with unnotched specimens were conducted as reported by Rink *et al.* [45]. The curve of F_M vs temperature can be explained as follows:

- at low temperature F_M decreases with temperature because this is the trend of the brittle strength of the material;

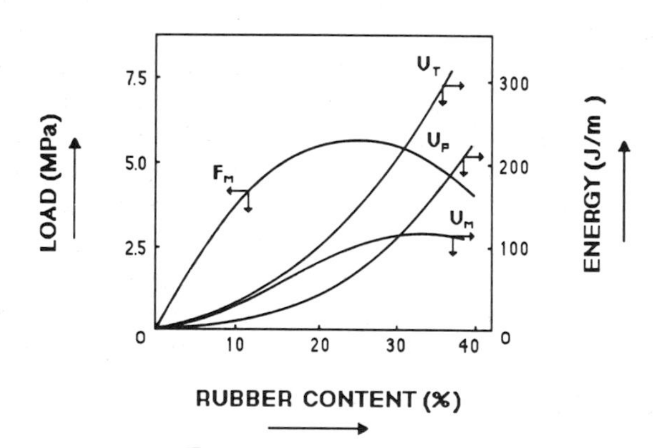

Fig. 4.10 Force F_M and energies U_M and U_T vs rubber content in impact tests on notched samples. (Reproduced from M. Rink, T. Riccō, W. Lubert and A. Pavan, *Journal of Applied Polymer Science*; published by Wiley Interscience, 1978.)

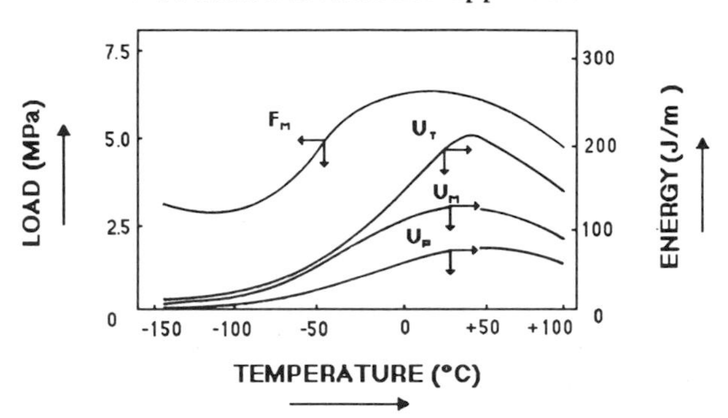

Fig. 4.11 Force F_M and energies U_M and U_T vs temperature in impact tests on notched samples. (Reproduced from M. Rink, T. Riccō, W. Lubert and A. Pavan, *Journal of Applied Polymer Science*; published by Wiley Interscience, 1978.)

- after a certain temperature the amount of plasticity at the notch reduces the stress intensity factor and the load for breaking increases;
- when the rupture is yield controlled F_M follows thc yield, decreasing with temperature.

The minimum of F_M could be taken as related to the DBTT of the material. The trend of load is similar to that reported by Wullaert [32] for Charpy tests of metals. In the case of energy, the introduction of rubbcr increases the total energy (impact strength) while the energy at maximum U_M remains constant after a certain amount of rubber content. Of course U_p, which is the difference between U_T and U_M, increases with rubber content. At low temperature all the energy terms are nearly constant. However, above the T_g value of the rubber particles, the dispersed particles become rubbery, promoting their toughening action.

All the energy terms after a certain temperature decrease; that may be attributed to a lesser orientation hardening of the crazed and yielded material at the tip of the notch of the advancing crack [45]. Advances in high speed tensile machines, even though they are very expensive in comparison with an impact pendulum or FW device, seem to promote the determination of properties at high strain rate [46]. It is possible to evaluate a material for a wide range of strain rates and temperatures and to create a useful map by plotting the main mechanical parameters vs both strain rate and temperature.

4.4 FRACTURE MECHANICS APPROACH

Fracture mechanics considers the fracture of solids initiated by defects, either natural or artificial. The main purpose of fracture mechanics is to determine material parameters which are independent of specimen geometry, similar to yield stress obtained in a regular tensile test, and to define rules or general criteria which are helpful for design against the failure of structures.

According to Williams [47] the mechanical systems can be divided into

- elastic systems and
- systems with energy dissipation.

For rubber toughened engineering polymers the elastic behaviour could be the result of high loading rates, low temperatures or the mechanical constraints (plane strain, large specimen size). This implies that they behave as both elastic systems and systems with energy dissipation depending on the test conditions as well as the particular material characteristics (matrix, toughening component, interphase and degree of anisotropy).

4.4.1 Fracture of elastic systems

With reference to Griffith's work [48], the fracture of an elastic system is the result of an energy balance among the following terms: U_1, the input energy, U_2, the dissipated energy, U_3, the stored energy, and U_4, the kinetic energy. Fracture occurs when the crack surface increases. The following relationship is the basis for fracture advancement [49]:

$$\frac{\mathrm{d}}{\mathrm{d}A}(U_1 - U_3) = \frac{\mathrm{d}}{\mathrm{d}A}U_2 - \frac{\mathrm{d}}{\mathrm{d}A}U_4 \tag{4.16}$$

where

$$\frac{\mathrm{d}}{\mathrm{d}A}(U_1 - U_3) = G \tag{4.17}$$

is the energy release rate,

$$\frac{\mathrm{d}U_2}{\mathrm{d}A} = R$$

is the fracture resistance and $\mathrm{d}U_4/\mathrm{d}A$ is the change of kinetic energy. At fracture initiation, the body is stationary and $U_4 = 0$, so

$$\frac{\mathrm{d}U_4}{\mathrm{d}A} > 0$$

and

$$G - R \geq 0. \tag{4.18}$$

In this case the system is unstable. If we assume that R is a unique material parameter related to the fracture resistance, G_c is the value of G corresponding to R (crack resistance of a material).

G for ideal conditions, i.e. when the force–deflection curve is linear, is determined by

$$G = \frac{\delta^2}{2BC^2}\frac{dC}{da} \tag{4.19}$$

where δ is the displacement, B the thickness, $C = \delta/F$ the compliance, F the load, dC/da the variation of compliance with crack advancement and G_c the value of G at crack instability.

Unfortunately, in many cases, the measurement of G_c is not as simple or is not possible. An alternative route which is based on the stress field around the crack tip may be used (i.e. the Irwin approach). This approach, at least in the elastic case, has been outlined in Ref. 50, and it is equivalent to the energy method. Without entering the mathematical treatment of the stress field around a sharp notch the most important points that can be raised are as follows:

- the form of the stress field near the crack tip is identical for all loadings;
- the magnitude of the stress field can be characterized locally by the parameter called the 'stress intensity factor' K.

We can 'postulate' at this point that the fracture will propagate when K reaches some critical value K_c. It is possible to demonstrate that the following relationships between K and G hold:

$$G_c = \frac{K_c^2}{E} \tag{4.20}$$

for plane stress and

$$G_c = \frac{K_c^2}{E}(1 - \nu^2) \tag{4.21}$$

for plane strain, where E is the elastic modulus and ν Poisson's ratio. For a uniformly loaded infinite plate containing a crack of length $2a$ we have in mode I loading

$$K_{Ic} = \sigma_c(\pi a)^{1/2} \tag{4.22}$$

and for systems of finite dimension

$$K_{Ic} = Y\sigma_c a^{1/2} \tag{4.23}$$

where Y is a geometric correction factor, σ_c is the critical stress applied to the system at crack instability, and a the crack length.

Y for different geometries is reported elsewhere [51]. G_{Ic} and K_{Ic} are material parameters but for polymers we have to consider their dependence on loading rate (viscoelastic effect) and temperature. In connection with the former, it is reported that mechanical loss peaks (T_g of the rubbers and the matrix) also affect the K values derived from Charpy impact studies (see, for example, Ref. 52 and references cited therein).

4.4.2 Fracture of systems with energy dissipation

For the testing of materials with strong energy dissipation like most rubber toughened engineering polymers, one should follow the approach proposed by Williams [47].

The fracture of materials is divided into four classes, which can be distinguished by the relationship of a term characterizing the material performance and the geometrical conditions of the test specimen. The term has the following expression:

$$l = \frac{K_{Ic}^2}{2\pi(M\sigma_y)^2} \tag{4.24}$$

where l represents the size of yielding in front of an initial crack. M is a constraint factor given by

$$M = \frac{\sigma}{\sigma_y} = (1 - 2v)^{-1/2}. \tag{4.25}$$

In the case of plane stress ($M = 1$),

$$l = \frac{1}{2\pi} \frac{K_{Ic}^2}{\sigma_y^2}. \tag{4.26}$$

If $l \ll a$, W, B (Fig. 4.12) we have a class 1 situation, which is called elastic fracture mechanics and is covered by our previous G_{Ic}, K_{Ic} analysis. If

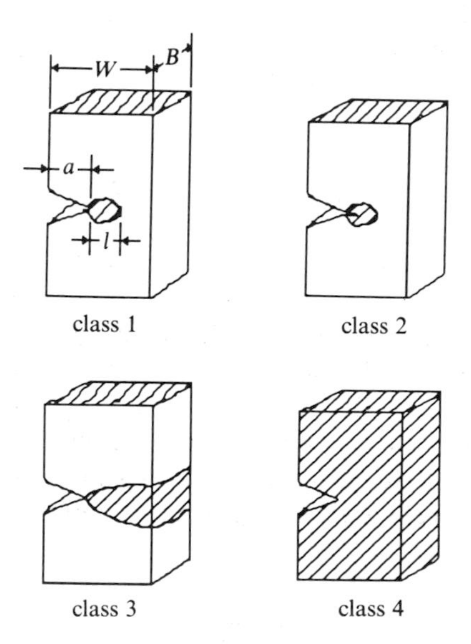

Fig. 4.12 Types of fracture. (Reproduced with permission from *Fracture Mechanics of Polymers* by J. G. Williams, published in 1984 by Ellis Horwood Limited, Chichester.)

$l < W - a$, but l is comparable with a or B, then class 2, as characterized by a significant yielding zone, exists. In class 3, i.e. $l > W - a$, the whole ligament has yielded. Finally, class 4 includes all the cases where dissipation of energy took place through the total body of the test specimen.

These assumptions have the advantage of being very simple for an engineering approach, even though there may be some questions about the physical basis. In the case of class 1, in the model proposed by Williams [47] the thickness effect is considered as well:

$$K_Q = K_{Ic1} + \frac{K_{Ic2}^2}{\pi B \sigma_y^2}(K_{Ic2} - K_{Ic1}). \tag{4.27}$$

For large thicknesses, $K_Q = K_{Ic1}$, the plane strain value.

For class 2 the load–displacement curve can have some degree of nonlinearity, which may be related to the width or a reduced thickness of the specimens. In the first case, as suggested by Hashemi and Williams [53], and later confirmed by Carpinteri, Marega and Savadori [54], we can define an apparent fracture toughness K_Q in the post-yield regime and relate it to the plane strain fracture toughness K_{Ic} through the equation

$$K_Q = (\pi a)^{1/2} F\left(\frac{a}{W}\right)\frac{2}{\pi}\sigma_y \cos^{-1}\left\{\exp\left[-\frac{\pi K_{Ic}^2}{8\sigma_y^2 a F(a/W)}\right]\right\}. \tag{4.28}$$

K_{Ic} is derived by curve fitting the experimental data for load–displacement curves, which exhibit semiductile behaviour up to brittle failure. In this case the plastic zone is modelled according to the Dugdale model (cohesive stress at the notch tip) [55]. The model describes the behaviour of polymers fairly well and can be written as

$$l_D = \frac{\pi}{16}\frac{K_{Ic}^2}{(M\sigma_y)^2} \approx 1.2l. \tag{4.29}$$

Figure 4.13 is an example of this approach for PA although examples of other materials can be found, for example PP (see Ref. 53). These size effects (width effects) arise from the coexistence of different structural crises and from the finiteness of the specimen. In fact, whereas brittle fracture is governed by a generalized force having physical dimensions $[F][L]^{-3/2}$ (stress intensity factor K_{Ic}), plastic collapse is governed by a generalized force having physical dimensions $[F][L]^{-2}$ (stress) [56]. As Fig. 4.14 shows, plastic collapse has the dominant effect on the separation mechanism when the dimensionless number

$$s = \frac{K_{Ic}}{\sigma_y W^{1/2}} \tag{4.30}$$

is bigger than 0.72. Brittle fracture is then inhibited by small structural sizes.

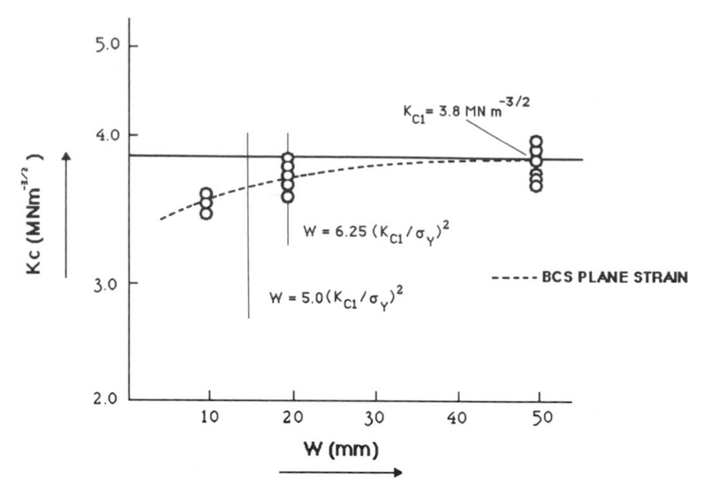

Fig. 4.13 The effect of specimen width on fracture toughness for single edge notched bend specimens of nylon (– – –, BCS plane strain). (Reproduced with permission from S. Hashemi and J. G. Williams, *Journal of Materials Science*; published by Chapman & Hall, 1984.)

This is particularly important for designing rubber toughened engineering polymers for structural applications, where they may have different trends of K_{Ic} and σ_y with rubber content, particle size etc. The brittle–ductile transition can be shifted not only by material parameters but also by the size effects.

Another approach for evaluating a fracture parameter for materials ranging from brittle to semiductile is the determination of the crack opening displacement (COD). Based on the Dugdale model (cohesive stress at the notch tip) [55], the COD δ_c is given by the following relationship:

$$\delta_c = \frac{8\sigma_y}{\pi E} a \log\left(\sec\frac{\pi\sigma_c}{2\sigma_y}\right) \tag{4.31}$$

with

$$G_{Ic} = \sigma_y \delta_c \tag{4.32}$$

where σ_y is the yield stress, a the crack length, σ_c the stress at break and E the tensile modulus. The crack opening displacement measurement is a more direct method than using load or energy values derived by measuring fracture parameters [57].

Class 3 represents a full yield failure resulting from an extensive amount of plasticity at the crack tip. The failure mode changes from semiductile to ductile tearing. This part of fracture mechanics, since the yield is the predominant phenomenon, is called post-yield fracture mechanics (PYFM) (the COD could be considered belonging to it). The physical approach proposed in this

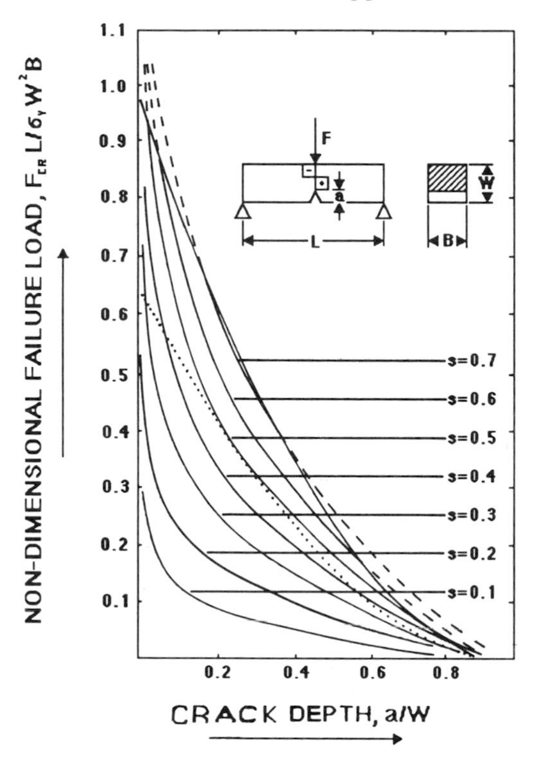

Fig. 4.14 Interaction between crack propagation and plastic hinge formation at the ligament: ---, plastic flow collapse (elastic–perfectly plastic material); ......, ultimate strength collapse (elastic–brittle material). $s = K_{Ic}/\sigma_y W^{1/2}$. (Reproduced with permission from A. Carpinteri, C. Marega and A. Savadori, *Journal of Materials Science*; published by Chapman & Hall, 1986.)

case is the J method. The J method is the energy per unit area necessary to create new fracture surfaces. Following Rice [58] it can be expressed as

$$J = -\frac{1}{B}\frac{dU}{da}\bigg|_{\delta} \tag{4.33}$$

where dU/da is the variation of energy under the load–deflection curve with crack length and B is the thickness. The definition is closely analogous to that of the strain energy release rate G. From F–δ records (where F is load per unit thickness and δ is the load-point displacement) of identical specimens with slightly different crack lengths the area between curves is $J\,da$. This result enabled Begley and Landes [59] to make the first experimental evaluation of J using the so-called 'multispecimen technique'. J can be obtained as a function of displacement and J_c is the value of J at the point of crack initiation.

4.4.3 Tests

(a) General criteria for K_c and G_c

The European Group of Fracture (EGF) is standardizing the K_c–G_c test following the main indication of ASTM E399 but with particular emphasis on the plastics problem and specifying that G_c is important for plastics [60]. The main indications are as follows.

- Three-point bend (SENB) and compact tension (CT) specimens are the recommended geometries with $0.45 < a/W < 0.55$ and $W = 2B$ initially.
- Particular care in sharpening the notch radius has to be taken into account.
- A crosshead speed of $10\,\mathrm{mm\,min^{-1}}$ is the basic condition but in all cases the loading time should be indicated.
- Speeds greater than $1\,\mathrm{m\,s^{-1}}$ or loading times less than 1 ms should be avoided because of dynamic effects.
- Displacements of loading application points have to be taken into consideration by using a displacement transducer, clip gauge etc., in particular for the G_c test. If there is some non-linearity in the diagram, Ref. 60 indicates how to elaborate the data or to reject them.
- The size criteria are

$$B,\ a,\ W - a > 2.5\left(\frac{K_{\mathrm{Ic}}}{\sigma_y}\right)^2 \tag{4.34}$$

  where σ_y is the tensile yield stress, conventionally taken at the maximum load.

G_{Ic} can be obtained in principle from the following relationship:

$$G_{\mathrm{Ic}} = \frac{(1 - \nu)^2}{E} K_{\mathrm{Ic}}^2 \tag{4.35}$$

for plane strain. The Young's modulus E for plastics must be obtained under the same frequency (time) and temperature conditions because of viscoelastic effects. Owing to many uncertainties in the calculations according to equation (4.35), it is preferable to determine G_c directly from the energy absorbed. G_c is calculated from the fracture energy, which is defined in Refs. 38 and 60, through the following expression:

$$G_{\mathrm{Ic}} = \frac{U}{BW\phi} \tag{4.36}$$

where ϕ is the energy calibration factor.

(b) General criteria for J_c

Although J can be experimentally obtained by using the multispecimen technique, significant effort has been dedicated to relating J to the work done or the area under the $F-\delta$ diagram. The results are standardized in ASTM E813.

Experimentally the value of J_c can be obtained using several specimens each loaded to a different amount of crack growth, having initial crack lengths which are > 0.5 of the specimen depth. Specimens are then cooled down and broken open to reveal the amount of crack extension. For each specimen the J value is calculated from the area under the load–deflection curve up to the unloading point using the relationship

$$J = \frac{2U}{Bb} \tag{4.37}$$

where B is the thickness and $b = W - a$, i.e. the uncracked ligament width. This approach has been proposed for several polymers by Hashemi and Williams [61].

Figure 4.15 shows a schematic representation of the test technique. An R curve is constructed and its intercept with the straight $J = 2\sigma\Delta a$ blunting

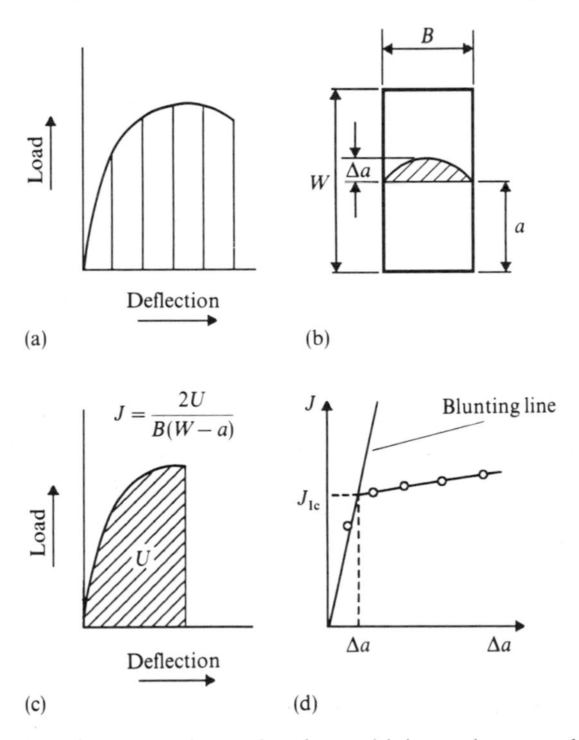

Fig. 4.15 Procedure for measuring J by the multiple-specimen method: (a) load–deflection curve of several specimens; (b) crack growth analysis; (c) J measurement of specific crack growth; (d) J plot and J_{Ic} evaluation. (Reproduced with permission from J. G. Williams and S. Hashemi, *Polymer Blends and Mixtures*; published by Kluwer Academic, 1985.)

line gives the value of J_c. It should be noted that the data-qualifying scheme and thus the determination of J_c are different in the related ASTM E813-81 and E813-87 standards, which may result in a large scatter in J_c, especially for toughened systems [62]. To overcome the difficulties related to the time and specimen consumption of the multiple-specimen R curve method a single-specimen technique is recommended. In the latter testing procedure, a specimen loaded at a constant rate is unloaded (to a given extent) and reloaded repeatedly and from the unloaded compliance the crack extension (Δa) is calculated. Hashemi and Williams [62] claimed that this single-specimen R curve method produces results comparable with those for the multiple-specimen method for toughened PAs as well. For the J_c value to be independent of the specimen size, ASTM has proposed the following requirements:

$$B,\ W,\ W-a \geq 25\,\frac{J_c}{\sigma_y}. \tag{4.38}$$

The slope of J vs Δa, $\mathrm{d}J/\mathrm{d}(\Delta a)$, is the tearing resistance which, according to Paris [63], determines the tearing stability of the crack growth in the specimen. It has become doubtful whether J_c, as a single fracture mechanics parameter related merely to the blunting, reflects properly the toughness of high impact plastics or not. It has been concluded recently that the energy dissipation processes are better and, in addition, are quantitatively involved in the J, T term [64], where J is the J integral and T is the tearing modulus given by [63]

$$T = \frac{\mathrm{d}J\,E}{\mathrm{d}(\Delta a)\sigma_y^2}. \tag{4.39}$$

The J, T approach seems to work well and correlates with experimental toughness values very well; this is also the case when the JT term is defined in impact tests [65].

(c) Application of fracture mechanics concepts to impact tests

The measurements of K_c–G_c and J_c values by impact tests can be attempted using the following guidelines.

K_{Ic}, G_{Ic}

Instrumented impact testing. K_{Ic} can be obtained from the maximum load provided that the influence of dynamic effects is very small and the load–time or load–deflection curves are linear (or almost linear) as pointed out in Ref. 60.

G_{Ic} is obtained from the measured K_{Ic} value and via E. Alternatively, it may be determined from the area under the load–time curve using the equation

$$G_{\mathrm{Ic}} = \frac{V_0 \int_0^t F \, \mathrm{d}\tau}{BW\phi} \tag{4.40}$$

where V_0 is the hammer speed, W the specimen width, F the force, ϕ the geometrical factor and B the thickness.

The COD is obtained from the relationship

$$\delta_{\mathrm{c}} = \frac{G_{\mathrm{Ic}}}{\sigma_{\mathrm{y}}}. \tag{4.41}$$

Energy measurement. Plati and Williams [38] have pointed out the difficulty of obtaining reliable data from instrumented impacts at high impact speeds. They proposed the following relationship for calculating G_{Ic} from the impact energy:

$$G_{\mathrm{Ic}} = \frac{U - U_{\mathrm{k}}}{BW\phi} \tag{4.42}$$

where U is the energy absorbed by the pendulum, U_{k} the kinetic energy of the specimen, B the thickness, W the width and ϕ the calibration factor. In a subsequent paper Newmann and Williams [66] extended this model to tensile impact. The determination of K_{Ic} and G_{Ic} has permitted the unification of the Charpy, Izod and tensile impact tests on notched specimens. The formula for energy is still valid for a blunt notch which is usually specified in standard methods.

The critical fracture energy in this case is called G_{b}, and can be linked theoretically [38] to G_{Ic} by the following relationship:

$$G_{\mathrm{b}} = \frac{2}{\pi} U'_{\mathrm{p}} \rho \frac{G_{\mathrm{Ic}}}{2} \tag{4.43}$$

where ρ is the root radius and U'_{p} is the energy per unit volume to yield, which is related to the radius of the plastic zone by

$$r_{\mathrm{p}} = \frac{\pi}{16} \frac{G_{\mathrm{Ic}}}{U'_{\mathrm{p}}}. \tag{4.44}$$

The approach through instrumented impact testing or energy measurement gives the same results [67] provided that the LEFM can be applied, the impact speed is low ($< 1\,\mathrm{m\,s^{-1}}$) [68] and reliable elastic constants (E, ν) are available. In this case the approach through K_{Ic} can also be followed. Williams and Hodgkinson [69] have shown that an increase in impact speed causes a strong increase in G_{c} owing to thermal blunting.

J_c

The determination of J_c through an impact test is not very easy. Two routes are proposed.

Instrumented impact testing. Server [70] for metals and Savadori [71] for polymers proposed a system (hammer and steel stop) for achieving Δa increments, enabling the R curve to be constructed.

Energy measurement [72]. The Charpy impact pendulum is modified by tapering the thickness of the supports. Crack growth may be controlled since the load-point displacement depends on the initial specimen length. In fact, if the specimen is short enough the crack may pass through the jaws, and hence total fracture is avoided. Instead, only a certain amount of crack growth occurs. The J value is calculated by the standard relationship

$$J = \frac{2U}{B(W-a)} \tag{4.45}$$

where U is the absorbed energy read on the pendulum dial, corrected for frictional losses, and a is evaluated by scanning electron microscopy (SEM) and/or optical microscopy.

(d) Application of fracture mechanism concepts to fatigue tests

The rate of crack growth per cycle $\mathrm{d}a/\mathrm{d}N$, often called the fatigue crack propagation (FCP) rate, is measured as a function of the upper and lower values of the stress intensity factor ΔK at the existing crack tip ($K_{max} - K_{min}$ and $R = K_{min}/K_{max}$). The recommended specimen geometry is CT. Over a restricted range of conditions, fatigue data can be fitted to the Paris law [73] of the form

$$\frac{\mathrm{d}a}{\mathrm{d}N} = A(\Delta K)^n \tag{4.46}$$

where A is a constant, ΔK is the stress intensity factor range ($\Delta K = K_{max} - K_{min}$) and n is an exponent which may vary between 4 and 12 for a certain range of polymers [74].

Figure 4.16 from Ref. 26 represents a schematic diagram of the Paris plot, and Fig. 4.17 from Ref. 26 is the application of this approach to PS, HIPS and ABS. It can be pointed out that the presence of rubber improves the fatigue crack propagation. Crack tip blunting due to toughening and eventually hysteresis heating is beneficial for FCP resistance, while a decrease in the E modulus is generally deleterious, since the creep process in the crack tip is normally superimposed on FCP [74].

It is well documented in the literature that the FCP response strongly depends on the mean stress level, especially for impact-modified plastics. It is

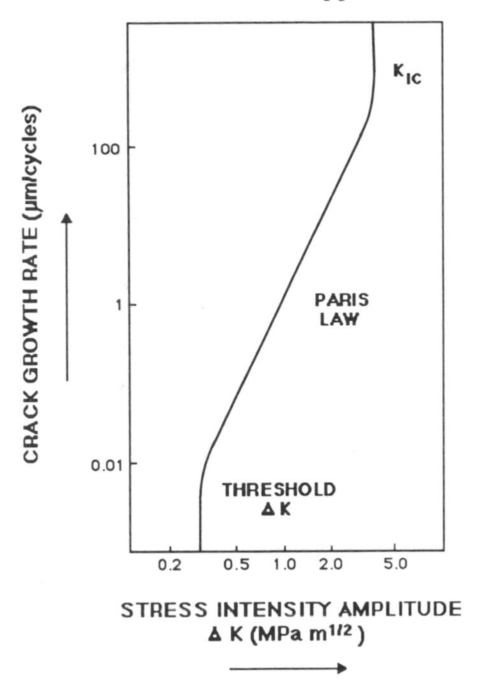

Fig. 4.16 Schematic diagram of relationship between fatigue crack growth rate and amplitude of stress intensity factor. (Reproduced with permission from C. V. Bucknall, *Polymer Blends and Mixtures*; published by Kluwer Academic, 1985.)

therefore a reasonable assumption that the overall FCP rate is composed of the cyclic fatigue rate and of the creep-related fatigue rate. Recent investigations showed, however, that the cyclic and creep-induced damages are not simply additive [75].

A still open and very interesting question exists as to whether there is a unified approach to the FCP for different materials or not. On the basis of recent contributions, the former statement may be valid [76].

A recent approach regards the crack and process zone as an entity (crack layer) and describes its propagation by the principle of irreversible thermodynamics [77]. The crack layer theory seems to work even for rubbery compounds [78].

4.4.4 Experimental data

The number of applications of FM to rubber toughened engineering polymers has been increasing in order to obtain reliable parameters useful for the design. On the other hand, the high ductility of many of them in a large temperature range does not allow an extensive use of LEFM.

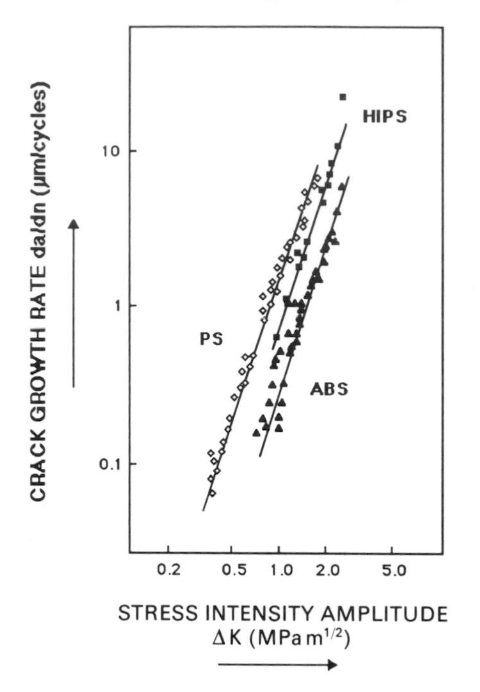

Fig. 4.17 Paris power law plots for PS, HIPS and ABS. (Reproduced with permission from C. V. Bucknall, *Polymer Blends and Mixtures*; published by Kluwer Academic, 1985.)

Nimmer *et al.* [79] pointed out that the LEFM approach can be applied to predict minimum failure loads of components if the material is tested at a temperature below its DBTT. This corresponds to a load–displacement curve of 'triangular' shape, i.e. catastrophic failure occurs (brittle fracture).

The epoxy resins, because of their intrinsic brittleness, even though they are rubber toughened, are among the materials analysed according to LEFM. With regard to the book *Fracture Behaviour of Polymers* by Kinloch and Young [80] and other references, one may conclude that G_{Ic} of epoxy resins is related to the volume fraction of rubber particles by a simple rule of mixtures [81, 82]:

$$G_{Ic} = G_{Ic}(\text{matrix})(1 - \psi_d) + \Delta G_{Ic} \qquad (4.47)$$

where $\Delta G_{Ic} \approx 4\Gamma_t \psi_d$ and Γ_t is the tearing energy of a rubber particle. Figure 4.18 from Bucknall and Yoshii [83] confirms this relationship, at least in a given range of rubber contents. Other microstructural factors, including particle size and particle distribution, are of secondary importance if the particle size is between 0.5 and 5 μm.

In addition Hwang *et al.* [84] and Karger-Kocsis and Friedrich [85] have shown that rubber improves not only impact strength but also the resistance

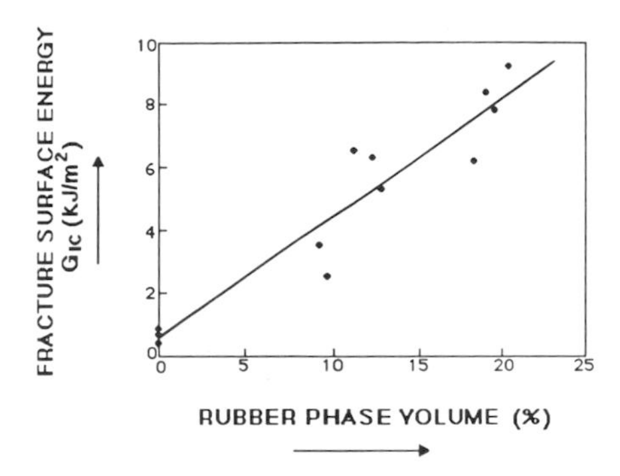

Fig. 4.18 Relationship between fracture surface energy and degree of rubber phase volume. (Reproduced from J. C. B. Bucknall and T. Yoshii, *British Polymer Journal*, 1978.)

against FCP. Based on fractographic analysis the relevant failure processes leading to a deceleration in FCP were determined and summarized in models [85]. FCP decreased further with increasing cycling frequency for toughened epoxies.

Recently considerable efforts have been undertaken to improve further the toughness performance of rubber modified epoxy by the help of computer modelling [86, 87]. A deeper understanding of the failure mechanisms and their interaction in such systems as in general for all rubber toughened engineering plastics would contribute considerably to reducing the experimental work of the trial-and-error type exercised so far.

As far as the interpretation of the fracture mechanical data is concerned, nowadays a clear trend can be observed. Researchers are trying to find correlations between fracture mechanical results and morphological or even molecular characteristics of polymers. Brostow and Macip [88] showed the link between the DBTT and the chain relaxation capability (CRC) and thus the free volume of polymers. This concept may serve as a basis of a more general approach regarding the interrelation between macroscopic fracture mechanical and microscopic molecular terms [89].

Recently several papers have treated the course of the fracture toughness as a function of crosslink [90, 91] density for toughened EP resins. Since the crosslink density is related to the CRC and thus to the free volume the above concept may be of paramount interest for the future.

The approach to describing the 'ductility region' of very ductile materials by fracture mechanics started with the work of Sridharan and Broutman on ABS [92]. The J method criterion using the Begley and Landes procedure [59] was applied and it was found that J_{Ic} defined in terms of the Rice contour integral [58] is reasonably independent of geometry. Much effort was undertaken to characterize different toughened polyamides (PAs) both at low and at high speed [72, 93–95].

One of the most promising conclusions [94] is that the K_{Ic} value determined through the relationship

$$K_{Ic}^2 = EJ_{Ic} \tag{4.48}$$

by using J_{Ic} values from ductile ruptures and K_{Ic} values from semibrittle and brittle ruptures can be combined for a general description of fracture by a unique parameter. Figure 4.19 reports these data in comparison with those for two other toughened polymers, ABS and PVC.

The FCP of rubber toughened block amide copolymer is lower than that of the parent material [95]. The slope of the da/dN vs ΔK plot (Fig. 4.20) becomes steeper as a result of toughening. Hysteretic heating has a marked influence on FCP response and the improvement of FCP resistance is lower at high ΔK levels.

Other data concerning rubber toughened PA are those of glass-filled PA. The introduction of glass fibres modifies the fracture toughness. According to Friedrich [96], the fracture toughness of the material may be obtained from the semiempirical relationship

$$K_{cc} = M_K K_{cm} = (a_K + n_K R_K) K_{cm} \tag{4.49}$$

where K_{cc} and K_{cm} are the fracture toughness of composite and matrix respectively, M_K is the microstructural efficiency factor, a_K is the matrix stress

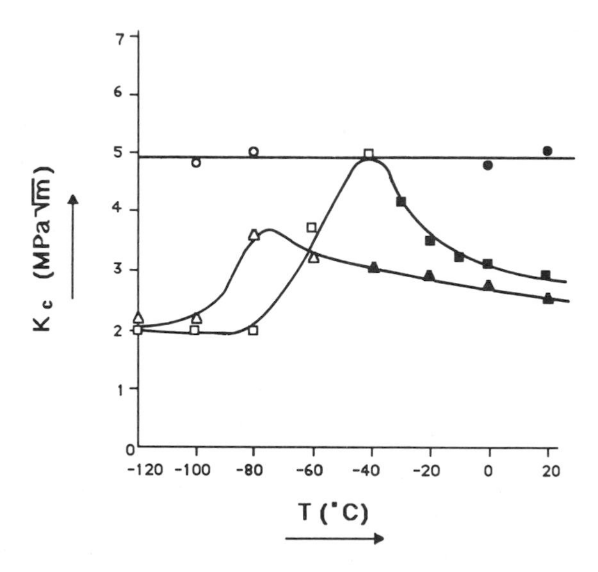

Fig. 4.19 K_c as a function of temperature for three toughened polymers (○, ●, nylon; □, ■, PVC; △, ▲, ABS): ●, ■, ▲, converted from J_c values ($K^2 = EJ_c$); ○, □, △, from LEFM tests. (Reproduced with permission from J. G. Williams and S. Hashemi, *Polymer Blends and Mixtures*; published by Kluwer Academic, 1985.)

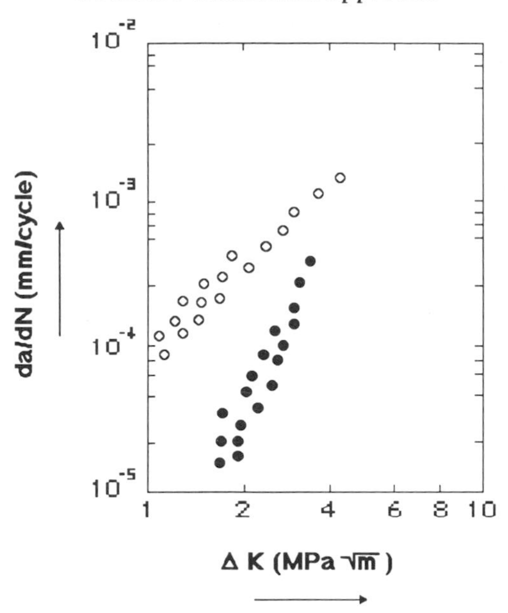

Fig. 4.20 The effect of rubber toughening on FCP response of block amide copolymer ($M_n = 24\,000\,\text{g mol}^{-1}$) ($f = 5$ Hz, $R = 0.1$; ○, initial block amide copolymer with $M_n = 17\,000\,\text{g mol}^{-1}$; ●, rubber toughened copolymer). (Reproduced from Y. Giraud and J. Coletto, *Makromolekulare Chemie, Macromolecular Symposia*; published by Huthig and Wepf, 1989.)

condition factor, n_K is the energy absorption ratio and R_K is the reinforcing effectiveness parameter. Figure 4.21 presents a schematic diagram of the relationship of the principal ΔK_{cc} vs R_K curves, indicating upgrading and deterioration of a composite's fracture toughness as a function of the ductility of the matrix.

The extension of this approach to the J method [97] could be the following:

$$\frac{J_{cc}}{J_{cm}} = \frac{(a_K + n_K R_K)^2}{R_K(E_f/E_m)(1 - \Psi_f)} \tag{4.50}$$

found with PP–elastomer–PE blends. The J method approach seems to be widely used even though some aspects concerning the exact definition of the crack blunting line [98, 99] and the evaluation of ductile tearing instability [100, 101] are still under discussion. Recently, Lee, Lu and Chang [99] proposed the hysteresis method for overcoming this controversy. Wert, Saxena and Ernst [102], exploring the applicability of the J method to plain and toughened polycarbonate, pointed out that a suitable method to deter-

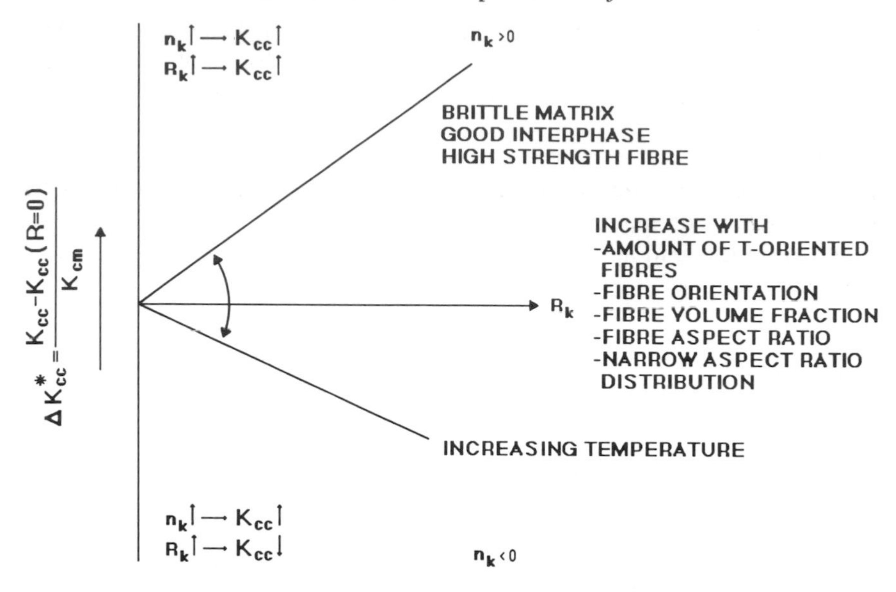

Fig. 4.21 Schematic representation of microstructural efficiency vs R curves indicating upgrading and deterioration in composites' fracture toughness. (Reproduced from J. Karger-Kocsis, *Application of Fracture Mechanics to Composite Materials*; published by Elsevier, 1989.)

mine J_{Ic} for these polymers does not exist. They proposed as a fracture parameter J_M, which can characterize the material's resistance in the early stages of crack growth. J_M, which seems to be size independent, can be obtained from load–deflection diagrams as follows:

$$J_M = G + \int_0^{\delta_p} \frac{\gamma_1}{bB} F \, d\delta_p. \tag{4.51}$$

where G is the elastic strain energy release rate, b the remaining ligament, δ_p the the plastic part of the deflection, F the load and $\gamma_1 = \gamma - \beta/n_\sigma$; γ and β are functions of only a/W; n_σ is the strain hardening coefficient.

Concluding this analysis, it is worth mentioning that another approach for evaluating the fracture toughness of very ductile materials is the essential work of fracture. The plane stress specific work of fracture (U_e) is obtained by extrapolating the ligament area of double edge notched (DEN) specimens to zero. The formula is

$$U_s = U_e + \beta_s U'_p b \tag{4.52}$$

where b is the ligament length, β_s is a shape factor and U'_p is the plastic work of fracture per unit volume. Using DEN specimens and different ligament sizes it is possible to obtain $U_s = U_c$ which seems very close to J_{Ic} [103–105]. Paton and Hashemi [106] have also shown that U_e of PC is independent of specimen length, W, sharpness of the initial notch and specimen geometry. The advantage of this method is that it can be used on samples which could not be studied by standard J integral procedures because of stringent size requirements.

Another approach within the concept of the essential work of fracture applies to the generalized locus method which determines the resistance to crack growth, including crack initiation, considering the locus of characteristic points on the load against load-point displacement curves of three-point bending specimens which differ in their a/W ratio [107].

4.5 FRACTOGRAPHY

With reference to Roulin-Moloney's book *Fractography and Failure Mechanisms of Polymers and Composites* [108], fractography is a term generally adopted for the science studying fracture surfaces in order to determine the sources of the fracture and the relation between the mechanical parameters, crack propagation mode and the microstructure of the material. For the methods of measurement of toughness of material it is an important means for understanding the micromechanism of crack advance in relation to the load and energy measured and the nature of material tested. In particular, with the fractography approach one can

- understand where the fracture has initiated, which can be a very useful 'exercise' for a 'post mortem' evaluation of broken objects,
- identify the amount of plastic and elastic deformation on and below fracture surfaces, quantifying also the volume of material involved in the failure mechanism, and
- define 'status', quantity, size, distribution, adhesion of rubber particles and the fundamental mechanisms (voids, crazing and shear yielding) at the origin of material performances.

This evaluation has certain limitations since it depends on energy, rate of load and geometry of samples which together with temperature have a strong influence on crack speed and on the role of the rubber phase. The main techniques for fractography studies at present are as follows.

- Optical microscopy is simple, inexpensive and extremely useful for an overall view of the fracture surface and for the evaluation of microtomed thin sections perpendicular to the fracture surface.
- SEM provides a large depth of field at high magnification, as compared with optical microscopy.
- Transmission electron microscopy (TEM) can be used, through replicas or thin sections and high resolution, to establish the relationship between the features observed on the fracture surface and the detailed microstructure of the base material.

- Image analysis offers the possibility of quantifying many features on the fracture surfaces.
- Scanning secondary ion mass spectrometry (SIMS) had rather limited application until recently but it has great potential for obtaining a surface composition map.
- Scanning acoustic microscopy (SAM) offers two unique features: firstly the elastic properties of the material can be imaged; secondly, it is possible to form images of surface features. Analysis by acoustic microscopy includes cracks, voids and delaminations, non-destructive testing (NDT) of adhesion and glue bands etc.

The fracture surface of broken specimens of plain polymers (amorphous, semicrystalline, thermoset or thermoplastic) as observed through optical or electron microscopy can be divided into the following zones [109].

- Mirror zones are close to the crack tip or defect where the fracture starts. They result from craze rupture in front of the crack tip and, because of the presence of a thin layer of highly oriented polymer with different refractive index from that of the bulk, interference colour fringes are frequently observed. It has been shown that the size of this mirror area increases with molecular weight in the case of PBT [110] and with temperature and time in the case of epoxy resins [111], and it can be related to applied critical stress according to the Dugdale model. This area is substituted by a 'stress whitened zone' in rubber toughened polymers.
- Mist regions are smooth and not necessarily confined to the vicinity of the fracture origin but can be observed elsewhere on the fracture surfaces.
- Hackle regions are rough and associated with a stage of fracture propagation where a large amount of strain energy is absorbed through both plastic deformation and the generation of new fracture surface areas. They tend to appear in areas where the stress field is changing rapidly or when the stress state changes from one of plane strain to plane stress. In Ref. 109 an example from a specimen PC after Izod impact is reported.

Other features of fracture surfaces are 'Wallner lines' formed when stress waves reflected from the specimen boundaries interact with the propagating crack front. In the case of rubber toughened polymers, the fracture surface features are modified by the presence of relevant plastic deformed zones just at fracture onset. In the case of toughened epoxy resins, it has been shown [108] that cavities form around fractured rubber particles or by the stretching of the rubber particles, and the growth of the resultant voids is the main mechanism. Analysis by SEM and by optical microscopy of thin sections (10–20 μm) perpendicular to fracture surfaces suggests that cavitation of the rubber particles and formation of shear bands are the main mechanisms of toughening. Hobbs, Bopp and Watkins [112] investigated the first zone of the impact surface of PA 6.6 and LDPE grafted with maleic anhydride.

Optical and electron (SEM, TEM) microscopy studies were used and particle size distributions were obtained using an image analyser. Samples were prepared using several techniques including direct examination, extraction, etching, petrographic thinning and replication [113]. X-ray scattering experiments were performed in order to calculate the apparent void sizes as a function of depth below fracture surfaces. The same stress whitening of the fracture surface was noted on all the test specimens. In the low impact samples this region was localized just behind the notch tip while in the high impact specimens the entire surface as well as an appreciable volume of subsurface material was whitened. Small shear lips and somewhat greater lateral contraction of the sample sides were visible in the high impact specimens.

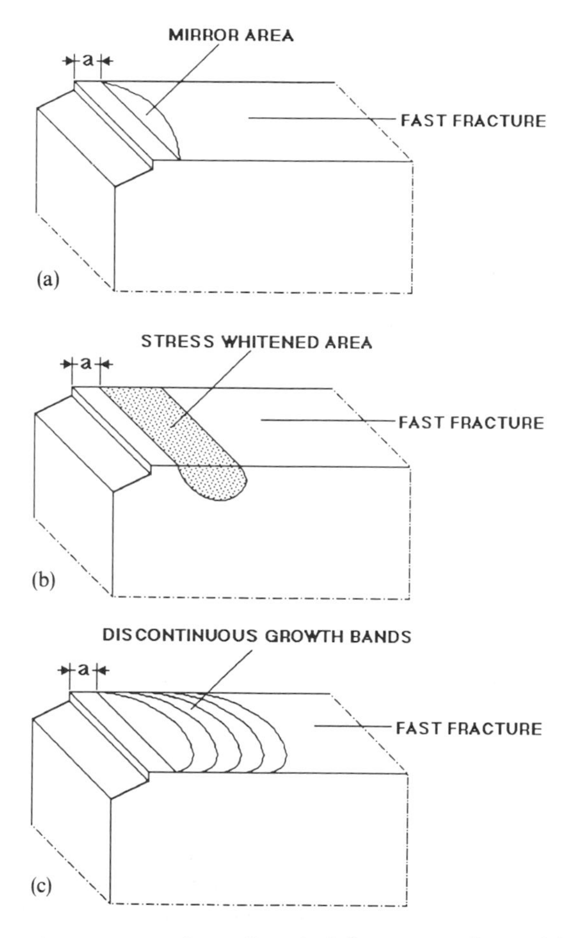

Fig. 4.22 Schematic representation of typical fracture surfaces: (a) homogeneous polymer with mirror area; (b) rubber toughened polymers; (c) fatigue specimen.

The results of this analysis have been the following.

- The primary deformation mechanism is shear flow, although X-ray studies show that dilatation failure processes occur simultaneously.
- Subsurface deformation and debonding processes contribute substantially to energy dissipation.
- A numerical model through image analysis has been developed for estimating the interaction between the dispersed impact modifier particles for predicting the shift in DBTT.

In the case of fatigue fracture, it is possible to show that there are fatigue striations as in metals [114]. Furthermore, the striation width was found to increase with increasing crack length and then with ΔK consistently with the trends found in metals, even if the dependence on ΔK is different from a power of 2 (4–20 in polymers). It has been found for PMMA, PS, PC and epoxy resin that there is an excellent one-to-one agreement between striation width and macroscopic growth rate. In the case of semicrystalline polymers, the crystalline texture could disturb the crack advance even if for nylon, PE and PVDF striation markings parallel to the advancing front have been found and associated with the macroscopic growth rate. Fracture surfaces of ABS and HIPS–PPO specimens have shown strong whitening with intensity increasing with ΔK. At low ΔK, bands were observed perpendicular to the direction of crack growth, as found with other polymers.

Figure 4.22 shows schematic diagrams of three broken specimens and the localized 'processing' zones, after which the fracture becomes unstable.

4.6 ENGINEERING DESIGN REQUIREMENTS

The design criteria are essentially determined by

- the level of deformation accepted in the product and
- the ability of the material to remain intact under service.

The first criterion depends on the deformation of the product in a particular application; for example, plastic body panels would not be used on a motor car if they underwent a significant degree of deformation. On the other hand, a large amount of deformation is expected from the bumpers made from toughened plastics. The second criterion has many aspects and may be evaluated according to the specific application requirements. An exhaustive approach to this evaluation would imply consideration of the following fracture and failure modes:

- impact loading;
- fatigue crack propagation;
- creep failure.

Rubber toughened engineering plastics are used where the predominant characteristic required is the impact resistance at room and low temperatures. Fatigue and creep failure are also important but mainly when glass-filled materials are used. With reference to previous sections, there are many tests for evaluating this mechanical behaviour, but the proper choice of fracture mechanical characterization for design purposes depends on the material's response under applied conditions. FM, unlike the traditional methods, can help to describe the behaviour of the material and part in a more rational way. For this purpose the scheme reported in Fig. 4.23 taken from Ref. 115 seems very useful, if the requirements are high stiffness or high strength or the material is applied under particular conditions (low temperature, high strain rate or high constraint) and LEFM with its parameters K_{Ic} and G_{Ic} is the right approach. If the material has low stiffness and low strength, then the fracture approach through PYFM whose parameters are J_{Ic} and δ_c is recommended. Analytical solutions for K_{Ic} exist and they permit a quantitative answer in many cases.

Figure 4.24 illustrates a comparison between analytical predictions of failure and the experimental measurements at a low temperature for a rubber

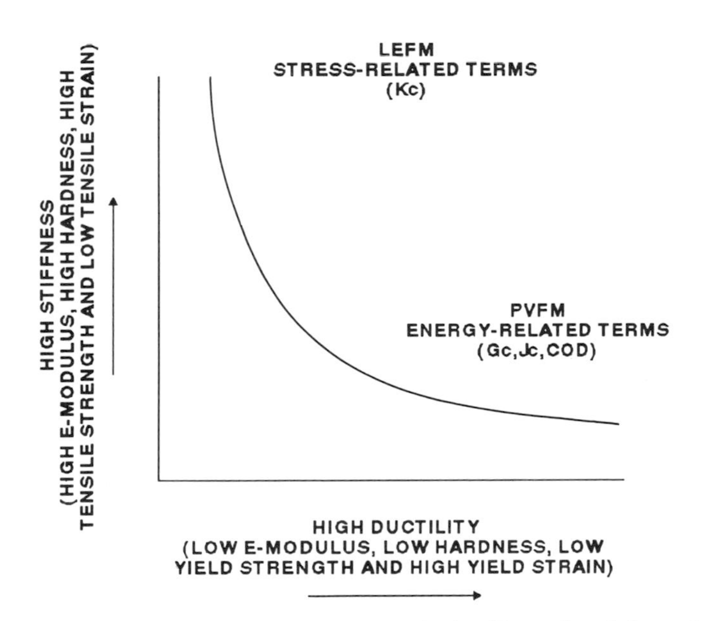

Fig. 4.23 Schematic diagram of FM design criteria. (Reproduced from J. Karger-Kocsis, *Encyclopedia of Composites*; reprinted with permission by VCH Publishers, © 1990.)

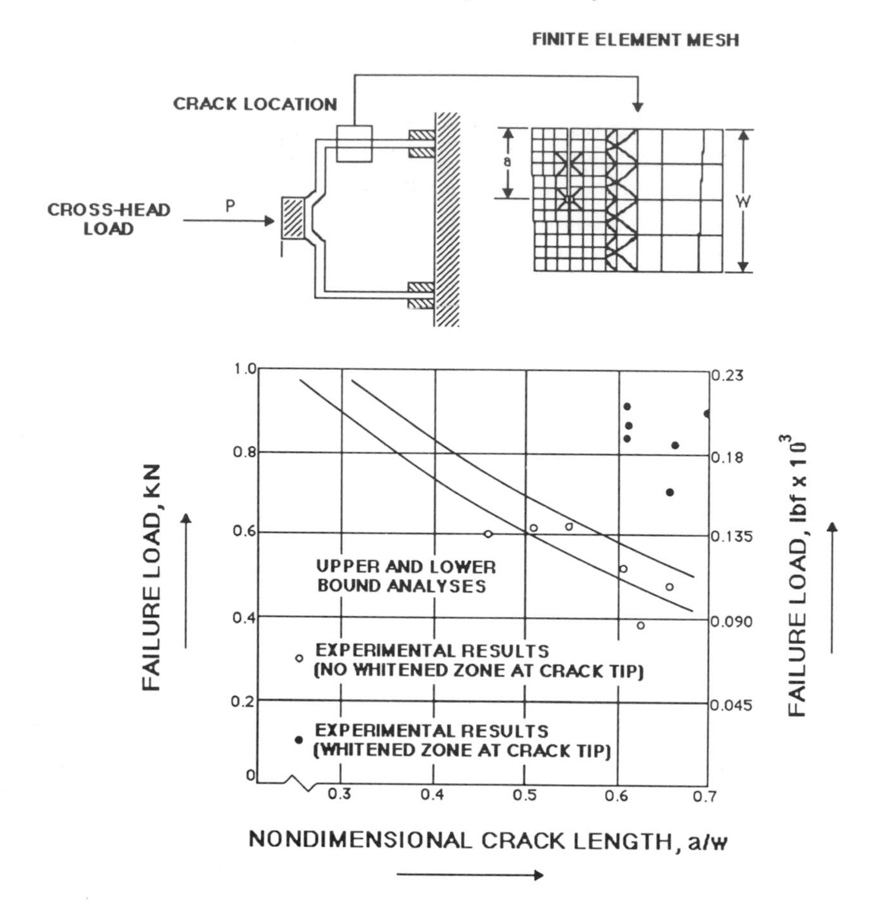

Fig. 4.24 Comparison of predicted and measured loads during the low temperature impact of cracked specimens: ○, experimental results, no whitened zone at crack tip; ●, experimental results, whitened zone at crack tip. (Reproduced with permission from *Engineered Materials Handbook*, Vol. 2, *Engineering Plastics*, ASM International, Metals Park, OH, 1988.)

toughened engineering polymer [116]. The material has a K_{Ic} value of about 3.5 MPa m$^{1/2}$ at $-50\,^{\circ}$C. The full lines in Fig. 4.24 represent the bounds of prediction. In the case of the fatigue loading of glass-filled polymers [115] it is possible to divide the response curve into two regions, crack initiation and propagation (Fig. 4.25). For short glass fibre and long glass fibre composite reinforced injection moulded articles crack propagation is of basic importance. In this case therefore the threshold value derived from static and dynamic fatigue measurements can be used for construction purposes.

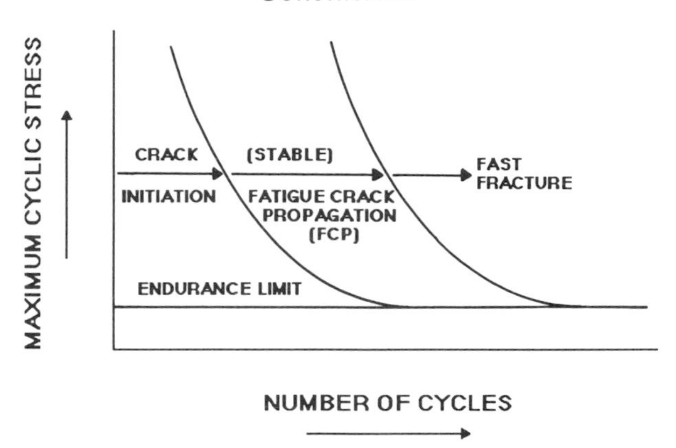

Fig. 4.25 Schematic diagram of maximum cyclic stress vs number of cycles. (Reproduced from J. Karger-Kocsis, *Encyclopedia of Composites*; reprinted with permission by VCH Publishers, © 1990.)

4.7 CONCLUSIONS

- The methods of measuring the toughness of rubber toughened engineering polymers are making progress: K_{Ic}, G_{Ic} are established parameters for brittle fracture; the BCS model and J method seem very promising even if their applicability to polymeric materials is still under discussion. In particular, the exact definition of the crack blunting line, the evaluation of ductile tearing, the geometry effect (size) on brittle–ductile behaviour and finally the physical significance of those 'ductile parameters' for so heterogeneous a material and the possibility of their use for design against catastrophic failure have to be a matter of future work.
- The impact strengths measured by Izod, Charpy or FW tests (instrumented or not) currently represent the easiest and cheapest ways of evaluating the mechanical strength of polymeric materials at high strain rate. Even if they are valid as criteria for quality control or used as a basis of comparison or an indicator in material development, in particular using the DBTT, unfortunately they cannot be used for design calculations or as a starting point for the formulation of theories in materials science.
- Traditional Izod, Charpy and tensile tests (on a notched specimen) can be rationalized by applying LEFM provided that the fracture is brittle: ductile or semiductile failures are more difficult to analyse.
- FW tests have to be used for practical purposes. Specimen thickness, size and the clamp systems can influence the fracture patterns and fracture energy [117]. In the case of brittle fracture, a mechanical model can be attempted but with ductile failure, owing to the interaction of the indenter (size, shape) with fracture energy and the DBTT, only an empirical approach on a comparative basis is possible.

- Fractography and dilatometry are effective means by which knowledge can be gained regarding the relationship between the material microstructure, fracture phenomena and mechanical parameters.

Good material design requires the use of the above methods.

APPENDIX 4.A SYMBOLS AND ABBREVIATIONS

4.A.1 Symbols

a	crack length
a_K	matrix stress condition factor
A	crack area
A_y	yield constant of given material
$b = W - a$	ligament width
B	thickness of specimen
C_p	heat capacity
C	compliance
C_G	modulus–yield stress coefficient
e	strain
E	Young's modulus
E_c	Young's modulus of continuous phase
E_d	Young's modulus of dispersed phase
E_f	Young's modulus of fibres
E_m	Young's modulus of matrix
F	load, force
F_{cr}	critical load
F_M	maximum load
G	energy release rate per unit crack area
G_b	apparent G for blunt crack
G_c	value of G at fracture
G_M	shear modulus
ΔH	activation enthalpy
I	moment of inertia of specimen
J	rate of change of energy with crack area
J''	loss compliance
J_c	value of J at fracture
J_{cc}	J integral of composite
J_{cm}	J integral of matrix
k	Boltzmann's constant
K	stress intensity factor
K_c	value of K at fracture
K_{c1}	plane strain value of K_c
K_{c2}	plane stress value of K_c
K_{cc}	fracture toughness of composite
K_{cm}	fracture toughness of matrix

K_{Ic}	K_c for crack opening mode I
K_Q	value of K_c for which validity is yet to be established
l	size of yielding
l_D	size of plastic zone in Dugdale model
L	span in bending test
M	constraining ratio
M_e	equivalent mass of hammer
M_K	microstructural efficiency factor
n	fatigue exponent
n	strain hardening exponent
n_K	energy absorption ratio
N	number of cycles
P	pressure
r_c	radius of clamps
r_p	plastic radius
R	fracture resistance
R_K	reinforcing effectiveness parameter
s	brittleness number
t	time
t_r	rebound time
T	temperature
U	energy
U_0	energy absorbed during impact process
U_0	hammer energy
U_1	external work
U_2	irreversible work
U_3	potential energy
U_4	kinetic energy
U_e	essential work of fracture
U_k	kinetic energy of broken specimen
U_M	energy at maximum
U'_p	energy per unit volume to yield
U_T	total energy
v	activation volume
W	specimen width
Y	calibration factor
γ	surface energy
γ_1	$\gamma-\beta/n_\sigma$, where γ and β are functions of a/W and n_σ is the strain hardening coefficient
δ	displacement
δ_p	plastic part of displacement
μ	pressure coefficient
ν	Poisson's ratio
ρ	notch radius or density
σ	stress

$\sigma_1, \sigma_2, \sigma_3$	principal stresses
σ_B	brittle stress–strength
τ	time variable
τ_y	shear yield stress–strength
ϕ	energy calibration factor
ψ	volume fraction
Ψ_f	volume fraction of fibres

4.A.2 Abbreviations

COD	crack opening displacement
DBTT	ductile–brittle transition temperature
FATT	fracture appearance transition temperature
FCP	fatigue crack propagation
FM	fracture mechanics
FW	falling weight
LEFM	linear elastic fracture mechanics
NDT	nil ductility temperature
PYFM	post-yield fracture mechanics

REFERENCES

1. Echte, A. (1989) *Adv. Chem. Ser.*, **222**, 15.
2. Wu, S. (1990) *Polym. Eng. Sci.*, **30**, 753.
3. Turner, S. (1983) The mechanical testing of plastics, in *Mechanical Testing of Plastics*, Longman, London, p. 1.
4. Knott, J. F. (1979) Modes of failure, in *Fundamentals of Fracture Mechanics*, Butterworths, London, p. 1.
5. Brostow, W. (1986) Impact strength: determination and prediction, in *Failure of Plastics*, Hanser, Munich, p. 196.
6. McCrum, N. G., Buckley, C. P. and Bucknall, C. B. (1988) Viscoelasticity, in *Principles of Polymer Engineering*, Oxford University Press, New York, p. 101.
7. Macchetta, A., Pavan, A. and Savadori, A. (1989) *Mater. Des.*, **10**, 293.
8. Casiraghi, F. (1983) *Polym. Eng. Sci.*, **23**, 902.
9. Dickie, R. A. (1978) Mechanical properties (small deformations) of multiphase polymer blends, in *Polymer Blends*, Vol. 1 (eds D. R. Paul and S. Newmann), Academic Press, New York, p. 353.
10. Hwang, J. F., Manson, J. A., Hertzberg, R. W., Miller, G. A. and Sperling, L. H. (1989) *Polym. Eng. Sci.*, **29**, 1466.
11. Takayanagi, M., Harima, H. and Iwata, Y. (1963) *J. Soc. Mater. Sci. Jpn.*, **12**, 389.
12. Hashin, Z. and Shtrikman, S. (1963) *J. Mech. Phys. Solids*, **11**, 167.
13. Kerner, E. H. (1956) *Proc. Phys. Soc. B*, **69**, 808.
14. Hsu, W. Y. and Wu, S. (1993) *Polym. Eng. Sci.*, **33**, 293.
15. Bucknall, C. B. (1978) *Adv. Polym. Sci.*, **27**, 121.
16. Oxborough, R. J. and Bowden, P. B. (1974) *Philos. Mag.*, **30**, 171.
17. Bucknall, C. B. (1978) Fracture phenomena in polymer blends, in *Polymer Blends*, Vol. 2 (eds D. R. Paul and S. Newmann), Academic Press, New York, p. 91.
18. Truss, R. W. and Chadwick, G. A. (1976) *J. Mater. Sci.*, **11**, 1385.
19. Heikens, D., Sioerdsma, S. D. and Coumans, W. J. (1981) *J. Mater. Sci.*, **16**, 429.

20. Coumans, W. J. and Heikens, D. (1980) *Polymer*, **21**, 957.
21. Naqui, S. I. and Robinson, I. M. (1993) *J. Mater. Sci.*, **28**, 1421.
22. Brown, N. (1988) Creep, stress relaxation and yielding, in *Engineering Plastics*, Engineered Materials Handbook, Vol. 2, ASM International, Metals Park, OH, p. 728.
23. Sternstein, S. S. and Ongchin, L. (1969) *Polym. Prepr. Am. Chem. Soc.*, **10**(2), 1117.
24. Sultan, J. M. and McGarry, F. J. (1973) *J. Polym. Sci.*, **13**, 29.
25. Kausch, H. H. (1987) The role of chain scission in homogeneous deformation and fracture, in *Polymer Fracture*, Springer, Berlin, p. 251.
26. Bucknall, C. V. (1985) Fatigue of high impact polymers, in *Polymer Blends and Mixtures* (eds D. J. Walsh, J. S. Higgins and A. Macconachie), NATO Series, Martinus Nijhoff, Dordrecht, p. 363.
27. Hertzberg, R. W. and Manson, J. A. (1980) Introduction to fatigue, in *Fatigue of Engineering Plastics*, Academic Press, New York, p. 1.
28. Boukhili, R. and Gauvin, R. (1990) *J. Mater. Sci. Lett.*, **9**, 449.
29. Takemori, M. T. (1990) *Adv. Polym. Sci.*, **91–92**, 263.
30. Sauer, J. A. and Chen, C. C. (1983) *Adv. Polym. Sci.*, **52–53**, 169.
31. Ferry, J. D. (1961) Illustrative applied calculations, in *Viscoelastic Properties of Polymers*, Wiley, New York, p. 425.
32. Wullaert, R. A. (1970) *Impact Testing of Metals, ASTM Spec. Tech. Publ. 466*, p. 148.
33. Wu, S. (1985) *Polymer*, **26**, 1855.
34. Borgrave, R. J. M., Gaymans, R. J. and Eichenwald, H. M. (1989) *Polymer*, **30**, 78.
35. Wu, S. (1988) *J. Appl. Polym. Sci.*, **35**, 549.
36. Wu, S. (1992) *Polym. Int.*, **29**, 229.
37. Savadori, A. (1985) *Polym. Test.*, **5**, 209.
38. Plati, E. and Williams, J. G. (1975) *Polym. Eng. Sci.*, **15**, 470.
39. Williams, J. G. (1984) Impact testing and dynamic effects, in *Fracture Mechanics of Polymers*, Ellis Horwood, Chichester, Chapter 8, p. 237.
40. Lubert, W., Rink, M. and Pavan, A. (1976) *J. Appl. Polym. Sci.*, **20**, 1107.
41. Casiraghi, T., Castiglioni, G. and Airoldi, G. (1982) *Plast. Rubber Process. Appl.*, **2**, 353.
42. Braglia, R. and Casiraghi, T. (1984) *J. Mater. Sci.*, **19**, 2643.
43. Dunn, C. M. R. and Williams, M. J. (1980) *Plast. Rubber Mater. Appl.*, 90.
44. Bucknall, C. B. (1977) *Toughened Plastics*, Applied Science, Barking, p. 243.
45. Rink, M., Riccō, T., Lubert, W. and Pavan, A. (1978) *J. Appl. Polym. Sci.*, **22**, 429.
46. Bardeneir, R. and Cordes, M. (1989) Evaluation of the mechanical properties of polymers using advanced testing technology. First International Conference on Sandwich Construction, Royal Institute of Technology, Stockholm.
47. Williams, J. G. (1984) Systems with energy dissipation, in *Fracture Mechanics of Polymers*, Ellis Horwood, Chichester, Chapter 5, p. 93.
48. Griffith, A. A. (1920) *Philos. Trans. R. Soc. London, Ser. A*, **221**, 163.
49. Williams, J. G. (1984) Energy of elastic systems, in *Fracture Mechanics of Polymers*, Ellis Horwood, Chichester, Chapter 2, p. 21.
50. Williams, J. G. (1984) Elastic stresses around crack tips, in *Fracture Mechanics of Polymers*, Ellis Horwood, Chichester, Chapter 3, p. 41.
51. Brown, W. F. and Srawly, J. E. (1966) *ASTM Spec. Tech. Publ. 410*.
52. Karger-Kocsis, J., Kiss, L. and Kuleznev, V. N. (1982) *Acta Polym.*, **33**, 14.
53. Hashemi, S. and Williams, J. G. (1984) *J. Mater. Sci.*, **19**, 3746.
54. Carpinteri, A., Marega, C. and Savadori, A. (1986) *J. Mater. Sci.*, **21**, 4173.
55. Dugdale, D. S. (1960) *J. Mech. Phys. Solids*, **8**, 100.
56. Carpinteri, A. (1982) *Eng. Fract. Mech.*, **16**, 467.

57. Knott, J. F. (1979) General yielding fracture mechanics, in *Fundamentals of Fracture Mechanics*, Butterworths, London, p. 151.
58. Rice, J. R. (1968) *J. Appl. Mech.*, **35**, 379.
59. Begley, J. A. and Landes, J. D. (1972) The *J* integral as a fracture criterion, in *Fracture Toughness, ASTM Spec. Tech. Publ. 514.*
60. Williams, J. G. and Cawood, M. J. (1990) *Polym. Test.*, **9**, 15.
61. Hashemi, S. and Williams, J. J. (1986) *Polym. Eng. Sci.*, **26**, 760.
62. Hashemi, S. and Williams, J. J. (1991) *J. Mater. Sci.*, **26**, 621.
63. Paris, P. C. (1979) *ASTM Spec. Tech. Publ. 668.*
64. Will, P. and Michel, B. (1989) *Int. J. Fatigue*, **11**, 125.
65. Grellmann, W., Seidler, S. and Nezbedova, E. (1991) *Macromol. Chem. Symp.*, **41**, 195.
66. Newmann, L. V. and Williams, J. G. (1980) *Polym. Eng. Sci.*, **20**, 572.
67. Hodgkinson, J., Savadori, A. and Williams, J. G. (1983) *J. Mater. Sci.*, **18**, 2319.
68. Adams, G. C., Bender, R. G., Cronch, B. A. and Williams, J. G. (1990) *Polym. Eng. Sci.*, **30**, 241.
69. Williams, J. G. and Hodgkinson, J. (1981) *Proc. R. Soc. London, Ser. A*, **375**, 231.
70. Server, W. L. (1979) *Elastic Plastic Fracture, ASTM Spec. Tech. Publ. 668*, p. 493.
71. Savadori, A. (1984) *Polym. Test.*, **4**(1), 73.
72. Sunderland, P., Kausch, H. H., Schmid, W. and Arber, W. (1988) *Makromol. Chem., Macromol. Symp.*, **16**, 365.
73. Paris, P. C. and Erdogan, F. (1963) *J. Basic Eng.*, **85**(4), 528.
74. Friedrich, K. and Karger-Kocsis, J. (1990) Fracture and fatigue of unfilled and reinforced polyamides and polyester, in *Solid State Behaviour of Linear Polyesters and Polyamide* (eds J. M. Schultz and S. Fakinov), Prentice-Hall, Englewood Cliffs, NJ, Chapter 5, p. 249.
75. Clark, T. R., Hertzberg, R. W. and Manson, J. A. (1990) *J. Test. Eval.*, **18**, 319.
76. Chow, C. L. and Lu, T. J. (1990) *Mater. Sci. Lett.*, **9**, 1427.
77. Kadota, K. and Chudnovsky, A. (1992) *Polym. Eng. Sci.*, **32**, 1097.
78. Haglan, H. and Moet, A. (1969) *Int. J. Fract.*, **40**, 285.
79. Nimmer, R., Takemori, M., McGuire, J., Weins, K. and Morelli, T. (1988) Annu. 46th Technical Conf., Society of Plastics Engineers, ANTEC, p. 1503.
80. Kinloch, A. J. and Young, R. J. (1983) Toughened multiphase plastics, in *Fracture Behaviour of Polymers*, Applied Science, London.
81. Kunz-Douglas, S., Beaumont, P. W. R. and Ashby, M. F. (1980) *J. Mater. Sci.*, **15**, 1109.
82. Kunz, S. and Beaumont, P. W. R. (1981) *J. Mater. Sci.*, **16**, 3141.
83. Bucknall, J. C. B. and Yoshii, T. (1978) *Br. Polym. J.*, **10**, 53.
84. Hwang, J. F., Manson, J. A., Hertzberg, R. W., Miller, G. A. and Sperling, L. H. (1989) *Polym. Eng. Sci.*, **29**, 1477.
85. Karger-Kocsis, J. and Friedrich, K. (1992) *Colloid Polym. Sci.*, **270**, 549.
86. Huang, Y. and Kinloch, A. J. (1992) *J. Mater. Sci.*, **27**, 2753.
87. Huang, Y. and Kinloch, A. J. (1992) *J. Mater. Sci.*, **27**, 2763.
88. Brostow, W. and Macip, M. A. (1989) *Macromolecules*, **22**, 2761.
89. Brostow, W. (1991) *Makromol. Chem., Macromol. Symp.*, **41**, 119.
90. Iijima, T., Yoshioka, N. and Tomoi, M. (1992) *Eur. Polym. J.*, **28**, 573.
91. Karger-Kocsis, J. and Friedrich, K. (1993) *Compos. Sci. Technol.*, **48**, 263.
92. Sridharan, N. S. and Broutman, L. J. (1982) *Polym. Eng. Sci.*, **22**, 760.
93. Williams, J. G. and Hashemi, S. (1985) Fracture toughness evaluation of blends and mixtures and the use of the *J* method, in *Polymer Blends and Mixtures* (eds D. J. Walsh, J. S. Higgins and A. Macconachie), Martinus Nijhoff, Dordrecht, p. 289.

94. Giraud, Y. and Coletto, J. (1989) *Makromol. Chem., Macromol. Symp.*, **23**, 225.
95. Riccō, T. and Pavan, A. (1992) *Angew. Makromol. Chem.*, **201**, 23.
96. Friedrich, K. (1985) *Compos. Sci. Technol.*, **22**, 43.
97. Karger-Kocsis, J. (1989) Microstructure and fracture mechanical performance of short-fibre reinforced thermoplastics, in *Application of Fracture Mechanics to Composite Materials* (ed. K. Friedrich), Elsevier, Amsterdam, p. 189.
98. Narishawa, L. and Takemori, M. T. (1989) *Polym. Eng. Sci.*, **29**, 671.
99. Lee, C., Lu, M. and Chang, F. (1993) *J. Appl. Polym. Sci.*, **47**, 1867.
100. Narishawa, L. and Takemori, M.T. (1988) *Polym. Eng. Sci.*, **28**, 1462.
101. Dekkers, M. E. J. and Hobbs, S. Y. (1987) *Polym. Eng. Sci.*, **27** 1164.
102. Wert, J., Saxena, A. and Ernst, H. A. (1990) *J. Test Eval.*, **18**, 1.
103. Saleemi, A. S. and Nairn, J. A. (1990) *Polym. Eng. Sci.*, **30**, 211.
104. Mai, Y. W. (1989) *Polym. Commun.*, **30**, 330.
105. Mai, Y. W. and Powell, P. (1991) *J. Polym. Sci., Phys.*, **29**, 785.
106. Paton, C. A. and Hashemi, S. (1992) *J. Mater. Sci.*, **27**, 2279.
107. Joe, C. R. and Kim, B. H. (1990) *J. Mater. Sci.*, **25**, 1991.
108. Roulin-Moloney, A. C. (1989) *Fractography and Failure Mechanisms of Polymers and Composites*, Applied Science, London.
109. So, P. K. (1988) Fractography, in *Engineering Plastics*, Engineered Materials Handbook, Vol. 2, ASM International, Metals Park, OH, p. 805.
110. Casiraghi, T. (1978) *Polym. Eng. Sci.*, **18**, 833.
111. Cantwell, W. J. and Roulin-Moloney, A. C. (1989) Fractography and failure mechanism of unfilled and particulate filled epoxy resins, in *Fractography and Failure Mechanisms of Polymers and Composites*, Applied Science, London, p. 233.
112. Hobbs, S. Y., Bopp, R. C. and Watkins, V. H. (1983) *Polym. Eng. Sci.*, **23**, 380.
113. Hobbs, J. J. (1980) *J. Macromol. Sci. Rev., Macromol. Chem. C*, **19**, 221.
114. Hertzberg, R. W. and Manson, J. A. (1980) Fatigue fracture micromechanisms, in *Fatigue of Engineering Plastics*, Academic Press, New York, p. 146.
115. Karger-Kocsis, J. (1991) Structure and fracture mechanism of injection-molded composites, in *Encyclopedia of Composites*, Vol. 5 (ed. S. M. Lee), VCH, New York, p. 337.
116. Nimmer, R. (1988) Impact loading, in *Engineering Plastics*, Engineered Materials Handbook, Vol. 2, ASM International, Metals Park, OH, p. 679.
117. Turner, S., Read, P. E. and Money, M. (1984) *Plast. Rubber Process. Appl.*, **4**, 369.

5

Toughening agents for engineering polymers

H. Keskkula and D. R. Paul

5.1 INTRODUCTION

Most plastics designed for engineering applications are multiphase materials that contain elastomeric impact modifiers and often other polymers. The addition of toughening agents usually increases the overall ductility of the polymer over a wide temperature range but also improves resistance to notch sensitivity and toughness of thick sections, reduces water absorption, etc. As more knowledge of miscibility, compatibility and interaction of polymers has become available in recent years, more varied approaches to improved polymer blends have become apparent. Simultaneous with this, there has been an explosion of activity in the field of toughened engineering plastics. Many new tough polymer blends based on polyamides, poly(phenylene oxide) (PPO), polyesters and polycarbonates (PCs) have been introduced in large volume applications, particularly in the automotive industry [1].

The manufacture of engineering polymer blends grew tremendously in the 1980s. Before 1980, blend technology leading to engineering polymers was essentially unknown on a commercial basis [2–5], except, of course, for blends of PPO with high impact polystyrene that were introduced in the late 1960s by General Electric Company under the trademark Noryl®. Recently, a number of multicomponent engineering polymers that contain impact modifiers have been introduced. The world-wide patent activity of compositions useful for such applications has been explosive. Hundreds of patents on the modification of polyamides alone have been issued during the 1980s. It is clear that the growth of rubber toughened engineering plastics will continue at a very high rate involving many combinations of polymers. Table 5.1 shows the growth of engineering polymers in the 1980s.

Alloys and blends involving combinations of various engineering polymers designed for improved heat deformation characteristics, rigidity, solvent resistance and toughness will be the main area for significant growth. Such

Table 5.1 US sales of engineering polymers*

Polymer	Sales for the following years ($\times 10^6$ tons)[†]			Percentage increase since 1981
	1981	*1985*	*1989*	
Thermoplastic polyesters	0.56	0.65	0.96	71
ABS	0.44	0.48	0.57	30
Polycarbonate	0.11	0.13	0.28	155
Nylon	0.13	0.18	0.27	108
Epoxy	0.15	0.16	0.22	47
PPO and copolymers	0.06	0.08	0.08	33
Polyacetal	0.04	0.05	0.06	50

* Data from *Modern Plastics*, January issues.
† Tons are long tons; 1 ton ≈ 1000 kg.

blends are expected to account for approximately 160 000 tons of engineering thermoplastic consumption in the US for 1990 [6]. As seen from Table 5.1, this is a significant percentage of the combined US sales of nylon, PC and PPO (630 000 tons in 1989).

In this chapter, the general synthetic methods and description of impact modifiers used in engineering polymers will be reviewed, as well as some specific uses in various polymer blends. However, because of the very large number of possibilities, the impact modifiers for combinations of engineering polymers, e.g. nylon–PPO [2, 7], PC–polyester [2], will not be covered in detail, since many impact modifiers described for a specific polymer are often used for such blends.

5.2 BACKGROUND

5.2.1 Historical

Acrylonitrile, butadiene, styrene or ABS materials, developed in the 1950s, were the first family of rubber toughened polymers that were useful for engineering applications. ABS polymers have an excellent balance of strength, rigidity and toughness, but they are deficient in some critical properties such as heat deformation temperature and resistance to solvent attack. These deficiencies have limited their use in many applications. Styrene–acrylonitrile grafted rubbers such as those that give ductility to the SAN matrix in ABS are widely used as components to enhance toughness in a number of modern engineering polymer systems.

Blends of poly(phenylene oxide) and high impact polystyrene (HIPS) were introduced in the late 1960s by the General Electric Company under the Noryl® trademark. These products were useful in many applications that required high heat deformation temperature, mechanical strength and toughness. Because of the thermodynamic miscibility of PPO and PS, a large number of tailor-made products with varying impact strength and heat distortion temperature were possible by adjusting the ratio of PPO to polystyrene in the matrix [8]. Addition of flame retardants, fiber reinforcement, etc. opened additional applications for these plastics.

Polycarbonate was introduced in the late 1950s and found rapid commercial acceptance because of its transparency, high strength and high heat deformation temperature. However, its high melt viscosity, poor impact strength of thick sections, and high cost motivated development of improved compositions with better-balanced properties. ABS–PC blends were found to give a desirable combination of properties useful for many applications. These blends have steadily grown and today they are one of the principal engineering polymer alloys.

All three of these first engineering polymer blends were based on rubber modified styrenic polymers. This technology provided an excellent basis for the synthesis of starting materials for blends of engineering polymers.

5.2.2 Methods of manufacture of toughened polymers

(a) Melt blending

The first commercially significant polymer blends in the late 1940s were prepared by extrusion blending. PVC modification serves as an excellent example of this approach. Melt blending technology continues to be used for producing specialty formulations of HIPS, but most often those of ABS. With the advent of block copolymers and thermoplastic elastomers, more formulations were developed by melt blending with styrenic polymers. The development of PPO–HIPS blends by GE was entirely based on the melt blending approach. At the present time many of the multiphase engineering polymers are manufactured continuously by melt blending often in twin-screw extruders.

(b) In situ polymerization

The polymerization of styrene in the presence of dissolved butadiene based rubber with shearing agitation is used for the manufacture of high impact polystyrene [9, 10]. It is also used for the synthesis of some varieties of ABS like those that are used in blends with PC. Emulsion polymerization of SAN copolymers in the presence of rubber seed latex is the basis of a major portion of commercial ABS polymers, however. In this process the grafted rubber particle size is controlled by the size of the seed rubber particle. The small (0.05–0.2 μm) graft rubber particles from the emulsion process are desirable for the modification of many engineering polymers. *In situ* polymerization is also practiced in the rubber toughening of epoxy resins using low molecular weight carboxy-terminated nitrile rubbers (CTBN).

5.2.3 Requirements of toughening agents

The various toughening agents or impact modifiers described in the literature are manufactured by many synthetic approaches and involve numerous chemical compositions. There are several important criteria, however, that must be kept in mind to obtain tough engineering polymers successfully, regardless of the type of elastomer used. These requirements are important for obtaining commercially viable blends useful for demanding engineering applications.

It is recognized that the toughened engineering composites are all multiphase systems that may contain several separate polymer domains in addition to the discrete elastomer particles that enhance blend toughness [11–13]. Rubber particles need to adhere to the matrix for satisfactory stress transfer in most instances. An exception, however, is observed in blends of PC with polyethylene, where the non-adhering polyethylene domains cause a change in the mechanism of deformation of the blends that gives rise to improved impact strength of thick sections and specimens with sharp notches. Satisfactory adhesion is often obtained by the formation of chemical bonds between the matrix and the rubber phase. Physical interactions may also be used to provide adequate adhesion. In the modern view of polymer–polymer interfaces, interpenetration of the two polymers occurs to an extent related to their thermodynamic interaction that provides a mechanism for interfacial adhesion [14]. The adhering rubber particles often need to be quite small, typically less than 0.1–0.2 μm, and uniformly distributed. Wu has suggested that in modified nylon 6,6 there is a critical interparticle distance below which ductile compositions are observed [15–17]. Later Borggreve, Gaymans and coworkers [18–20] reached similar conclusions in studying the toughening of nylon 6. However, particles can also be too small for effective toughening [21, 22] and, in general, some optimum range of sizes is needed. Of course, there is often a critical concentration of toughening agent that must be exceeded. Another requirement for rubber toughened polymers is that the rubber phase morphology must not change during melt fabrication processes, i.e. rubber particle size and distribution should remain unaltered. This is usually assured by crosslinking the rubber phase, such as in the core–shell impact modifiers. The use of chemically modified gum rubber involves more variables for achieving a stable, controlled morphology. To achieve improved impact strength at low temperatures the glass temperature of the elastomer must be well below the desired use temperature.

In addition to the requirements summarized above, it is desirable that commercial toughening agents be thermally stable, easily dispersible and stable to oxidation and UV light.

5.2.4 Frequently toughened polymers

Engineering polymers that are most frequently rubber toughened include polyamides (Chapter 7), polyesters (Chapter 8), epoxy resins (Chapter 6), poly(phenylene oxide) [8], polycarbonates (Chapter 8) and polyacetals [23, 24]. Toughening of polyimides (Chapter 10) and polysulfones and polyarylether ketones is discussed in Chapter 9. Combinations of engineering polymers have been found to be useful in applications requiring high strength, high heat deformation temperatures, solvent resistance and toughness. Rubber modified PA–PPO [25, 26], PBT–PC, PA–PC, PET–PBT and PET–PC blends serve as examples of this fast-growing segment of engineering plastics.

Recent patent literature reveals that numerous elastomers, particularly those that contain some acid or polar functionality, are effective in toughening a variety of engineering polymers. The voluminous patent by Epstein [27] on rubber modification of polyamides serves as an excellent example. His claimed invention requires elastomer particles to be from 0.01 to 1 μm in diameter, and to adhere to the polyamide matrix, and to have a tensile modulus in the range from 0.007 to 138 MPa. Within this broad range of requirements for the rubber toughening component, a very large variety of elastomers is described in 168 examples and 55 claims [27].

5.3 GUM ELASTOMERS

A large number of unvulcanized, or gum, elastomers and rubbers are available for potential use in blends with engineering polymers. There are some limitations, however, that make some elastomers less desirable. Because of the high temperatures encountered in compounding and fabrication, thermal stability of the elastomer is of great importance. For this consideration, elastomers containing double bonds in the main polymer chain, e.g. polybutadiene or butadiene containing copolymers, are less desirable than those that do not, e.g. ethylene–propylene copolymers. Parts made from such materials may also be subjected to high use temperatures, UV light and oxygen. In these cases, blends containing butadiene based elastomers are subject to thermal or oxidative degradation and consequent loss of strength and discoloration. However, appropriate use of antioxidants permits polybutadiene rubbers to be used for impact modification of some engineering polymers. The morphological location of the rubber particles in the part may also help to alleviate the stability problem. Polycarbonate–ABS and PPO–HIPS blends serve as examples where polybutadiene based rubbers are used. Engineering polymers that require low temperature toughness benefit from the use of butadiene based elastomers, because their glass transition temperatures are significantly lower than most saturated elastomers such as those based on *n*-butyl acrylate or ethylene–propylene.

5.3.1 Butadiene based elastomers

A large number of elastomers based on diene monomers are available [28]. In addition to polybutadiene (PBD), elastomers are manufactured from isoprene and chloroprene, as well as a variety of copolymers with styrene, acrylonitrile, etc. The total US 1989 production of synthetic rubbers was about 2.3×10^6 tons and those based on butadiene constituted about 60% of the total [29, 30]. The predominance of butadiene elastomers has been reported world wide as well [31]. Gum PBD and copolymers are manufactured by a number of processes that give characteristic features to each. Free radical emulsion polymerization was first developed during World War II as a part of the Government Synthetic Rubber Program. GR-S, a butadiene–styrene (77:23) copolymer, was one of the first successful synthetic rubbers. It was polymerized with a so-called 'hot' recipe and the polymerization was carried to only about 60% conversion since higher conversions would lead to crosslinked rubbers. Subsequently, so-called 'cold' recipes were employed that permitted significant reduction of undesirable side reactions such as gel formation. By the use of certain organic hydroperoxides, polymerization was possible at around 0 °C. Sodium-initiated polymerization of butadiene was practiced in Germany in the 1930s, but this approach caused a high level of 1,2 enchainment of butadiene and, therefore, gave a high glass transition temperature (T_g). Sodium initiation, however, was a precursor to the development of the Ziegler–Natta and butyl lithium processes in the 1950s that are now used extensively to manufacture PBD and other diene based copolymers. Polybutadienes with extremes of microstructures ranging from 1,2 to the *cis*-1,4 addition are possible [28].

Hydrocarbon soluble gum rubbers based on butadiene have been used exclusively for the manufacture of HIPS and certain ABS polymers. In this process 5–15% of rubber is dissolved in styrene or styrene–acrylonitrile. The rubber solution is polymerized thermally or by peroxide initiators while applying shearing agitation. Agitation is particularly important during the early phases of polymerization when phase inversion takes place and the size of the rubber droplets is established. In the final stages of polymerization, the temperature is increased and the rubber particles become crosslinked during devolatilization. Rubber particles that are formed in the mass process have a complex structure and contain numerous PS or SAN occlusions. The presence of occlusions in the particles produces a large increase of rubber phase volume relative to the actual amount of rubber used. Thus, mass-made HIPS and ABS make very efficient use of the rubber [9, 10, 14, 32].

Butadiene based elastomers were reported as impact modifiers for polyamides already in the 1960s by Murdock, Nelan and Segall [33]. They used a butadiene–methyl methacrylate–methacrylic acid rubber for toughening a mixed polyamide matrix.

(a) Telechelic polymers

Low molecular weight butadiene–acrylonitrile copolymers with carboxy, hydroxy or amine end groups are frequently used for rubber toughening epoxy

Table 5.2 Characteristics of the reactive liquid rubbers [36]

Rubber	*Acrylonitrile (%)*	$\bar{M}_n$	*Viscosity η (MPa s) at 25 °C*	T_g *(°C)*	*End group*
CTB 162	0	4200	60 000	− 83	Carboxy
CTBN 8	18	3600	150 000	− 60	Carboxy
CTBN 13	26	3600	540 000	− 42	Carboxy
HTBN 34	17	3600	140 000	− 60	Primary hydroxy
VTBNX 23	16	3600	250 000	− 60	Secondary hydroxy + vinyl
ATBN	0–18	–	–	–	Aromatic amine

resins [34, 35]. Table 5.2 shows the composition of some liquid rubbers available from B. F. Goodrich. They are telechelic, solvent-free polymers with a number-average molecular weight in the range 3600–4200. The synthesis of the telechelic low molecular weight difunctional carboxy-terminated butadiene–acrylonitrile copolymers (CTBN) has been reviewed by Riew [37]. Free radical polymerization is typically used. Anionic polymerization may also be used in the absence of polar monomers. The resulting polymers have a narrow molecular weight distribution, but large concentrations of organometallic catalysts are needed to produce low molecular weight polymers. In free radical initiated processes that give broader molecular weight distributions, typically initiators which contain carboxy groups are used. Several reviews describe the chemistry, morphology and physical properties of epoxy resins containing carboxy-terminated rubber particles [14, 34].

While most of these polymers are used for toughening epoxy resins, bismaleimide resins are toughened by incorporation of CTBN [38, 39]. Maleimide resin prepolymers are produced by the reaction of maleic anhydride with an aromatic diamine followed by the incorporation of the CTBN rubbers, molding and then curing at 200–250 °C to obtain the final toughened resin.

The use of amine-terminated rubbers (ATBN) in polyimides is reported also [40]. Here, another diamine, 1,4-piperazine, is added to the bismaleimide prepolymer. A rigid polymer of the following structure results:

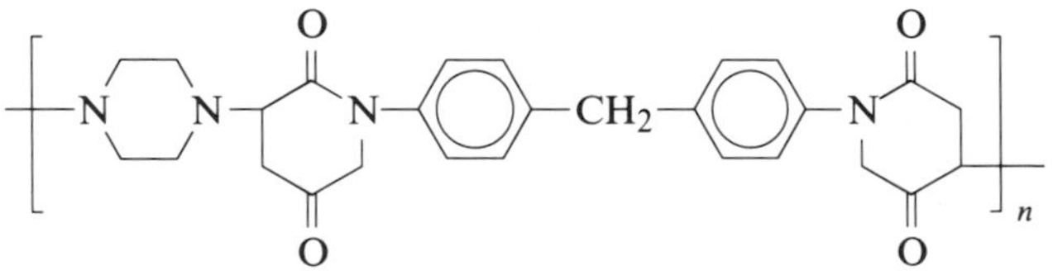

To this polymer ATBN rubber is added, and the reaction is completed in a two-step process yielding a random multiblock structure of rigid polyimide sequences alternating with flexible ATBN blocks. In a similar system the chemical incorporation of ATBN into a polyimide yielded films with improved tear resistance and showed a two-phase morphology, typical of rubber toughened polymers [41].

5.3.2 Ethylene based elastomers [42]

(a) Ethylene–propylene copolymers

Ethylene–propylene (EP) random copolymers, also referred to as EPDM or EPM, possess many desirable features that make them attractive as toughening agents for engineering polymers, e.g. excellent thermal and UV stability and satisfactory glass transition temperature of about −50 °C. Chemical modification of EP copolymers or EP–diene monomer (EPDM) terpolymers by, for example, maleic anhydride, to enhance interaction or to cause reactivity between the elastomer and the matrix may be accomplished using either melt extrusion or solution processes.

EPM and EPDM elastomers are synthesized by Ziegler–Natta processes. These products and processes have been reviewed in some detail [43]. In connection with their use as modifiers for engineering polymers, several aspects of their synthesis should be mentioned. In order to produce an elastomer that is not crystalline or has a low degree of crystallinity and that has a sufficiently low glass temperature, the content of propylene is usually between 25 and 50 mol.%. The use of most functional monomers is not possible in EP polymerization, as they deactivate the catalyst. In fact, monomers must be free of all active hydrogen compounds; <10 ppm is sufficient to stop the polymerization [43]. EPDM is a terpolymer that contains a small amount of a non-conjugated diene monomer, such as 5-ethylidene-2-norbornene, 1,4-hexadiene or dicyclopentadiene. They are chosen because they interfere with EP polymerization less than the conjugated dienes and provide a double bond away from the main chain. This latter point is important, so that ozone or oxygen attack does not result in the reduction of molecular weight of the elastomer or formation of gel. Furthermore, the off-main-chain double bond provides excellent sites for grafting with reactive monomers and for cross-linking. It is also important to note that melt processing of EPM polymers is strongly dependent on molecular weight and molecular weight distribution. In general, branched polymers with a broad molecular weight distribution are more resilient, swell more and extrude faster.

Typically, EPM and EPDM are used in engineering polymer blends together with maleic anhydride modified elastomers. Particularly, functionalized ethylene–propylene rubbers are used in rubber toughening of polyamides to control the rubber phase adhesion, particle size and uniformity of dispersion [15, 44]. EPM rubbers may also be functionalized by reacting with glycidyl methacrylate. Such polymers are used to toughen polyamides and

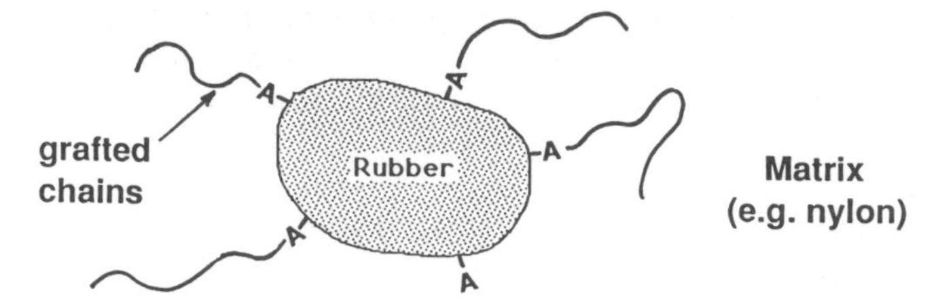

Fig. 5.1 Schematic of a maleic anhydride modified rubber particle grafted with nylon (rubber is EPDM-g-MA or SEBS-g-MA). (Reprinted from D. R. Paul, *Mechanical Behaviour of Materials – VI*, Copyright 1991, Page 287, with kind permission from Pergamon Press Ltd, Headington Hill Hall, Oxford OX3 0BW, UK.)

polyesters [45, 46]. Often, combinations of functionalized elastomers are used with commodity EPM and EPDM [19, 44]. A schematic illustration in Fig. 5.1 shows grafting of matrix chains to the rubber through reactive sites for adhesion and control of the morphology of the rubber phase.

EPDM rubbers may also be grafted with a variety of monomers. EPDM-g-SAN is available from Uniroyal Chemical under the trade name Royaltuf® 372 [47]. It is useful for impact modification of PC blends with PBT and PET [48–50] and for the recycling of PET beverage bottles that contain a polyethylene base [50].

(b) Ethylene copolymers

Ethylene can be copolymerized with a variety of comonomers by using a high pressure, free radical process. Many polar or functional monomers, e.g. vinyl carboxylic acids, esters and anhydrides, carbon monoxide, sulfonates [27, 51–53], are suitable as minor comonomers to provide sites for chemical reactions or adhesive bonding when blended with appropriate matrix polymers. A long list of ethylene copolymers and terpolymers suitable for blending with polyamides is given in a 1965 reference [52] and later by Epstein [27].

Typical copolymerization of ethylene and acrylic or methacrylic acid involves the use of about 3–6 mol.% acid that is randomly incorporated into the copolymer. Usually the polymerization is limited to low conversion to obtain gel-free and homogeneous copolymers. Ethylene copolymers and terpolymers containing vinyl acetate are also widely produced including terpolymers with monomers such as glycidyl methacrylate and half esters of maleic acid [54, 55]. In addition to the blends with polyamides, ethylene copolymers are used for impact modification of polybutylene terephthalate [45].

(c) Ionomers [56]

Most commercially available ionomers useful for toughening engineering polymers are based on ethylene and acrylic or methacrylic acid copolymers.

They contain less than about 10 mol.% of acid monomer, are partially neutralized with zinc or sodium cations and are available from DuPont under the trade name Surlyn®. In these polymers, the ionic groups interact to form ion-rich domains that act as physical crosslinks and have a strong influence on the mechanical properties of the ionomer. The structure and properties of the ionomers are influenced by the type of the polymer backbone, ionic content (amount of carboxy functionality and degree of neutralization) and the type of cation. It is clear that a variety of ionomers are possible. In addition to their uses in blends with engineering polymers, they are useful in a number of packaging applications, sporting goods, footwear etc. Some toughened polyamide blends reported in the patent literature contain pure ionomers [27, 55] or are used in combination with other elastomeric impact modifiers [27].

5.3.3 Polyurethanes

The composition and properties of polyurethane elastomers can be varied over a wide range by an appropriate choice of soft and hard segments. Typically, the soft segments are based on long chain polyester or polyether diols and the hard segments are aromatic diisocyanates connected by chain extenders. Hydroquinone bis(2-hydroxyethyl) ether and 1,4-butanediol are often used as chain extenders. The choice of diol is most important in determining the T_g of the elastomer. For instance, poly(butylene adipate) diol gives $T_g = -40\,°C$, while poly(tetramethylene oxide) diol results in an elastomer with a T_g of $-80\,°C$. An excellent review of thermoplastic polyurethane elastomers is reported by Meckel, Coyert and Wieder [57].

The choice of monomers, monomer ratios, sequence of addition, and the method of polymerization are all important in controlling the nature of the elastomer. Most conveniently, polyurethane elastomers are synthesized in a single-pass extrusion process, yielding elastomer granules directly. However, it should be noted that the thermal history of the process is a critical variable, since there is a phase separation between the soft and hard segments that can affect phase morphology and crystallinity. A large number of references are given by Meckel, Coyert and Wieder [57] on the synthetic and morphological aspects of polyurethanes.

The use of thermoplastic polyurethane elastomers in blends with engineering polymers is clearly suggested by their polar nature and is related to a large number of variables that may affect the performance of the blend. Some of these variables are (a) chemical and physical interaction with the matrix, (b) elastomer particle size and size distribution, (c) elastomer micromorphology, e.g. crystallite size distribution, and (d) effect of thermal history on the elastomer phase morphology. The choice of polyurethane elastomers for the modification of engineering polymers depends to a large extent on the intended application. The facts that polyurethanes, especially those based on polyesters, have a propensity for hydrolysis at elevated temperatures and are attacked by a number of organic solvents limit their use. Nevertheless, a large number of blends of thermoplastic polyurethane (TPU) elastomers with engineering polymers have been reported. Poly(oxymethylene) or acetal resins are effectively toughened by the incorporation of TPU elastomers [24]. Toughened acetal resins are available from DuPont under the trade name of Delrin® 100 ST and 500T [23]. Other toughened polymers are also available from BASF and Hoechst-Celanese. They have an excellent balance of mechanical properties and are used in a variety of applications such as appliances, gears and outdoor equipment. Other polyurethane blends with polycarbonate [58] and with ABS or related styrenic polymers have been reported [57]. Ternary blends of TPU–polycarbonate–ABS or TPU–polycarbonate–PBT are also described in the patent literature [57]. These ternary blends are primarily targeted for the transportation applications where improved processing characteristics and solvent resistance are needed.

5.4 EMULSION MADE ELASTOMERS

Emulsion polymerization has been widely used since the 1940s to manufacture elastomers. As described earlier, butadiene–styrene rubbers were produced in large quantities under the Government Synthetic Rubber Program. In this section a distinction is made between such gum rubbers and the elastomers obtained by high conversion processes that usually yield crosslinked rubber particles or microgels if based on butadiene as a major monomer. Such polymers, first synthesized in the late 1940s, were designated as GRS 2004 and GRS 2003. They were crosslinked polybutadiene and 70:30 butadiene–styrene copolymer latexes respectively. In addition to foam rubber, they were used for impact modification of styrenic polymers [59]. The development of high conversion butadiene–styrene copolymers for latex paints by the Dow Chemical Company took place in the late 1940s. These copolymers were also used in blends to toughen polystyrene [59, 60]. The manufacture of Cycolac® brand by ABS Borg-Warner was one of the most successful emulsion graft polymerization processes developed to date where GRS 2004, a latex with a broad particle size distribution, was used as a seed for the polymerization of styrene–acrylonitrile [61]. A substantial amount of SAN was chemically bound to the PBD latex particles, while some SAN copolymer was also produced independently. The resulting ABS composition and properties are affected by the emulsion polymerization variables such as the amount of monomers, the amount and type of initiator, emulsifier concentration, reaction temperature and time, and chain transfer agents. Such polymers can be blended with additional SAN to alter the rubber concentration and the properties of the ABS.

5.4.1 Core–shell graft polymers

The grafted rubber particles in Cycolac ABS involve sequential emulsion polymerization steps like those used to form core–shell type modifiers to be discussed here; however, their formation may be considered as a result of *in situ* polymerization of ABS. Core–shell modifiers available at the present time for toughening engineering polymers, however, are specifically designed for this purpose and have minimal ungrafted shell material. The elastomer core is usually a copolymer based on either butadiene or *n*-butyl acrylate. The rubber core is typically crosslinked to give final products with permanent rubber particle morphology that is not affected by melt blending with the matrix polymer. Butadiene based core elastomers have a lower T_g in the range from -60 to $-85\,^{\circ}C$, while those based on *n*-butyl acrylate have a higher T_g of $-46\,^{\circ}C$ and above. Often T_g values of commercially available impact modifiers may be somewhat higher because of added comonomers. Typically, butadiene based graft copolymers give blends with better low temperature toughness than those based on acrylate elastomers. Of course, the acrylate based core–shell modifiers have better thermal and UV stability.

The latex particle size for impact modification is of crucial importance. Toughening of styrenic copolymers requires that some particles be relatively large, e.g. > 0.2 μm, while rubber modification of engineering polymers usually is effective with smaller particles in the range of < 0.2 μm. Typically, emulsion polymer particles range from about 0.1 to 0.3 μm in diameter, although much smaller particles may be produced, where the micelle size is the ultimate limiting factor. Also larger particles above 1 μm are possible through various agglomeration techniques. They include chemical approaches, pressure agglomeration, addition of solvents, electrolytes and water soluble polymers. At zero gravity in space, however, it is possible to grow uniform latex particles to a very large size, so far up to 30 μm (J. M. Vanderhoff, personal communication).

The formation of the grafted shell can be accomplished by using a number of different approaches. For instance, the graft monomer(s) may be added in a single shot or in multiple shots of constant or varying composition. Thus, multiple shell layers of different type or composition can be produced. A number of commercially important methyl methacrylate grafted core–shell polymers by Rohm and Haas are manufactured by such processes [30, 62]. Another approach involves the continuous addition of monomers [63]. Each of these approaches will lead to somewhat different types of graft structures and micromorphologies as shown schematically in Fig. 5.2.

The amount of polymer grafted is an important variable that affects the nature of the graft layer. Typically, the larger the amount of the monomer to be polymerized, the lower is the fraction that is chemically bound to the rubber particle. At the same time, however, the graft layer is thicker and usually more uniform. For example, BL-65, a Sumitomo-Naugatuck product, has 50% rubber and 50% SAN copolymer, of which 40% is chemically bound to the rubber and 60% is not [64]. Graft rubbers that have a high level of non-rubbery component are usually easy to handle as a free-flowing powder or a granulated material. A graft polymer that contains from about 70% to 80% rubber has a significantly different shell that may be visualized in some cases as

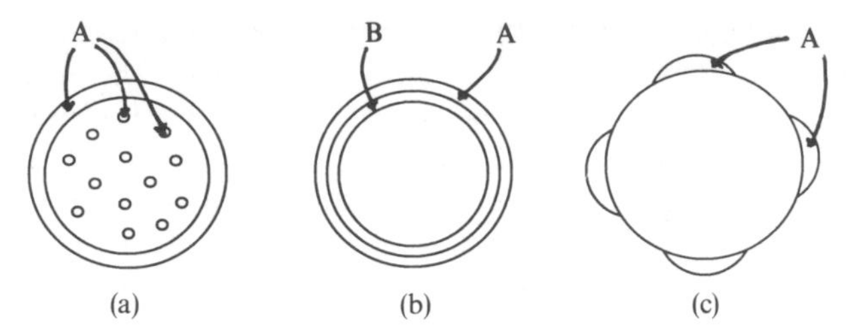

Fig. 5.2 Schematic representation of core–shell particles with varied structures: (a) particle with a uniform shell of polymer A, core contains some polymer A inclusions formed during the grafting step; (b) particle with a dual grafted shell [30]; (c) particle with a partial shell of polymer A [65].

shown in Fig. 5.2(c) [65]. Some of these polymers may tend to clump during storage and their effectiveness as an impact modifier may be quite different than those with thicker graft layers. However, some of the core–shell modifiers from Rohm and Haas have a high rubber content but are available in granules that can be dispersed to the ultimate emulsion particles by melt blending.

As discussed above, there are many important variables in emulsion polymerization of core–shell impact modifiers. It is important to point out that these variables may all be independently controlled in designing an engineering polymer blend, giving a wider latitude of variables for optimization of blend properties in critical engineering applications. For example, particle size can be set independently of the chemical formulation used to ensure adhesion to the matrix, which cannot be done straightforwardly in elastomers that have no fixed particle geometry.

Core–shell impact modifiers for engineering polymers often have a methyl methacrylate copolymer outer shell [30, 66]. This shell composition is quite useful since PMMA is thermodynamically miscible with or is wetted by a large number of polymers [11, 14, 32]. Consequently, such core–shell latex particles disperse readily during melt blending and in the solid state adhere to

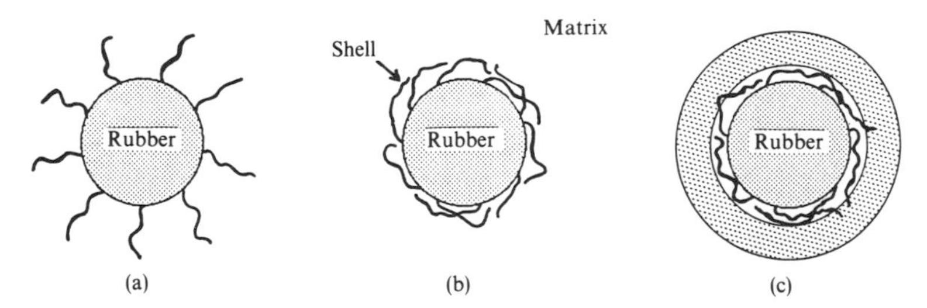

Fig. 5.3 PMMA grafted core–shell modifiers showing physical coupling: (a) miscibility of PMMA with SAN or PVC matrix; (b) wetting–adhesion of PMMA with polycarbonate; (c) wetting agent (e.g. PC) that wets or adheres to PMMA and the matrix (e.g. certain polyesters). (Reprinted from D. R. Paul, *Mechanical Behaviour of Materials – VI*, Copyright 1991, Page 287, with kind permission from Pergamon Press Ltd, Headington Hill Hall, Oxford OX3 0BW, UK.)

many matrix polymers. Apparently, because of a satisfactory physical interaction between the MMA based shell and a number of matrix polymers, they have been widely used as toughening agents. A schematic illustration in Fig. 5.3 shows a number of ways that the PMMA shell is useful in blends with several types of polymers. It indicates that effective toughening is obtained through its miscibility with the SAN or PVC matrix [32], or through wetting or adhesion with polycarbonate. Furthermore, polycarbonate may act as a wetting agent for polyesters as shown in Fig. 5.3(c). As an example, a PMMA grafted *n*-butyl acrylate rubber has been used by itself [66] or in combination with other impact modifiers for improving toughness, resistance to heat aging [67] and stress crack resistance [68] of polycarbonate. Several polyesters, such as PBT and PET, may be toughened to 800–1000 $J\,m^{-1}$ notched Izod impact levels with the use of PMMA based core–shell polymers. Also, blends of polyesters, e.g. PET–PBT, and blends of polyester–polycarbonate are similarly toughened [30]. These graft copolymers have also been found to be useful as impact modifiers for styrenic polymers [63]. Furthermore, they are effective in mixed matrix blends. Even blends that contain mixed immiscible components, e.g. ABS, HIPS, PC and polyesters, retain toughness in blends that contain about 20% of appropriate MMA grafted elastomer particles [69]. These results indicate that some MMA type core–shell particles may be effective for certain scrap polymer recycling in the form of 'compatibilized' blends.

5.4.2 Core–shell graft polymers with reactive sites

The use of MMA grafted rubber particles for toughening of a variety of polymers often does not require chemical bonding with the matrix [27]. Toughening of polyamides, however, seems to require such strong interaction. For this purpose, core–shell grafted rubbers with reactive sites have been developed. Figure 5.4 schematically shows two ways such particles can be used for modifying polyamides. In principle, reactive monomers, containing carboxy or other functional groups, can be introduced in the last stage polymeri-

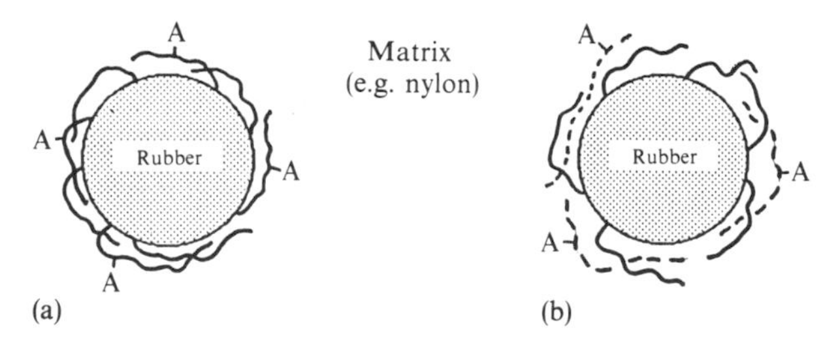

Fig. 5.4 Core–shell modifier attachment to a matrix polymer: (a) reaction of functionalized grafted chains with the matrix; (b) reaction of matrix with functionalized additive, e.g. styrene–maleic anhydride or styrene–acrylic acid copolymer, that is miscible with the shell. (Reprinted from D. R. Paul, *Mechanical Behaviour of Materials – VI*, Copyright 1991, Page 287, with kind permission from Pergamon Press Ltd, Headington Hill Hall, Oxford OX3 0BW, UK.)

zation [70]. This process leads to a product that has reactive sites on the surface of the grafted rubber particles. Alternatively, the impact modifiers may be blended with a reactive polymer that is miscible with the PMMA shell as shown in Fig. 5.4. During melt blending with polyamides the functional sites react with the amine end groups or amide linkages in the polyamide to provide chemical coupling between the phases [71]. This reaction appears necessary for an effective toughening of polyamides [72].

Mixed matrices of PPO and nylon 6,6 are also rubber modified in a similar way. In these cases, some compatibilization of the two matrix components is required in order to obtain the proper dispersion and morphology as well as good interfacial adhesion between these phases. Appropriate processing with reactive components, such as maleic anhydride, has been used for this purpose [73].

5.5 BLOCK COPOLYMERS

5.5.1 Butadiene–styrene block copolymers [74]

Styrene based block copolymers made by anionic polymerization are frequently used for toughening polymers. A number of books and reviews have been published on the synthesis, morphology and properties of these polymers [75–77].

By controlling the initiator concentration and the amount and the sequence of monomer addition, a large variety of structures are possible. Block sequence and size may be changed to obtain polymers of controlled phase structure, chemical nature, and physical properties. Tapered block polymers have a copolymer sequence between the distinct homopolymer block segments. So-called 'star' structures are formed by using multifunctional terminators–initiators [77].

The soft block is typically based on polybutadiene or polyisoprene. These blocks may be subsequently hydrogenated to eliminate substantially double bonds and to render the polymer more stable [78]. Structures resembling ethylene–propylene or ethylene–butene can be made in this way by controlling the microstructure of the diene block during its formation. Styrene is most often used as the monomer to produce the hard block. Other, non-polar monomers may also be used such as α-methyl styrene to increase the softening temperature of the rigid domain [77]. Polar and ionic monomers, however, cannot be used in anionic polymerization. Marginally, MMA may be used, but the low polymerization temperatures required render such block copolymers economically unattractive.

It should be pointed out that, in addition to the anionic mechanism, a number of other polymerization methods have been used for the synthesis of graft and block copolymers including cationic, free radical, irradiation and mechanochemical processes. The mechanochemical process has been of considerable interest since most blends are made by mechanical mixing in an extruder. An interesting review of graft and block copolymers by Ceresa described their synthesis by mechanochemical processes [75].

5.5.2 Specialty graft and block copolymers

(a) Grafted rubbers

Butadiene based elastomers can be readily grafted with a variety of monomers, utilizing solution or emulsion processes [9]. For illustration, Dean [79] has reported grafting EPDM with a styrene–methyl methacrylate–methacrylic acid terpolymer. This approach involves the peroxidation of the rubber first, followed by a solution polymerization of the three monomers in the presence of the peroxidized rubber. The resulting graft copolymer was found to be a toughening agent for nylon 6. The grafted component was also determined to be thermodynamically miscible with the polyamide.

Another approach involves grafting a preformed ethylene–ethyl acrylate copolymer with nylon 6 during the polymerization of ε-caprolactam in the presence of sebacic acid as a polymerization terminator [53]. The resulting graft rubber has a good balance of tensile strength, rigidity and impact strength.

(b) Block copolymers

Polyamide–polyether block copolymers with ester linkages have been reported to be useful materials or components in blends [80]. Thermoplastic elastomers based on polyamide with polyether, or polyester, blocks are available from Atochem and General Electric. A variety of compositions are possible. Such block copolymers from Atochem are available under the trade name Pebax and have some unique properties that illustrate the potential of such an approach. In addition, they are useful in blends with various inorganic and organic additives. They produce desirable blends with polyamides, PVC and even polyolefins. Also, novel PET and PBT blends with a polyesteramide impact modifier have been reported [81].

5.5.3 Hydrogenated block copolymers [78]

Elimination of double bonds in the main polymer chain greatly enhances thermal, oxidative and UV stability. Hydrogenation of styrene–butadiene block copolymers readily accomplishes this. It is possible to hydrogenate such polymers completely, including the benzene ring [82]. However, in most cases only the elastomeric block segment is hydrogenated. Depending on the microstructure of the polybutadiene segment, i.e. percentage 1,2 enchainment of butadiene, a variety of styrene–ethylene–butene-1-styrene (SEBS) block copolymers are obtained. The center block should be substantially amorphous and have a relatively low T_g. Commercial systems are usually optimized for the best compromise and have glass transition temperatures in the range from about −55 to −60 °C [78]. Of course, non-hydrogenated PBD segments would have a lower T_g of about −80 °C.

Partial hydrogenation of a polyisoprene in heptane solution was first described in 1963 using a triisobutylaluminum and tetraisopropyl titanate catalyst [83]. Details of the hydrogenation of high 1,2 addition polybutadienes have also been reported [84]. Hydrogenation of butadiene in SBS block copolymers to a high conversion for elimination of aliphatic double bonds was first described in 1972 [85]. Here, nickel acetylacetonate–triethylaluminum catalysis was used to hydrogenate 98% of the aliphatic unsaturation. Block copolymers are also hydrogenated in a solvent mixture of *n*-hexane and cyclohexane by use of cobalt naphthenate and triethylaluminum as catalyst at a hydrogen pressure of 7 kgf cm^{-2} (690 kPa) and 50 °C for 5 h [86]. Complete hydrogenation of double bonds may not be achieved or desired. The unreacted double bonds may be useful as reactive sites for further chemical modification. However, incomplete hydrogenation reduces the stability of the block copolymer. Commercially available SEBS, e.g. Kraton® 1652 (low molecular weight) and 1651 (high molecular weight), are essentially completely hydrogenated, however. As in all elastomers, carefully selected antioxidants are added to provide protection against oxidative and thermal degradation. Styrene containing block copolymers and this derivatized modification have been used extensively for toughening a large group of engineering polymers including polyamides, polyesters, PPO and polycarbonate and in blended matrix systems [76, 87–90].

5.5.4 Functionalized block copolymers

Styrene–butadiene–styrene and styrene–EB–styrene type block copolymers are frequently used as starting materials for the synthesis of functionalized rubber modifiers. Because they are not chemically crosslinked, chemical reactions in solution are readily carried out. However, performing these chemical reactions in an extruder offers a faster and economically attractive alternative. There are a number of chemical reactions of SBS and SEBS block copolymers described in the patent literature [91]. As a commercial example, Shell Chemical Company has introduced a maleic anhydride functionalized version of an SEBS block copolymer, Kraton® FG1901X, as an impact modifier for polyamides and polyesters. The use of this functionalized SEBS gives significantly reduced rubber particle size in polyamides in comparison with the blends with unmodified SEBS. A large increase in the notched Izod impact is achieved for nylon 6, nylon 6,6, PBT and PET by blending with Kraton® FG1901X or in combination with unmodified SEBS polymers [21, 70, 87, 88]. Effective use of block copolymers is also found in multicomponent blends with PPO, polypropylene and polyamides [76, 91]. Polystyrene segments in SEBS can also be functionalized by lithiation, followed by reaction with carbon dioxide [92]. SEBS copolymers with carboxy-modified polystyrene segments and double-bond-free rubbery center blocks give very thermally stable toughening agents for impact modification of polyamides and polyesters [92].

5.5.5 Morphology

The morphology of block copolymers has received considerable theoretical and technological attention since their discovery [32, 75, 93]. The microphase separation of the component blocks is influenced by the composition, architecture, sequence length and the average molecular weight of the block copolymer [93]. The hard domains act as physical crosslink sites that can be reversibly disrupted and reformed by heat and, thus, they are called 'thermoplastic' elastomers. In blends with other polymers, block copolymer domains are formed as a result of macrophase separation and they vary considerably in size and shape. Figure 5.5 shows a schematic of how block copolymer may be dispersed in matrices such as PS or PPO from the microscale to macrodomains. This is compared with an actual photomicrograph of an SEBS domain in polycarbonate. High molecular weight block copolymers disperse in an approximately spherical form up to several microns in diameter, while those of the lower molecular weight block copolymers assume a stringy type morphology. Photomicrographs in Fig. 5.6 for blends of two types of SBS copolymers, Kraton 1101 ($\bar{M}_w = 93\,000$) and Kraton 1102 ($\bar{M}_w = 65\,000$), with high impact

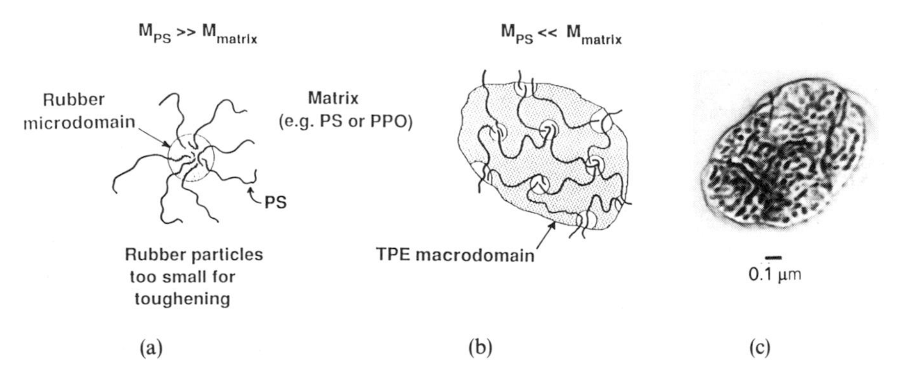

Fig. 5.5 Schematic of SBS block copolymer particle morphology in several matrix polymers: (a) very small rubber particles when $M_{PS} \gg M_{matrix}$ (polystyrene or PPO); (b) large rubber particles when $M_{PS} \ll M_{matrix}$; (c) STEM photomicrograph of an SEBS particle dispersed in polycarbonate [89]. (Parts (a) and (b) reprinted from D. R. Paul, *Mechanical Behaviour of Materials – VI*, Copyright 1991, Page 287, with kind permission from Pergamon Press Ltd, Headington Hill Hall, Oxford OX3 0BW, UK; part (c) reproduced with permission from D. Gilmore and M. Modic, *Society of Plastics Engineers, Technical Papers*, **47**, 1371, 1989.)

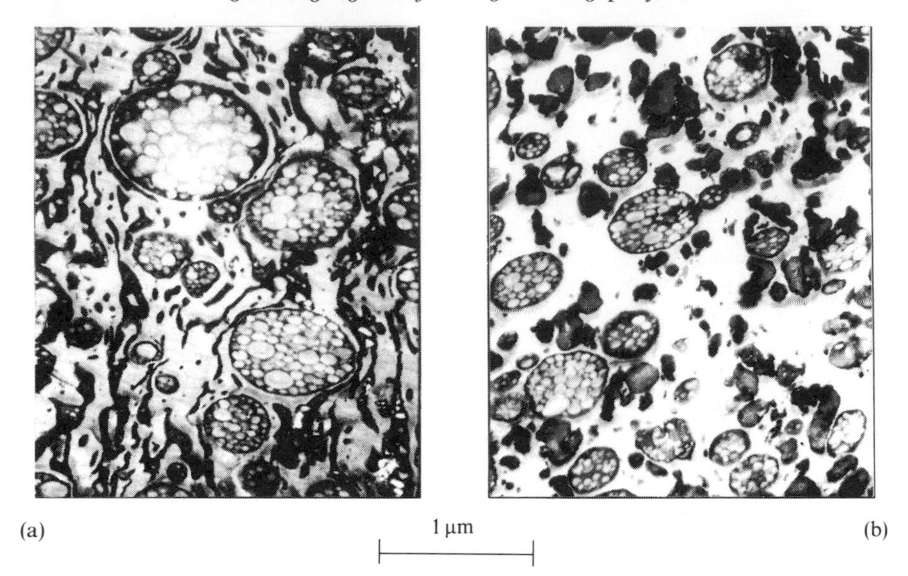

Fig. 5.6 Dispersion of 15% SBS block copolymer in high impact polystyrene (transmission electron microscopy of ultrathin OsO_4 stained sections): (a) Kraton 1102 ($\bar{M}_w = 65\,000$, 29% styrene); (b) Kraton 1101 ($\bar{M}_w = 93\,000$, 29% styrene). Complex particles with polystyrene occlusions are from HIPS. (Reprinted with permission from Ref. 94. Copyright 1982 MMI Press.)

polystyrene illustrate the two types of morphologies [94]. The high molecular weight SBS is dispersed as solid particles, while those from the lower molecular weight SBS show highly elongated irregular morphologies.

5.6 MALEIC ANHYDRIDE MODIFIED POLYMERS

5.6.1 Maleic anhydride reactions

Maleic anhydride (MA) is frequently used to functionalize elastomers as mentioned above. The free radical chemistry of MA has been reviewed by a number of researchers [95, 96]. As early as the 1930s, Farmer and Wheeler applied for a patent to modify rubbers with MA [97]. They discovered that heating a natural rubber solution in the presence of MA and benzoyl peroxide for several hours at 70–100 °C led to new elastomers that were significantly altered in their physical behavior. Gel-free products resulted using a low concentration of MA and benzoyl peroxide. Typically, MA does not homopolymerize but may copolymerize to form an alternating copolymer for example with styrene. Importantly, it is capable of reacting with polymeric double bonds and free radicals. For instance, an 'ENE' reaction [71, 92, 96] involves an allylic hydrogen reacting with the double bond of MA as shown in Fig. 5.7. This scheme depends on the unsaturation in the base polymer for

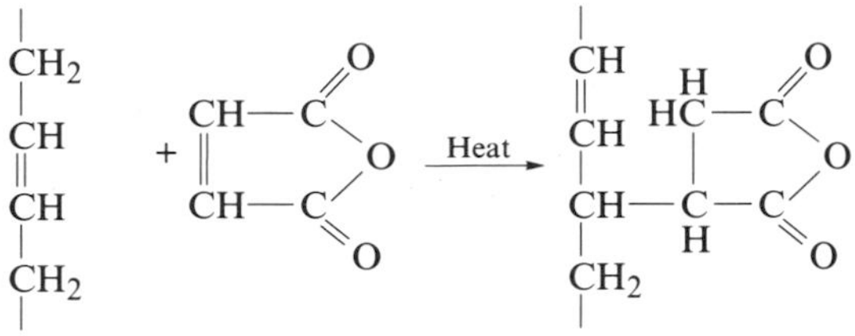

Fig. 5.7 Maleic anhydride reaction with elastomers containing main chain or side chain unsaturation [91, 98, 99].

reaction sites and does not require an initiator. Accordingly, some remaining unsaturation may be desirable in the hydrogenated butadiene based polymers to obtain MA attachment by this reaction to the polymer. EPDM rubbers, where the unsaturation is in the side chain, also react readily with MA by the same mechanism.

Saturated hydrocarbon polymers can also be functionalized with MA. Exposing the polymer to sufficiently high temperatures or the use of organic peroxides to form polymeric free radicals will cause the addition of MA to the polymer chain by the scheme shown in Fig. 5.8. There are many references, particularly in the patent literature, that show the capability of MA to react with saturated and unsaturated hydrocarbon elastomers [7, 27, 91, 98, 100–103]. For instance, when a low molecular weight saturated EPM rubber was reacted with maleic anhydride in a dry hexane solution containing 5% dicumyl peroxide at temperatures of about 150 °C for 60 min, the reaction product contained 1.7% combined maleic anhydride [101]. Commercial SEBS block copolymers can be reacted with maleic anhydride in solution at 81 °C with benzoyl peroxide as initiator, or in an extruder at about 260 °C with 2,5-dimethyl-2,5-di(*t*-butylperoxy)hexane as initiator. Increasing the initiator concentration from 0.01% to 0.1% caused an increase of maleic anhydride addition from 0.2% to 1.6% [91]. Hydrogenated butadiene polymers that contain residual unsaturation of the order of 0.5–20% are readily reacted with maleic anhydride [98, 102]. The maleic anhydride adduct is formed by the 'ENE' type reaction illustrated in Fig. 5.7. The hydrogenated polymer and MA are dissolved in toluene, and the solution, containing 5% MA based on the weight of hydrogenated polymer, is heated at 200 °C for 24 h. This reaction

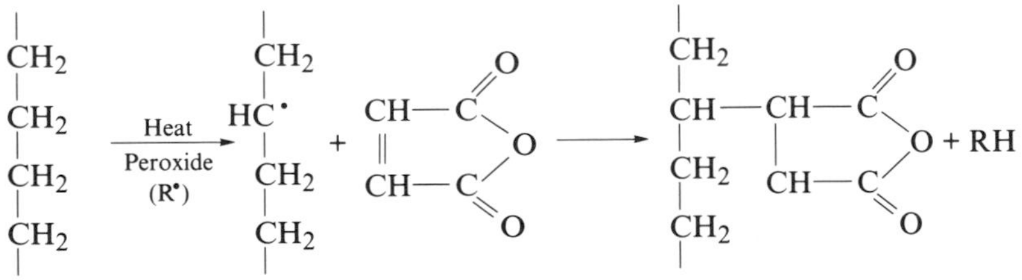

Fig. 5.8 Maleic anhydride reaction with saturated hydrocarbon elastomers.

Table 5.3 Commercially available maleated toughening agents used in engineering polymers

Elastomer	*Source*	*Trade name*	*Reference*
EPM	Exxon*	Exxelor VA-1801	[106]
	Exxon	Exxelor VA-1803	[106]
EPDM	Uniroyal	Royaltuf 465A	[44, 86]
Hydrogenated SBS	Shell Development	Kraton® FG1901X	[88]
E–alkyl acrylate–MA terpolymer	CdF Chemie	Lotadcr 4700; HX-8040; HX-8140	[51]
E–MMA–monoethyl maleate	DuPont	Vamac G	[51]
Thermoplastic elastomer	DSM	Keltaflex N35	[19]
Thermoplastic elastomer	Akzo-Enka	Arnited	[19]

* Experimental maleated (< % MA) EPM rubbers from Exxon are designated XX 1201, 1301 and 1601 [19].

may be carried out in an extruder as well. The hydrogenated polymer is extruded through a twin-screw extruder at 280 °C and 2% molten MA is added through an injection port to the barrel, and the unreacted MA removed under reduced pressure in another port [98]. Peroxide-initiated MA modification of EPM rubbers in an extruder serves as an example of a general approach to render polymers useful for interaction in blends with polyamides. Peroxides are beneficially used in extrusion modification even with elastomers that contain some unsaturation. It has been shown [91, 104] that improved grafting of MA is obtained with the addition of peroxides. Borggreve, Gaymans and Luttmer [20] demonstrated that MA reacts with hydrocarbon rubbers that do not contain double bonds. They successfully modified EPDM, EPM and linear polyethylene in a twin-screw extruder in the presence of bis(*t*-butylperoxyisopropyl) benzene. The reacted polymers had 0.5%, 0.4% and 0.2% combined MA respectively. Even in the absence of peroxides, low density polyethylene was modified by heating with maleic anhydride for 45 min at 275 °C, resulting in blends with polyamides that showed significantly improved homogeneity and properties in comparison with blends with unmodified polyethylene [105].

Commercially available elastomers that have been functionalized with maleic anhydride are listed in Table 5.3.

5.6.2 Analysis of MA-modified elastomers

The amount of maleic anhydride reacted with elastomers may be determined by a number of methods [71]. Oxygen analysis of the modified elastomer readily gives the combined maleic anhydride content. IR spectroscopy and titration are also used. It is important to distinguish bound and unbound anhydride which can be done by appropriate removal of unreacted MA. IR spectra of the starting and the reacted elastomers may be compared at several absorption bands, e.g. at 1785 and 1852 cm^{-1} typical of anhydrides, to determine the amount of chemically bound MA. IR results have been found to be in good agreement with values determined by titration of polymer solutions in THF with 0.1 M tetrabutyl ammonium hydroxide in methanol [99]. Colorimetric titration with potassium methoxide and phenolphthalein as indicator is also used [91]. Another titration method involves the analysis of MA-modified EPM where the soluble fraction of the reaction product is refluxed in water-saturated xylene for 1 h and then titrated with a hot solution of 0.05 N ethanolic KOH, using 1% thymol blue as an indicator [107]. Typically, viscosity and gel content of the reaction product are also determined. The inherent viscosity (0.1 g per 100 ml of solvent) may be determined in tetrachloroethylene at 30 °C [99]. Gel formation during the reaction of elastomers with maleic anhydride in the presence of peroxides often takes place [107]. It is important to minimize and/or control gel levels for maleated products designed for impact modification of engineering polymers.

5.6.3 Analysis of polyamide–maleic anhydride containing polymer blends

The reaction of amine end groups of polyamides with maleated elastomers is illustrated in Fig. 5.9. The extent of this reaction may be assessed by a number of methods. One approach involves the determination of the graft copolymer and the free nylon or maleated elastomer using a coacervation method. The blend is first dissolved in *m*-cresol and then diluted with cyclohexane. On centrifugation, the upper cyclohexane layer will contain all of the unreacted, free elastomer. Free rubber is obtained from this fraction by extracting the material by refluxing in toluene for 48 h. Details of this process are described by Hergenrother and Ambrose [108]. Polyamide in the graft copolymer was obtained from nitrogen analyses.

IR spectroscopy is used to follow the disappearance of the characteristic absorption bands of MA at 1785 and 1852 cm^{-1} indicating that grafting has taken place. The extent of anhydride groups in the starting elastomer is determined first by examining a film of known thickness. For elastomers containing vicinal carboxy or alkoxycarbonyl groups, the characteristic IR band lies at 1695 cm^{-1} [99].

The presence of polyamide in the graft polymer is shown by IR absorption at 1667 cm^{-1} to indicate amide carbonyl, 1563 cm^{-1} for —NH bending and 3333 cm^{-1} for —NH stretching. The overall polyamide content in the graft may be determined by Kjeldahl analysis for nitrogen content.

The graft copolymer containing polyamide can be characterized by heating a sample in sulfuric or hydrochloric acid to above 200 °C. Under these

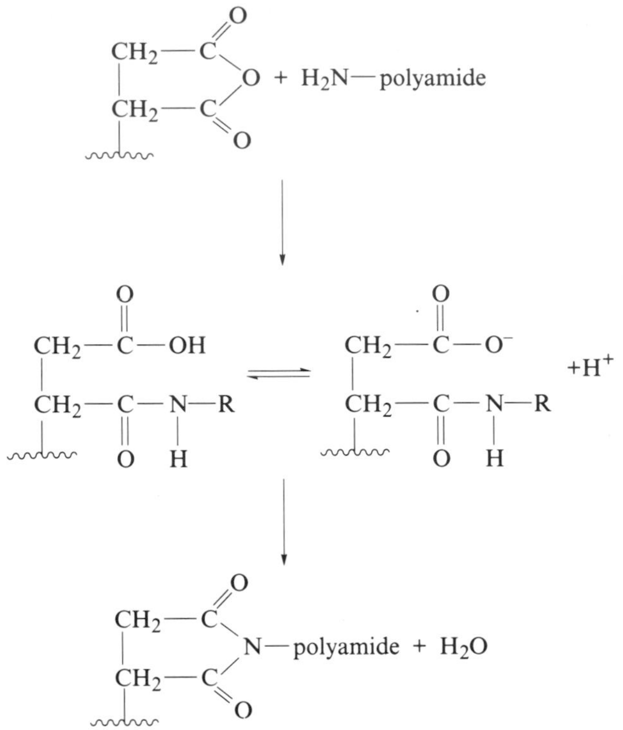

$$\text{(succinic anhydride ring: } CH_2{-}C(=O){-}O{-}C(=O){-}CH_2\text{, grafted to chain)} + H_2N{-}\text{polyamide}$$

$$\downarrow$$

$$CH_2{-}C(=O){-}OH \,/\, CH_2{-}C(=O){-}N(H){-}R \rightleftharpoons CH_2{-}C(=O){-}O^- \,/\, CH_2{-}C(=O){-}N(H){-}R \; + H^+$$

$$\downarrow$$

$$\text{(succinimide ring: } CH_2{-}C(=O){-}N{-}C(=O){-}CH_2\text{)}{-}\text{polyamide} + H_2O$$

Fig. 5.9 Reaction between maleic anhydride and an amine end group in a polyamide.

conditions, PA degrades to the starting lactams which are volatile [99]. Determination of the change in molecular weight due to grafting may be determined by GPC using *m*-cresol at 100°C on porous polystyrene packed columns [99]. Dynamic mechanical properties also provide a good indication that grafting has taken place, particularly in the 100–150°C temperature range. Copolymer grafted to PA will have a high modulus in comparison with the ungrafted elastomer.

5.6.4 Uses of maleic anhydride modified ethylene–propylene rubbers in blends

A major fraction of the recent literature on impact modification of polyamides uses MA functionalized EPM or EPDM rubbers [7, 19, 101]. Contributions by Wu [15–17] and Borrgreve and coworkers [19, 20] have been most significant in developing an improved understanding of the requirements for toughening polyamides including rubber particle size and concentration [19, 20] and the criteria for matrix–rubber particle dispersion [20] (Chapters

2 and 7). Borggreve, Gaymans and Luttmer concluded that the amount of MA bound to the elastomer, within the range 0.1–0.8 wt%, is not a critical variable whereas obtaining a uniform rubber particle dispersion is most important [20]. Wu [17] has used these results to propose that the distance between rubber particles, i.e. interparticle distance or matrix ligament thickness, is a key parameter (Chapters 2 and 7). A number of polyamide blends with MA functionalized ethylene–propylene polymers have been reported in the patent literature [99, 101]. Several commercially available toughened nylons are based on these polymers.

5.6.5 Uses of maleic anhydride modified block copolymers in blends

As discussed earlier, hydrogenated and functionalized block copolymers, such as SEBS and SEBS-g-MA, are useful toughening agents for engineering polymers. They are stabilized with a combination of antioxidants to provide protection during melt blending, fabrication and long-term use at elevated temperatures. A number of commercially available maleated elastomers are listed in Table 5.3.

While commercially available SEBS-g-MA polymers are functionalized in the midblock, functionalized benzene rings in SEBS block copolymers are also produced to provide excellent thermal stability. Such carboxy-modified polymers have been found to be useful at about 20% concentration for toughening polyamides, such as nylon 6,6, PBT and PET [92]. A large number of SBS, SEBS and SEBS-g-MA combinations have been used for the impact modification of a variety of engineering polymer blends. While each of these polymers is often an effective toughener by itself, some synergistic combinations have been reported.

Blends of SEBS (Kraton G1652) with nylon 6 yield particles, ~5 μm, that are too large and do not adequately adhere to the matrix for toughening [21]. On the other hand, blends with a maleated version, SEBS-g-MA, result in particles, ~0.05 μm, that are too small for toughening nylon 6. Particle size varies continuously between these limits when nylon 6 is blended with both SEBS and SEBS-g-MA [21, 76]. Toughening goes through a maximum as the ratio of the two block copolymers is changed at constant total elastomer content, in the region where optimum particle sizes are achieved. The situation with nylon 6,6 is more complex [109].

A combination of block copolymers has also been used in mixed matrix engineering polymers, e.g. nylon and PPO. By using a 50:50 mixture of a high molecular weight SEBS (molecular weight 174 000; 33% styrene) and low molecular weight SEBS-g-MA a balance of toughness, high flexural modulus and excellent solvent resistance of the blend was achieved [110]. Alternatively, PPO, PA, SBS and maleic anhydride may be extrusion blended together to obtain compositions with useful combinations of toughness and flexural stiffness [26].

5.7 FUTURE TRENDS

The rapid growth of toughened engineering polymers designed for a large-scale automotive and other uses requiring high temperature resistance, toughness and chemical resistance is undoubtedly going to continue for the foreseeable future. This assures an excellent growth opportunity for a variety of toughening agents. These demanding applications, however, require toughening agents with improved performance. Elastomers with low T_g are needed to impart toughness at low use temperatures while thermal and oxidative resistances are needed to survive the high temperatures required for processing these materials. In addition, these elastomers must be dispersed within the matrix to an appropriate morphology (or size scale) and adequately coupled to the matrix. These two issues are often interrelated and specific to the particular matrix material. Previous discussion indicated that numerous possibilities exist for attachment of the elastomer phase to the matrix, but clearly there are many additional possibilities, both physical and chemical. The control of elastomer phase morphology is of utmost importance and requires further study to answer questions related to obtaining and maintaining an optimum rubber particle size or producing a system with a dual particle size distribution.

Core–shell emulsion made elastomers offer considerable promise for continued growth and possibilities for new products. As pointed out earlier, the composition, particularly that of the surface, and particle size are determined during the polymerization, thus providing well-defined tougheners for many engineering polymers. Since their surface layer can be varied over a wide range of compositions, from styrene and methyl methacrylate copolymers to copolymers containing functional groups, they can provide either physical or chemical coupling to a large variety of matrix polymers. Polyesters, polycarbonates, polyacetals and blends of these polymers have been toughened with standard modifiers that do not contain reactive functionality, whereas polyamides seem to require functionalized core–shell modifiers [30]. A main advantage of core–shell modifiers is that control of particle size or morphology and adhesion to the matrix can be decoupled. Another approach for controlling rubber phase morphology is achieved by the use of 'artificial latexes'. Polymers may be emulsified by dissolving them in a solvent and dispersing the solution in water containing a surfactant, followed by stripping away the solvent. This method permits the use of rubbers that are not normally synthesized by emulsion polymerization.

Functionalized elastomers appear to be in a strong position for growth since many large volume engineering polymer blends contain potentially reactive sites. Typically, maleic anhydride functionalized EPM rubbers and SEBS block copolymers have been used since they readily react with the end groups of polyamides. Other engineering thermoplastics are also capable of reacting with functionalized modifiers. In many instances, however, combinations of functionalized and unmodified elastomers are likely to provide even better results than a single modifier [21, 76, 86]. Functionalized elastomer modifiers alone may be too finely dispersed in a matrix, and the addition of an unmodified elastomer permits the adjustment of the particle morphology to an optimum level. Undoubtedly, other new elastomeric modifiers may be found to be beneficial in some future engineering polymer blends. For instance silicone elastomers [111] and fluoroelastomers [112] have been demonstrated to work in several toughened systems. The problems of high cost and reactivity may inhibit extensive use of such specialty elastomers, however.

Finally, continued efforts will be required to produce a better understanding of the various toughening mechanisms that are applicable to engineering polymers. It appears that a multimechanism response to deformation leads to some of the exceptionally tough polymers. Shear yielding and dilatation are the principal responses to stress. While crazing may be still the predominant dilatational process for engineering polymers, rubber particle cavitation and interfacial separation may provide a significant contribution as well. Understanding these issues should provide help in choosing toughening agents with appropriate composition, crosslink density and molecular architecture. A multimechanism response has been well described by Takemori [113] showing, indeed, that there is competition between crazing and shear yielding during fatigue of several engineering polymers.

In all too many cases the concepts and materials used for toughening engineering polymers to date have been borrowed from commodity materials such as PVC, HIPS and ABS. It is evident that the future lies in new ideas and modifiers that are specifically designed for the diverse array of engineering polymers.

REFERENCES

1. Juran, R. (ed.) (1989) *Modern Plastics Encyclopedia*, McGraw-Hill, New York.
2. Wigotski, V. (1988) *Plast. Eng.* (November), 25.
3. Clagett, D. C. (1986) Engineering plastics, in *Encyclopedia of Polymer Science and Engineering*, Vol. 6, 2nd edn, Wiley, New York, pp. 94–131.
4. Bakker, M. (1980) Engineering plastics, in *Kirk-Othmer Encyclopedia of Chemical Technology*, Vol. 9, 3rd edn, Wiley, New York, pp. 118–37.
5. Utracki, L. A. (1987) *Soc. Plast. Eng., Tech. Pap.*, **33**, 1339.
6. Anonymous (1990) *Mod. Plast.*, **67**(1), 53.
7. Nishio, T., Sanada, T. and Hosoda, S. (1987) European Patent Appl. 0237187 (to Sumitomo Chem. Co. Ltd), September 16, 1987.
8. Cooper, G. D., Lee, G. F., Katchman, A. and Shank, C. P. (1981) *Mater. Technol.*, Spring, 13.
9. Bucknall, C. B. (1977) *Toughened Plastics*, Applied Science, London.
10. Soderquist, M. E. and Dion, R. P. (1989) Styrene polymers, in *Encyclopedia of Polymer Science and Engineering*, Vol. 16, 2nd edn (ed. E. R. Moore), Wiley, New York.
11. Hobbs, S. Y., Dekkers, M. E. J. and Watkins, V. H. (1988) *Polymer*, **29**, 1598.
12. Hobbs, S. Y., Dekkers, M. E. J. and Watkins, V. H. (1989) *J. Mater. Sci.*, **24**, 2025.
13. Hobbs, S. Y., Dekkers, M. E. J. and Watkins, V. H. (1988) *J. Mater. Sci.*, **23**, 1219.
14. Paul, D. R., Barlow, J. W. and Keskkula, H. (1988) Polymer blends, in *Encyclo-*

pedia of Polymer Science and Engineering, Vol. 12, 2nd edn, Wiley, New York, pp. 339–461.
15. Wu, S. (1985) *Polymer*, **26**, 1855.
16. Wu, S. (1987) *Polym. Eng. Sci.*, **27**, 335.
17. Wu, S. (1988) *J. Appl. Polym. Sci.*, **35**, 549.
18. Borggreve, R. J. M. and Gaymans, R. J. (1989) *Polymer*, **30**, 63.
19. Borggreve, R. J. M., Gaymans, R. J. and Schuijer, J. (1989) *Polymer*, **30**, 71.
20. Borggreve, R. J. M., Gaymans, R. J. and Luttmer, A. R. (1988) *Makromol. Chem., Macromol. Symp.*, **16**, 195.
21. Oshinski, A. J., Keskkula, H. and Paul, D. R. (1992) *Polymer*, **33**, 268.
22. Oostenbrink, A. J., Molenaar, L. J. and Gaymans, R. J. (1990) Polyamide–rubber blends: influence of very small rubber particle sizes on impact strength. 6th Annual Meeting of Polymer Processing Society, Nice, April 18–20, 1990, poster.
23. Flexman, E. A., Huang, D. D. and Snyder, H. L. (1988) *Polym. Prepr.*, **29**(2), 189.
24. Flexman, E. A., Jr (1989) US Patent 4,804,716 (to DuPont), February 14, 1989.
25. Hobbs, S. Y., Dekkers, M. E. J. and Watkins, V. H. (1989) Proc. 2nd Topical Conf. Engineering Tech., AIChE, San Francisco, CA.
26. Shibuya, N., Sobajima, Y. and Sano, H. (1988) US Patent 4,743,651 (to Mitsubishi Petrochemical), May 10, 1988.
27. Epstein, B. N. (1979) US Patent 4,174,358 (to DuPont), November 13, 1979.
28. Tate, D. P. and Bethea, T. W. (1985) In *Encyclopedia of Polymer Science and Engineering*, Vol. 2, 2nd edn, Wiley, New York, p. 537.
29. Reich, M. (1990) *Chem. Eng. News*, **68** (April 9), 15.
30. DeVries, J. (1989) Proc. 1st International Congress on Compatibilizers and Reactive Polymer Alloying (Compalloy '89), New Orleans, LA, p. 97.
31. Greek, B. F. (1988) *Chem. Eng. News*, **66** (March 21), 25.
32. Paul, D. R. and Newman, S. (eds) (1978) *Polymer Blends*, Vols 1 and 2, Academic Press, New York.
33. Murdock, J. D., Nelan, N. and Segall, G. H. (1966) US Patent 3,274,289 (to Canadian Industries Ltd), September 20, 1966.
34. Drake, R. and Siebert, A. (1975) *SAMPE Q.*, **6**(4), Paper AB-11.
35. Siebert, A. R. (1987) *Makromol. Chem., Macromol. Symp.*, **7**, 115.
36. Suspene, L. and Pascault, J. P. (1989) *Soc. Plast. Eng., Tech. Pap.*, **47**, 604.
37. Riew, C. K. (1981) *Rubber Chem. Technol.*, **54**(2), 374.
38. Takeda, S. and Kakiuchi, H. (1988) *J. Appl. Polym. Sci.*, **35**, 1351.
39. Scholle, K. F. M. G. J. and Winter, H. (1988) 33rd Int. SAMPE Symp., March 7, 1988, pp. 1109–20.
40. Maglio, G., Palumbo, R. and Vitagliano, V. M. (1989) *Polymer*, **30**, 1175.
41. Ezzell, S. A., St. Clair, A. K. and Hinkly, J. A. (1987) *Polymer*, **28**, 1779.
42. Doak, K. W. (1986) Ethylene polymers, in *Encyclopedia of Polymer Science and Engineering*, Vol. 6, 2nd edn, Wiley, New York, pp. 383–501.
43. Ver Strate, G. (1986) In *Encyclopedia of Polymer Science and Engineering*, Vol. 6, 2nd edn, Wiley, New York, pp. 522–64.
44. Roura, M. J. (1984) US Patent 4,478,978 (to DuPont), October 23, 1984.
45. Olivier, E. J. (1990) US Patent 4,948,842 (to Copolymer Rubber and Chemical Corp.), August 14, 1990.
46. Epstein, B. N. (1979) US Patent 4,172,859 (to DuPont), October 30, 1979.
47. Uniroyal Chemical, *Technical Booklet on Royaltuf® Modified EPDM*.
48. Wefer, J. M. (1985) US Patent 4,493,921 (to Uniroyal), January 15, 1985.
49. Wefer, J. M. (1989) US Patent 4,814,381 (to Uniroyal), March 21, 1989.
50. Wefer, J. M. (1990) US Patent 4,895,899 (to Uniroyal), January 23, 1990.
51. Lavengood, R. E., Patel, R. and Padwa, A. R. (1987) European Patent Appl. 0220155 (to Monsanto Co.), April 29, 1987.
52. Br. Patent, 998,439 (to DuPont), July 14, 1965.

53. Seven, M. K. and Bellet, R. J. (1969) US Patent 3,456,059 (to Allied Chemical), September 2, 1969.
54. Hofmann, H., Leibner, H. and Braun, D. (1973) DDR Patent 95690, February 12, 1973.
55. Roura, M. J. (1982) US Patent 4,346,194 (to DuPont), August 24, 1982.
56. Lundberg, R. D. (1984) Ionomers, in *Kirk-Othmer Encyclopedia of Chemical Technology*, Supplement Vol., 3rd edn, Wiley, New York, pp. 546–73.
57. Meckel, W., Coyert, W. and Wieder, W. (1987) in *Thermoplastic Elastomers* (eds N. R. Legge, G. Holden and M. E. Schroeder), Hanser, New York, Chapter 2.
58. Goldblum, K. G. (1962) US Patent 3,431,224 (to General Electric).
59. Keskkula, H. (1989) *Adv. Chem. Ser.*, **222**, 289–99.
60. Thompson, S. J. (1980) *The S/B Latex Story*, Pendell, Midland, MI.
61. Calvert, W. C. (1966) US Patent 3,238,275 (to Borg-Warner), March 1, 1966.
62. Goldman, T. D. (1984) US Patent 4,443,585 (to Rohm and Haas), April 17, 1984.
63. Keskkula, H., Maass, D. A. and McGreedy, K. M. (1984) US Patent 4,460,744 (to Dow), July 17, 1984.
64. Kim, H., Keskkula, H. and Paul, D. R. (1990) *Polymer*, **31**, 869.
65. Moritani, M., Inoue, T., Motegi, M., Kawai, H. and Kato, K. (1971) in *Colloidal and Morphological Behavior of Block and Graft Copolymers* (ed. G. E. Molau), Plenum, New York, pp. 33–46.
66. Witman, M. W. (1983) US Patent 4,378,449 (to Mobay), March 29, 1983.
67. Liu, P. Y. (1981) US Patent 4,260,693 (to General Electric), April 7, 1981.
68. Boutni, O. M. and Liu, P. Y. (1987) US Patent 4,656,225 (to General Electric), April 7, 1987.
69. Keskkula, H., Cheng, T. -W. and Paul, D. R. (1989) *Soc. Plast. Eng., Tech. Pap.*, **47**, 1816.
70. Owens, F. H. and Clovis, J. S. (1972) US Patent 3,668,274 (to Rohm and Haas), June 6, 1972.
71. Lawson, D. F., Hergenrother, W. L. and Matlock, M. G. (1990) *J. Appl. Polym. Sci.*, **39**, 2331.
72. Gelles, R. (1987) Society of Plastics Engineers, Chicago Section, Engineering Polymers and Composites, September 23, 1987, pp. 75–83.
73. Bates, G. M., Chambers, G. R. and Ting, S.-P. (1987) US Patent 4,681,915 (to General Electric), July 21, 1987.
74. Legge, N. R., Holden, G. and Schroder, M. E. (eds) (1987) *Thermoplastic Elastomers*, Hanser, New York.
75. Ceresa, R. J. (1964) Block and graft copolymers, in *Encyclopedia of Polymer Science and Technology*, Vol. 2, 1st edn, Wiley, New York, pp. 485–528.
76. Modic, M. J., Gilmore, D. W. and Kirkpatrick, J. P. (1989) Proc. 1st International Congress on Compatibilizers and Reactive Polymer Alloying (Compalloy '89), New Orleans, LA, p. 195.
77. Holden, G. and Legge, N. R. (1987) In *Thermoplastic Elastomers* (eds N. R. Legge, G. Holden and M. E. Schroder), Hanser, New York, pp. 47–65.
78. Gergen, W. P., Lutz, R. G. and Davison, S. (1987) In *Thermoplastic Elastomers* (eds N. R. Legge, G. Holden and M. E. Schroder), Hanser, New York, pp. 508–40.
79. Dean, B. D. (1986) US Patent 4,593,066 (to Atlantic Richfield Co.), January 3, 1986.
80. Deleens, G. (1987) In *Thermoplastic Elastomers* (eds N. R. Legge, G. Holden and M. E. Schroder), Hanser, New York, pp. 215–30.
81. Chen, A. T. (1987) *Soc. Plast. Eng., Tech. Pap.*, **33**, 1395.
82. Falk, J. C. and Van Fleet, J. (1978) *Plast. Rubber Mater. Appl.*, **15**(11), 123.
83. Breslow, D. S. and Matlack, A. S. (1963) US Patent 3,113,986 (to Hercules), December 10, 1963.

84. Halasa, A. F., Carlson, D. W. and Hall, J. E. (1980) US Patent 4,226,952 (to Firestone), October 7, 1980.
85. Wald, M. M. and Quam, M. G. (1972) US Patent 3,700,633 (to Shell), October 24, 1972.
86. Saito, K., Nishimura, Y. and Izawa, S. (1987) European Patent 219,973 (to Asahi), April 29, 1987.
87. Kirkpatrick, J. P. and Preston, D. T. (1988) *Elastomerics*, **120** (October), 30.
88. Gelles, R., Modic, M. and Kirkpatrick, J. (1988) *Soc. Plast. Eng., Tech. Pap.*, **46**, 513.
89. Gilmore, D. and Modic, M. (1989) *Soc. Plast. Eng., Tech. Pap.*, **47**, 1371.
90. Gilmore, D. W. and Modic, M. J. (1989) *Plast. Eng.* (April), 51.
91. Gelles, R., Lutz, R. G. and Gergen, W. P. (1988) European Patent Appl. 0261748 (A2) (to Shell), March 30, 1988.
92. Gergen, W. P., Lutz, R. G. and Martin, M. K. (1987) European Patent Appl. 0215501 (A2) (to Shell), March 25, 1987.
93. Molau, G. E. (ed.) (1971) *Colloidal and Morphological Behavior of Block and Graft Copolymers*, Plenum, New York.
94. Keskkula, H. (1982) in *Polymer Compatibility and Incompatibility, Principles and Practice* (ed. E. Solc), Hardwood, New York, pp. 323–54.
95. Fettes, E. M. (ed.) (1964) *Chemical Reactions of Polymers*, Wiley, New York, pp. 203–19.
96. Trivedi, B. C. and Culbertson, B. M. (1982) *Maleic Anhydride*, Plenum, New York, pp. 172–3.
97. Farmer, E. H. and Wheeler, J. (1940) US Patent 2,227,777 (to British Rubber Producers Research Association), January 6, 1940.
98. Hergenrother, W. L., Matlock, M. G. and Ambrose, R. J. (1985) US Patent 4,508,874 (to Firestone), April 2, 1985.
99. Hammer, C. F. (1976) US Patent 3,972,961 (to DuPont), August 3, 1976.
100. Coran, A. Y. and Patel, R. (1983) *Rubber Chem. Technol.*, **56**, 1045.
101. Olivier, E. J. (1986) US Patent 4,594,386 (to Copolymer Rubber and Chemical Corp.), June 10, 1986.
102. Hergenrother, W. L., Matlock, M. G. and Ambrose, R. J. (1984) US Patent 4,427,828 (to Firestone), June 24, 1984.
103. Gaylord, N. G., Mehta, M. and Kumar, V. (1983) In *Modification of Polymers* (eds C. E. Carraher, Jr and J. A. Moore), *Polymer Science and Technology*, Vol. 21, Plenum, New York, p. 171.
104. Phadke, S. V. (1988) US Patent 4,757,112 (to Copolymer Rubber and Chemical Corp.), July 12, 1988.
105. Brunson, M. O. and McGillen, W. D. (1969) US Patent 3,484,403 (to Eastman Kodak), December 16, 1969.
106. Exxon, *Technical Booklet on Exxelor Modifiers.*
107. Gaylord, N. G., Mehta, M. and Mehta, R. (1988) *Am. Chem. Soc. Symp. Ser.*, **364**, 438.
108. Hergenrother, W. L. and Ambrose, R. J. (1974) *J. Polym. Sci., Polym. Chem. Ed.*, **12**, 2613.
109. Yamaguchi, N., Nambu, J., Toyoshima, Y., Mashita, K. and Ohmae, T. (1989) 5th Annual Meeting, PPS, Kyoto, April 11, 1989.
110. Modic, M. J. and Gelles, R. (1988) European Patent 255,184 (to Shell), February 3, 1988.
111. Yorkgitis, E. M., Tran, C., Eiss, N. S., Hu, T. Y., Yilgor, I., Wilkes, G. I. and McGrath, J. E. (1984) *Adv. Chem. Ser.*, **208**, 137.
112. Mijovic, J., Pearce, E. M. and Nir, Z. (1984) *Adv. Chem. Ser.*, **208**, 293.
113. Takemori, M. T. (1990) Crazing in polymers, in *Advances in Polymer Science*, Vol. 2, Springer, Berlin, p. 263.

6

Rubber modified epoxy resins

S. J. Shaw

6.1 INTRODUCTION

Epoxy resins are nowadays used extensively as adhesives and matrix resins for fibre reinforced composite materials where advantage is taken of favourable properties such as high modulus, low creep and reasonable elevated temperature performance. However, such characteristics in an epoxy require moderate to high levels of crosslinking which can and usually does result in brittle behaviour.

Although several approaches exist for enhancing the toughness of epoxy resins, possibly the most successful has concerned the use of reactive liquid elastomers. Numerous investigations have shown that the addition of such materials can, if used correctly, result in substantial toughness improvements whilst minimizing the effect on other important properties.

In this account an attempt will be made to discuss the various factors which can be regarded as important for the successful utilization of rubber modification as a toughness enhancement technique. In addition, the effects of various service-related parameters such as rate and temperature on the mechanical–fracture properties of toughened epoxies, in bulk, adhesive joint and composite form, will be considered together with a discussion of the various mechanisms which have been proposed to account for the high levels of toughness enhancement achieved.

6.2 FORMULATION AND CHEMISTRY

6.2.1 Epoxy resin chemistry

Epoxy resins are essentially compounds which contain, on average, more than one epoxide group per molecule. A wide variety of resin types have been developed over the years providing potential use in many applications under a variety of conditions. Without doubt the type that has been studied most with regards attempts at rubber modification has been the diglycidyl ether of bisphenol A epoxy (DGEBA), which has the structure

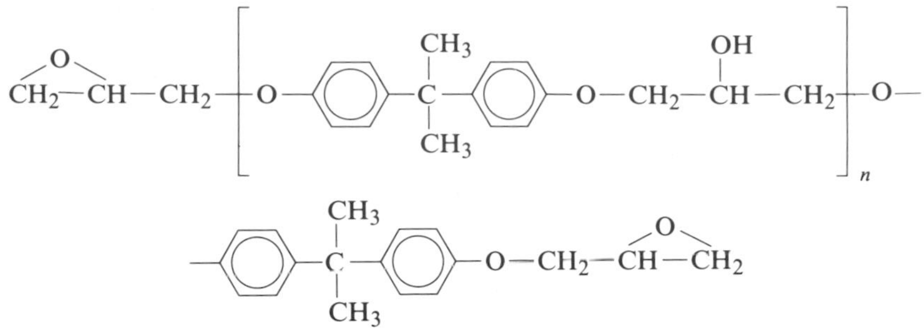

$$\overset{O}{\overbrace{CH_2-CH}}-CH_2-\left[O-C_6H_4-\underset{CH_3}{\overset{CH_3}{C}}-C_6H_4-O-CH_2-\underset{}{\overset{OH}{CH}}-CH_2\right]_n-O-C_6H_4-\underset{CH_3}{\overset{CH_3}{C}}-C_6H_4-O-CH_2-\overset{O}{\overbrace{CH-CH_2}}$$

with values $0<n<1$ being investigated predominantly.

Although epoxies having greater functionalities, e.g. resins based on the tetraglycidyl diamino diphenyl methane (TGDDM) system, have been studied with a view to rubber modification, the results obtained have been disappointing owing primarily to the comparatively high crosslink densities which these systems are capable of achieving on cure. Reasons for this behaviour are discussed later in this chapter.

In order to convert the relatively low molecular weight epoxy prepolymer to the crosslinked network necessary for successful utilization, crosslinking systems, generally referred to as curing agents or hardeners, are usually required. The types and number of curing agents which can be, and have been, employed are substantial. Although it is clearly beyond the scope of this review to itemize these various types and to discuss their relative attributes, it is perhaps of interest to mention those which have been employed in rubber modified formulations either in practical applications or as a means of studying behaviour. For a more detailed treatise concerning both epoxy resins and their curing agents, the reader is referred to various excellent accounts which exist in the literature [1, 2].

Possibly the two most common classes of epoxy curatives are amines and anhydrides, the former being particularly so with the majority of rubber modified formulations produced either commercially or for research purposes. A diverse range of amines can be and has been employed ranging from primary and secondary aliphatic and aromatic amines, which cure by addition mechanisms, through to tertiary amine based systems which cure by catalytic means. Although all these types have been employed at one time or another in rubber modified formulations, only a few can be regarded as having achieved prominence. Two such examples are piperidine and dicyandiamide, which have achieved fame for totally different reasons. Piperidine, a cycloaliphatic monosecondary amine, has the structure shown below:

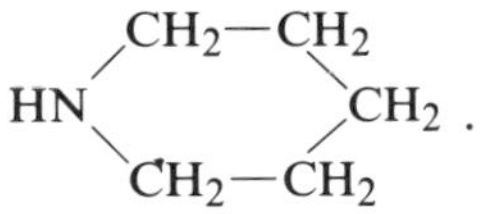

$$HN\begin{matrix} CH_2-CH_2 \\ \\ CH_2-CH_2 \end{matrix}CH_2 \, .$$

Because of the existence of only one secondary amine group, once this has reacted with an epoxide group a catalytic mechanism resulting from the remaining tertiary amine is the sole means of any further reaction and this is believed to be the mechanism by which this curing agent operates. Piperidine has been shown to provide certain substantial advantages in terms of speed of reaction and chemistry, which allows beneficial effects of rubber modification to be realized in many systems in a relatively simple way. Precise reasons for this are described below. This curing agent has therefore been employed in many investigations devoted primarily to fundamental study. As far as the author is aware, although one of the first curing agents employed in epoxy resin technology, it is rarely used in commercial applications in general and rubber modified formulations in particular.

Dicyandiamide, having the structure

$$H_2N-\underset{\displaystyle NH_2}{\underset{|}{C}}=N-CN\,,$$

is nowadays widely employed in many proprietary rubber modified epoxy formulations, particularly adhesive, where advantage is taken of its latent characteristics, exhibiting pot lives from days to years depending on storage temperature, when employed in one-part formulations. On heating to temperatures in excess of approximately 140 °C, rapid polymerization commences with acceptable properties being achieved in short periods of time.

The majority of curing agents employed in epoxy resin systems provide a substantial contribution to the properties of the crosslinked products. Thus choice of curing agent can therefore be of critical importance and needs to be considered carefully. This is particularly true for rubber modified formulations where curing agent can influence many factors capable of contributing to the final properties of the cured product. These can include its influence on epoxy–rubber compatibility and morphology, epoxy–rubber chemistry, rate of cure, epoxy crosslink density etc. The effects and importance of factors such as these will be considered later in this chapter.

6.2.2 Types of rubber modifiers

The various types of elastomeric materials which have been considered and studied with a view to rubber modification of epoxies are listed below:

- reactive butadiene–acrylonitrile rubbers;
- polysiloxanes [3];
- fluoroelastomers [4];
- acrylate elastomers [5, 6].

By far the most widely studied and indeed employed in the majority of proprietary rubber modified epoxy systems have been the butadiene–acrylonitrile rubbers. Consequently the majority of this section and indeed this chapter will be concerned primarily with these materials.

As the name implies, butadiene–acrylonitrile rubbers comprise a relatively low molecular weight backbone of butadiene and acrylonitrile groups with reactive groups in the terminal positions of the molecule as indicated below:

$$X-\left[(CH_2-CH=CH-CH_2)(CH_2-\underset{\substack{|\\CN}}{CH})\right]-X$$

In addition, rubbers having reactive groups pendent to the main polymer chain have been developed and studied. Several variants of these elastomers have been investigated with the major variables being threefold, namely (a) acrylonitrile content [7], (b) molecular weight [8] and (c) terminal functionality [9].

As will be discussed, a compatibility–incompatibility balance will essentially determine the success or otherwise of attempts at toughness elevation. This is predominantly controlled by the relative polarities of the resin and rubber components. With regard to the latter, this is greatly influenced by the acrylonitrile content, this portion of the molecule imposing a degree of polarity on an otherwise non-polar structure. As will be discussed, morphological characteristics and various properties have been related to rubber acrylonitrile content.

For reasons also related to compatibility–incompatibility and morphology requirements, elastomer molecular weight is also an important variable.

Good adhesion between the elastomeric and matrix components of a rubber modified thermoset has also been shown to be of vital importance. This is generally achieved by ensuring a means by which the rubber component can react with the resin, the provision of terminal reactivity usually being the preferred option. The type of rubber modifier employed therefore, at least in terms of terminal functionality, will depend on the chemical characteristics of the matrix polymer. In the case of epoxies several different functionalities have been studied including epoxy, phenol, vinyl, hydroxyl, mercaptan, amine and carboxyl, with at least reasonable success being achieved throughout [10]. Possibly the greatest benefit, however, has been provided by elastomers containing the carboxyl functionality (CTBNs) and the next section will discuss the detailed chemistry which would be expected in a formulation containing such an elastomer. The influence of elastomer structures of the types mentioned above on the mechanical–fracture properties of cured rubber modified epoxies will be considered later.

Investigations involving other elastomer types such as polysiloxanes and polyacrylates have been conducted. However, these materials are generally only considered when service conditions are of such an extreme nature as to prevent the use of the more effective butadiene–acrylonitrile systems. Polysiloxanes exhibit important advantages in having improved low temperature flexibility and excellent weatherability. Furthermore, the molecular structure of the butadiene–acrylonitriles renders them prone to thermal instability and thus unsuitable for long-term use in high temperature environments. The polysiloxanes, acrylates and fluoroelastomers can be considered more suitable for such conditions.

6.2.3 Epoxy–elastomer reaction mechanisms

As mentioned above, generally considered of fundamental importance is a reaction which should occur between the elastomer modifier and the epoxy resin requiring toughness enhancement, since a successful outcome generally guarantees an adequate bond between the elastomeric and epoxy phases. As will become apparent later in this chapter, when toughening mechanisms are discussed, this is considered vital for effective toughness enhancement.

Under certain conditions an esterification reaction will occur between an epoxy and a carboxyl-terminated elastomer as indicated in the following scheme:

$$\sim\!\overset{\overset{\displaystyle O}{\|}}{C}\!-\!OH + \overset{\;\;O\;\;}{\widehat{CH_2\!-\!CH}}\!-\!CH_2\!-\!O\!\sim$$

$$\downarrow$$

$$\sim\!\overset{\overset{\displaystyle O}{\|}}{C}\!-\!O\!-\!CH_2\!-\!\overset{\overset{\displaystyle OH}{|}}{CH}\!-\!CH_2\!-\!O\!\sim .$$

For such a reaction to occur to any significant extent, temperatures in excess of approximately 160 °C would be required. However, catalysts such as triphenylphosphine have shown success in allowing a quite considerable reduction in this reaction temperature.

With many, indeed the majority of, curing agents, selectivity problems exist where the esterification reaction is not encouraged. In these circumstances the normal epoxy cure reactions, promoted by the presence of the curing agent, occur to the almost total exclusion of the vitally important epoxide–carboxyl esterification reaction. Research has shown that in such cases, although phase separation and two-phase morphological characteristics can occur, toughness enhancement is generally disappointing, this being attributed to poor rubber–matrix bonding. In these circumstances this most important reaction is conducted as a preliminary process prior to the formulation and cure of the rubber modified epoxy, i.e. an epoxy–elastomer adduct is prepared producing in essence an epoxy-terminated butadiene–acrylonitrile elastomer. When added to the epoxy–curing agent mixture, the adduct is able to react essentially as an epoxide via the normal curing mechanism and thus problems associated with curing agent selectivity are not encountered. In some formulations, however, for example those containing piperidine as a curing agent, the formation of a preliminary adduct is not necessary owing to the curative favouring the epoxy–carboxyl esterification reaction. Studies have shown that not until esterification is virtually complete does the normal epoxy curing reaction occur.

With traditional proprietary amine-terminated butadiene–acrylonitrile (ATBN) rubbers, selectivity-associated problems do not exist, since the terminal secondary amine groups are capable of reaction with the epoxide groups via the usual mechanism. Modified ATBNs having primary amine-terminal functionality, which have recently been developed, would also be expected to be free from selectivity problems for similar reasons. In a similar manner, selectivity and reactivity problems would not be expected with butadiene–acrylonitrile elastomers having epoxy and mercaptan functionality, since both groups would be capable of reaction with an epoxy–curing agent formulation.

Attempts to employ hydroxyl-terminated elastomers for toughness enhancement have generally encountered difficulties in that the conditions required to promote the necessary hydroxyl–epoxide reaction generally lead to self-polymerization of the epoxy, the latter presumably occurring at the expense of the former and thus limiting the extent of both the elastomer–epoxy reaction and eventual adhesion between the elastomeric and epoxy matrix phases. A recent interesting approach to this problem has been described where the difficulty in chemically linking the elastomer to the epoxy has been overcome by employing a co-reactant, toluene diisocyanate (TDI), capable of reaction with both epoxide and hydroxyl functionalities, resulting in the generation of both urethane and oxazolidone groups between the elastomeric and epoxide components of the system [11].

6.2.4 Compatibility and morphology development

Most researchers would now agree that in order to provide an improvement in toughness whilst minimizing any deterioration in other important properties/parameters such as modulus and T_g, phase separation, followed by the gradual development of a two-phase morphology, is critically important. In order to achieve this outcome, the rubber must initially dissolve and become dispersed on a molecular level in the epoxy, but be encouraged to precipitate out when epoxy crosslinking occurs. Under these circumstances the required two-phase morphology will develop with the formation of rubbery particles dispersed in and bonded to the crosslinked epoxy matrix. Thus elements of both initial compatibility and eventual incompatibility must be regarded as vital elements of the system. This approach can be viewed in a more theoretical sense by considering interactional thermodynamics in general and solubility parameters in particular.

The level of compatibility exhibited by a mixture of two substances is controlled fundamentally by the Gibbs free energy of mixing, ΔG_m, as described in Chapter 2. By considering a cohesive energy density approach to molecular interaction, the level of compatibility between two substances can be estimated by use of solubility parameters, δ, the values of which determine the tendency of molecules to attract their own species in preference to those of dissimilar species. It is, however, important to recognize that this simplified approach only considers molecular interactions involving dispersion forces. In systems where polar interactions and hydrogen bonding play a prominent intermolecular role, anomalous effects would be expected.

Firstly, in order to develop what is now regarded as the optimum morphology (small elastomeric particles present in the crosslinked resin matrix) good initial compatibility between the epoxy and rubbery components is required. Thus, in thermodynamic terms, ΔG_m should be negative. This would be promoted by similar values of δ together with a low molar volume, V (i.e. low molecular weight). However, at some stage in the cure process, in order to avoid the continued intimate dispersion of the elastomeric component, a transition in the value of ΔG_m from negative to positive would be necessary so as to allow the required phase separation process. With most, if not all, rubber modified epoxy formulations, this transition is facilitated by the general molecular weight buildup which occurs during cure (resulting in an increased magnitude of V and hence ΔG_m).

Most of the investigations conducted on epoxies have indicated several factors likely to influence morphology. These include elastomer concentration [10], molecular weight and acrylonitrile content [7, 8], type and extent of curing agent employed and cure conditions [12, 13].

An increase in the amount of elastomer incorporated into an epoxy has been shown to increase the volume fraction of the elastomeric phase, as indeed would be expected. In addition it has been shown to provide increases in the size of the rubbery domains. Elastomer concentrations in excess of about 20 wt% have usually resulted in a phase inversion process where a transition in morphological characteristics occurs, resulting in epoxy particles embedded in an elastomer matrix [14]. Although numerous investigations have indicated a positive correlation between elastomer volume fraction and hence toughness, mechanical properties have been shown to deteriorate rapidly at elastomer concentrations in excess of this value.

Similarly, both the elastomer molecular weight and acrylonitrile content have been studied [7, 15]. Generally molecular weight reduction and increased acrylonitrile content have demonstrated similar effects with reductions in both elastomeric particle size and volume fraction with an increased tendency towards elastomer incorporation in the matrix.

The nature of the curing agent employed can influence the final properties of the modified epoxy in several ways [16–18]. In addition to exerting a major influence on the structure and hence properties of the crosslinked epoxy phase, it will also exert a substantial effect on the morphological characteristics of the cured product. This it can do by influencing the initial compatibility–incompatibility balance between the epoxy and elastomer. Of greater significance, however, will be the effect of curing agent reactivity. Highly reactive curing agents can impose what can be described as 'morphological constraint', where the rapid crosslinking reactions essentially promote the premature onset of gelation, which in turn interferes with and prevents full phase separation. This inevitably results in low volume fractions of elastomeric phase and hence relatively poor toughness characteristics.

Recent studies have indicated that cure conditions can have as significant an effect on the toughness of rubber modified epoxies (as will be discussed later in this chapter) as the formulation variables described above [13]. Once again the influence of these variables, particularly cure temperature, on morphology can be considered responsible. An increase in cure temperature would promote both an improvement in epoxy–elastomer compatibility as well as an increase in cure rate. Both factors would tend to retard phase separation resulting in relatively small values of elastomer volume fraction and particle size. On the other hand, increased cure temperature would also reduce formulation viscosity which would be expected to enhance phase separation, thereby having the reverse effect on morphology. In reality, therefore, the effect of cure temperature would be dependent on the relative importance of these three parameters, i.e. compatibility, viscosity and cure rate. Although some conflicting trends have therefore been understandably observed, particularly with regard to rubber volume fraction, an increase in cure temperature has usually been found to result in an increased rubber particle size [13].

At this stage in the discussion it would be of benefit to highlight briefly the morphological factors that are considered important for optimum toughening in rubber modified epoxies. Although this is still regarded as a somewhat controversial area, recent studies have focused on what would now seem to be the most critical parameters. Essentially the debate is centred on the relative importance of two morphological parameters, these being rubber volume fraction and particle size–particle size distribution. The early work of Sultan and McGarry suggested that the key morphological factor was particle size, with large particles (>1 μm diameter) promoting crazing within the epoxy matrix and smaller particles (0.01 μm) leading to shear deformation [16]. They proposed that craze generation provided the most efficient means of toughness enhancement and that, therefore, the formation of large particles was the key to successful rubber modification. Crazing is now, of course, recognized as not being a significant deformation mechanism in relatively highly crosslinked polymers, so in this respect alone the Sultan and McGarry theory has been largely discounted. However, in one major respect, the role of particle size and, in particular, particle size distribution has been and continues to be emphasized, particularly in respect of certain commercial rubber modified epoxy formulations, where a deliberate attempt is made to promote bimodal particle size morphology. This approach owes its existence to the work of Riew, Rowe and Siebert who found that, by incorporating bisphenol A into a DGEBA–

CTBN–piperidine system, substantial improvements in toughness could be achieved over and above those obtained from formulations not containing bisphenol A [10]. They noted in particular that this beneficial effect was related to a bimodal particle size morphology which they regarded as responsible for the improved toughness.

Although some workers view particle size and particle size distribution as providing significant contributions to toughness, most now recognize that the critical morphological factor is rubber volume fraction, with an increase in this parameter generally promoting increased toughness. Although this will partly depend on the quantity of elastomer initially added, the other factors previously described such as elastomer molecular weight, acrylonitrile content and curing agent reactivity will also have an important effect. It is particularly worthwhile to note that evidence from several sources has shown a rubber volume fraction exceeding the volume fraction of added rubber, thus providing an insight into the composite nature of the elastomeric domains. An anomalous effect such as this can only be resolved if it is assumed that the rubber particles have a not insignificant concentration of epoxy, either on the molecular level or as subinclusions dispersed in the elastomeric phase. As the chemistry outlined in section 2.3 will clearly testify, molecular integration of epoxy into the elastomeric phase will clearly occur and indeed it is possible to view the rubber particles as being composed of a copolymeric species derived from both rubber and epoxy. Thus the frequently observed rubber volume fraction effect is not unexpected.

To summarize, the two-phase morphology frequently observed in rubber modified epoxies plays a critical role in determining the extent of toughness enhancement. Many factors, including both formulation variables and cure conditions, have been shown to influence morphological parameters such as rubber phase volume fraction, particle size and size distribution. The following sections of this chapter will discuss the influence of these variables on the mechanical properties of rubber modified epoxies and, in certain cases, will attempt to relate these properties to the morphological characteristics discussed above.

6.3 TOUGHENING MECHANISMS

Several mechanisms have been proposed to account for the toughness enhancements obtained by the incorporation of elastomeric particles into glassy polymers. It is therefore of interest to consider each of the main mechanisms and to assess their relevance to rubber modified epoxies.

6.3.1 Rubber tear

A mechanism devoted solely to the deformation and tearing of the rubber particles present in a two-phase system was proposed initially by Mertz,

Claver and Baer in 1956 [19]. Its main proposal was that rubber particles quite simply hold the opposite faces of a propagating crack together and that the toughness of such a system is dependent primarily on the energy required to rupture the particles together with that required to fracture the glassy matrix. Although this mechanism has been regarded as irrelevant as regards toughened thermoplastics, such as high impact polystyrene and acrylonitrile–butadiene–styrene terpolymer (where crazing and shear yielding respectively are considered to apply), a number of fairly recent investigations using microscopy techniques have shown, in the case of rubber modified epoxies, clear evidencc of stretched rubber particles spanning loaded cracks. Indeed from such observations, Kunz-Douglas, Beaumont and Ashby proposed that the toughness enhancement provided by rubber particle incorporation was dependent primarily on the degree of elastic energy stored in the rubber particles during loading of the two-phase system [20]. In addition they developed a quantitative model based on these proposals which is depicted mathematically thus:

$$G_{Ic} = G_{Ice}(1 - V_p) + \left(1 - \frac{6}{\lambda_t^2 + \lambda_t + 4}\right) 4\gamma_t V_p \qquad (6.1)$$

where V_p is the volume fraction of the rubber particles, γ_t is the particle tear energy, λ_t is the rubber particle extension ratio at failure and G_{Ice} is the fracture energy of the epoxy matrix.

As clearly indicated by equation (6.1) this mechanism depicts toughness enhancement as being due solely to the properties of the elastomeric particles with no account being taken whatsoever of any interaction between elastomeric phase and matrix. Although, as mentioned previously, a number of studies have provided clear evidence of rubber particles spanning loaded cracks, thus suggesting a rubber tearing contribution to toughness enhancement, a number of significant inconsistencies between this mechanism and various experimental observations have been observed, throwing some considerable doubt on the universal applicability of the theory, for example the following.

- It does not explain the existence of stress whitening frequently observed with a wide range of rubber modified epoxy formulations under a variety of experimental conditions.
- It does not account for a yielding and plastic flow contribution to toughness. Numerous fractography studies have shown a pronounced increase in the extent of plastic deformation occurring in the matrix as a direct result of rubber particle incorporation, a factor which the model is unable to explain.
- The rate and temperature dependence of fracture toughness, observed in a number of studies, is generally opposite to the manner in which the tearing energy of elastomers varies with the same experimental variables.

Thus for the above reasons this rubber particle tear mechanism is not generally regarded as having universal applicability with rubber modified epoxies. However, under certain circumstances, where for example other energy absorption mechanisms such as matrix shear yielding are inoperative, it is likely to play a significant part.

6.3.2 Multiple crazing

The multiple crazing theory, due to Bucknall and Smith, proposes that toughness enhancement is attributed to the generation and efficient termination of crazes by rubber particles [21]. This process has been adequately demonstrated with thermoplastics such as high impact polystyrene. Craze initiation has been shown to occur at regions of high stress concentration which, with rubber particles embedded in a brittle matrix, is at the equatorial region normal to the applied stress direction. Craze termination occurs when the craze encounters another rubber particle. This stabilizes the craze and prevents it from growing into a larger crack-like structure. Thus a greater amount of energy can be absorbed by the system prior to eventual failure, thereby leading to an effective improvement in the toughness of the polymer.

The suggestion that crazing could be the dominant toughness enhancement mechanism in rubber modified epoxies has been proposed by a number of workers. As mentioned previously, Sultan and McGarry in their early work proposed that their improvements in toughness were due to the generation of crazes in the vicinity of rubber particles in an identical manner to that found in modified thermoplastics. In addition the frequently observed stress whitening phenomenon was attributed to the generation of crazes [16].

More recently the possibility of crazing as a toughening mechanism in rubber modified epoxies has been suggested by Bucknall and Yoshii in addition to shear deformation [18]. This proposal was based on evidence obtained from tensile creep experiments conducted on rubber modified epoxy specimens where increases in specimen volume were observed in addition to longitudinal extension. Bucknall and Yoshii suggested that the volume increase could be associated with a massive crazing within the specimen. In spite of these proposals it is now generally believed that crazing is the exception rather than the rule under most conditions in both unmodified and rubber modified epoxies. The above observations of both stress whitening and volume dilation have more recently been attributed, as will be discussed later, to void generation and growth within rubber particles in response to stress application. In addition recent studies with thermoplastics have indicated an apparent transition from a crazing to a shear yielding mechanism as the length of polymer chain between physical entanglements decreases, thus suggesting that with thermosetting polymers, where crosslink density is high, and thus chain length between crosslinks is short, crazing would be suppressed [22].

Although still a somewhat controversial issue, crazing is not nowadays generally believed to provide a major contribution to energy absorption and toughness enhancement in rubber modified epoxies.

6.3.3 Shear yielding

This theory was proposed by Newman and Strella in 1965 following work on acrylonitrile–butadiene–styrene thermoplastics [23]. Its main proposal is that shear deformation, occurring either as shear bands or as a more diffuse form of shear yielding, which, like crazing, is also initiated at stress concentrations resulting from the presence of rubber particles, is the main source of energy absorption and hence toughness enhancement.

Possibly the greatest difficulty with this mechanism has concerned its inability to account for stress whitening, owing to shear yielding being a constant volume deformation process. As a result it has been suggested that crazing and shear yielding occur simultaneously in many polymers, with the former accounting for the frequently observed stress whitening effect. Indeed, as previously described, a combined mechanism of this type has been proposed as the major source of toughness enhancement in rubber modified epoxies by Sultan and McGarry [16].

Although it is now generally recognized that crazing does not occur to any significant degree in epoxies, it is of interest to note that the toughening mechanism now regarded by many workers as being the most applicable to rubber modified epoxies is, to a large degree, based on a 'dual-mode' mechanism of this type. Based on rubber particle cavitation and matrix shear yielding, this mechanism will now be described.

6.3.4 Cavitation–shear yielding

A toughening mechanism developed and proposed independently by Kinloch, Shaw and Hunston [24] and by Pearson and Yee [25] is now generally recognized as the most consistent in terms of the experimental data and observations generated in recent years. Based on the concept of both rubber particle cavitation and matrix shear yielding, it is of interest to discuss this mechanism in some detail (also mentioned in Chapter 2).

Many investigations have demonstrated a positive correlation between toughness and the extent of plastic deformation found on fracture surfaces. Such a simple correlation has therefore suggested a mechanism based on yielding and plastic shear flow of the matrix as the primary source of energy absorption in rubber modified epoxies. Since greatly enhanced plastic deformation in the matrix has usually been found to accompany the incorporation of rubber particles, it is clearly necessary to focus attention initially on the particles and in particular the stress distribution which exists around them when located in the vicinity of a stressed crack tip.

Goodier first considered this effect in 1933 by studying the case of an isolated spherical particle embedded in an isotropic elastic matrix which is subjected to a uniform uniaxial tensile stress remote from the particle [26]. He found that, for a rubbery particle possessing a considerably lower shear modulus than the matrix, a maximum stress concentration of 1.9 occurred in the equatorial regions of the particle. Allowing for further factors such as particle–particle interaction, which would inevitably occur in practical systems, together with a triaxial as opposed to a uniaxial stress state, which particles would inevitably be subjected to, a stress concentration factor of approximately 1.6 would exist.

Now the interactions between a triaxial stress field at the tip of a loaded crack and the triaxiality associated with rubber particles experiencing the presence of the crack tip stress field could be envisaged as resulting in the initiation of two important processes.

Initially, the development of a triaxial stress would gradually promote dilation of the matrix. This, together with triaxial stresses inherent in the rubber particles owing to differential thermal contraction effects resulting from the initial cure process, would provide the conditions necessary for cavitation of the rubber particles. It is this cavitation process, rather than crazing of the epoxy matrix, which can be considered responsible for the stress whitening effects frequently observed with rubber modified epoxies.

In addition, the increasing presence of stress concentrations around the rubber particles during loading would promote both initiation and growth of shear yield deformation zones in the matrix. Owing to the large number of particles which would clearly exist, the degree of yielding generated would be substantially greater than would otherwise have occurred in an unmodified epoxy. However, since the particles would also act as sites for yield termination, yielding would remain localized in the vicinity of the crack tip.

It is reasonable to assume that both the cavitation and shear yielding processes would occur during the early stages of load application. In addition, once initiated, rubber particle cavitation would greatly enhance further shear yielding in the matrix. This would be partly attributable to an increased stress concentration which would inevitably accompany void formation. In addition, however, void formation would considerably reduce the level of constraint on the matrix adjacent to voided particles. This would essentially relieve the degree of triaxiality experienced by the matrix in the interparticle regions, and in turn lower yield stress and thus promote further extensive shear yielding. Crack tip blunting would, as a result, increase extensively resulting in the increased development of the plastic zone at the crack tip. Thus toughness would be considerably enhanced, as indeed has frequently been found in practice.

6.4 COMPARISON OF TOUGHENING METHODS

In addition to elastomeric modification, two other approaches to toughness enhancement in epoxies have been investigated, these being modification by high modulus particulates and thermoplastics, and it is of interest to consider briefly these two approaches and in particular to discuss their merits in comparison with elastomeric modification.

6.4.1 Particulate toughening

Several investigations have shown that the incorporation of particulate fillers such as silica and alumina trihydrate can enhance the toughness of crosslinked epoxies whilst contributing to a greatly enhanced modulus, a significant advantage over elastomeric modification where a reduction in modulus is inevitable [27–34]. Variables such as particle size and size distribution, particle surface chemistry and particulate volume fraction have all been investigated indicating the most important formulation variables required for optimum toughening.

The mechanism considered relevant is, however, somewhat different to that previously discussed for elastomeric modification and therefore worthy of brief attention.

Although a particulate reinforcement could be expected to impose stress concentrating effects on the epoxy matrix (as a result primarily to a substantial modulus differential between reinforcement and matrix) this is not generally considered significant. Instead a mechanism based essentially on the impeding characteristics of the particles, frequently referred to as the crack pinning mechanism, has been proposed [35–37]. The mechanism proposes that a propagating crack front, when encountering an inhomogeneity, becomes temporarily pinned at that point. An increase in load increases the degree of bowing between pinning points, resulting in both a new fracture surface and an increase in the length of the crack front. These processes will absorb energy and therefore fracture toughness will increase.

Although some fairly modest improvements in toughness have been observed with particular reinforcement, these have not generally been of the same magnitude as received from elastomeric incorporation. Thus, in a direct comparison, rubber modification would prevail. However, as mentioned above, a substantial modulus increase accompanying toughness enhancement is, of course, a major advantage. As a result several studies have recently been conducted exploring the possibilities of combining both particulate reinforcement and rubber modification. Much of this work, conducted in particular by Kinloch and coworkers, has shown the significant benefits which can be achieved by a combination of this type [38, 39].

6.4.2 Thermoplastic modification

Attempts at imparting significant improvements in toughness to high crosslink density epoxies such as the tetrafunctional TGDDM systems, by rubber modification, have generally resulted in failure. This has been attributed to the inability of the crosslinked epoxy matrix to undergo significant shear yielding in response to the presence of the rubber particles. This has therefore necessitated consideration of other means of toughness enhancement. One such approach which has received a relatively small degree of attention has concerned the use of thermoplastics.

Attempts at thermoplastic modification have been conducted employing primarily polyethersulphone and polyetherimide. Much of the work employing the former has been conducted by Bucknall and Partridge who attempted to study the relationship between the structure and mechanical properties of polyethersulphone modified epoxies [40]. Two types of epoxies were studied in their work, one being trifunctional and the other the current traditional aerospace resin, TGDDM. They found that morphology, studied using both dynamic mechanical spectroscopy and scanning electron microscopy, was dependent on both the type and concentration of the resin and the curing agent employed. In certain circumstances, two-phase morphological characteristics were observed, with a polyethersulphone-rich phase dispersed in the crosslinked epoxy matrix. Because of both the high modulus and T_g values of the polyethersulphone employed (2.8 GPa and 210 °C respectively), large concentrations of thermoplastic generally had a minor effect on these particularly important parameters. This of course represents a significant advantage over rubber modification where reductions in these parameters, particularly modulus, can occur. Unfortunately, however, polyethersulphone addition was found to have only a minor beneficial effect on fracture energy, where thermoplastic additions of up to 40 parts per hundred (phr) of resin in epoxy resulted in less than a 100% increase in fracture energy. Although this meagre improvement can be regarded as unexpected in view of the comparatively high fracture energy of polyethersulphone (approximately $2\,\mathrm{kJ\,m^{-2}}$) it is questionable whether significantly greater improvements in toughness would have been achieved with rubber modification of the high crosslink density resins employed.

Somewhat similar studies have been conducted by other workers including Diamont and Moulton [41] using polyetherimide as the thermoplastic modifier and McGrath and coworkers who employed hydroxyl- and amine-terminated polysulphones as modifiers for diglycidyl ether type epoxies [42, 43]. Although in both cases increases in toughness were obtained, as in the work described above the magnitude of the improvements was small in comparison with what could be expected from elastomeric modification.

To summarize therefore, although the use of tough, thermoplastic polymers offers a means by which the brittleness of crosslinked epoxies can be improved, in comparison with rubber modification the level of improvement is generally poor, particularly with relatively low T_g systems based on difunctional resins. With higher crosslink density systems such as those studied by Bucknall and Partridge, where rubber modification would probably provide a similar, relatively poor level of toughness enhancement to that provided by thermoplastics, the high modulus and T_g values of the latter would be regarded as highly advantageous.

6.5 BULK FRACTURE AND MECHANICAL PROPERTIES

6.5.1 Experimental techniques

Numerous experimental techniques have been employed to assess the properties of rubber modified epoxies and it is of interest to review briefly some of these characterization approaches before considering the properties that toughened epoxies can exhibit.

Since toughness enhancement is usually the main objective of elastomeric modification, experimental methods of quantifying improvements in toughness have received considerable attention. Most of the recent advances in the understanding of epoxy fracture have been through the application of fracture mechanics, whereby quantifiable expressions of toughness such as the critical stress intensity factor, K_{Ic}, and critical strain energy release rate, G_{Ic}, are frequently quoted. The reader is referred elsewhere for a detailed discussion [44–46] and to Chapter 1.

It is important to recognize that a cracked structure may be stressed in three different modes as indicated in Fig. 6.1. As shown these are denoted modes I, II and III. The cleavage or tensile opening mode, mode I, is technically the most important owing to it being that which most readily results in failure. In order to differentiate between the three loading modes, fracture energy, G_c, and fracture toughness, K_c, values usually employ the appropriate subscript, for example K_{Ic}, I denoting that the crack is stressed in a tensile opening mode.

Numerous test geometries have been employed to determine values of mode I fracture energy, G_{Ic}, and fracture toughness, K_{Ic}. The most common ones are illustrated in Fig. 6.2.

As discussed in Chapter 1, provided that linear elastic fracture mechanics requirements are met, the following simple relationships between K_{Ic} and G_{Ic} exist, thus allowing conversion of one parameter to another provided that modulus and Poisson's ratio data are available:

$$G_{Ic} = \frac{K_{Ic}^2}{E} \text{ for plane stress} \quad (6.2)$$

and

$$G_{Ic} = \frac{K_{Ic}^2(1 - \nu)}{E} \text{ for plane strain} \quad (6.3)$$

where E is Young's modulus and ν is Poisson's ratio.

Of all the mechanical characterization techniques employed in the study of modified epoxies, dynamic mechanical analysis (DMA) has provided possibly the most beneficial information. Capable of providing both mechanical data

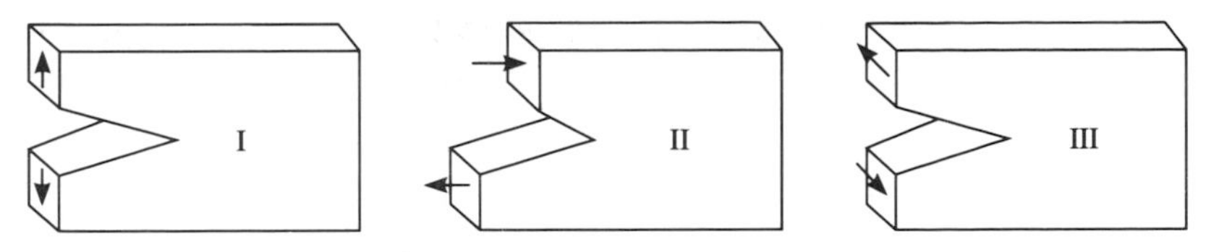

Fig. 6.1 Modes of loading: I, tensile opening; II, in-plane shear; III, tearing.

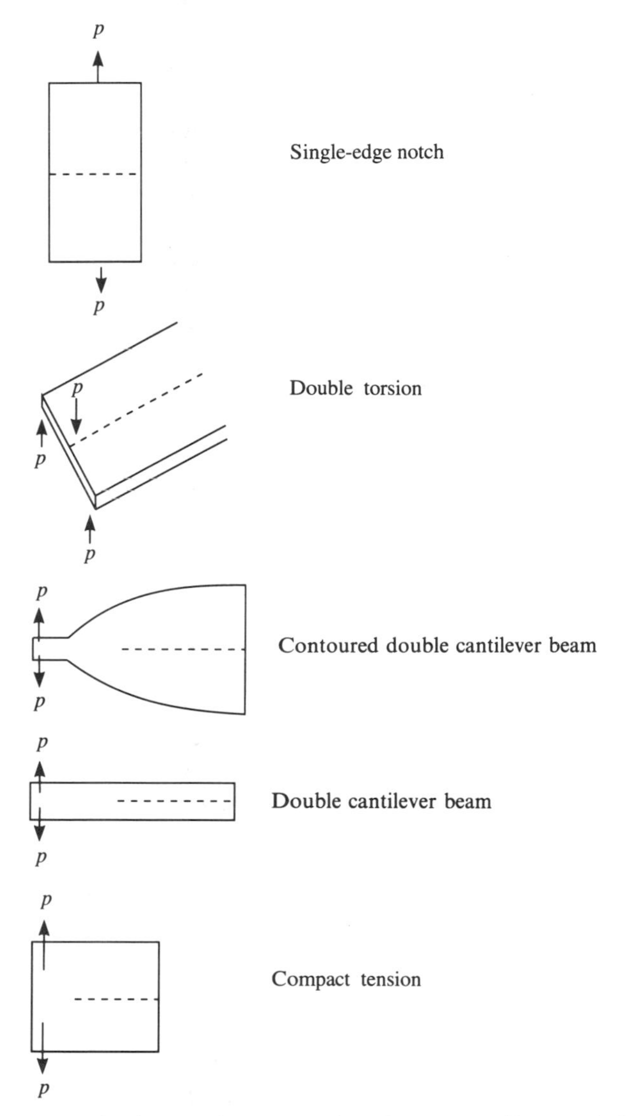

Fig. 6.2 Fracture mechanics specimens employed to assess fracture behaviour of epoxies.

and valuable information concerning vitally important morphology and phase separation behaviour characteristics, it has been employed extensively in many laboratories. A brief account of the information it can provide would therefore be useful. Essentially, dynamic mechanical analysis involves the measurement of a material's ability to store and dissipate mechanical energy when subjected to deformation. Because of its ability to measure viscoelastic behaviour, DMA is capable of detecting low-order molecular transitions of the type generally considered responsible for and vital to the properties of rubber modified epoxies. Although various approaches to this technique exist, that employed by the current author and colleagues has provided valuable information. Essentially the technique involves the application of a sinusoidal deformation to one end of an epoxy specimen (typically rectangular in shape) with the resultant torque imposed by the deformation measured at the other end. The information obtained is analysed to provide measures of elastic shear modulus, G^1, loss shear modulus, G^{11}, and $\tan \delta$, the so-called loss angle. Measurements are generally taken at various temperatures over a relevant temperature range, $-150\,°C$ to $+150\,°C$ being typical for rubber modified epoxies. Figure 6.3 shows typical dynamic mechanical data obtained from

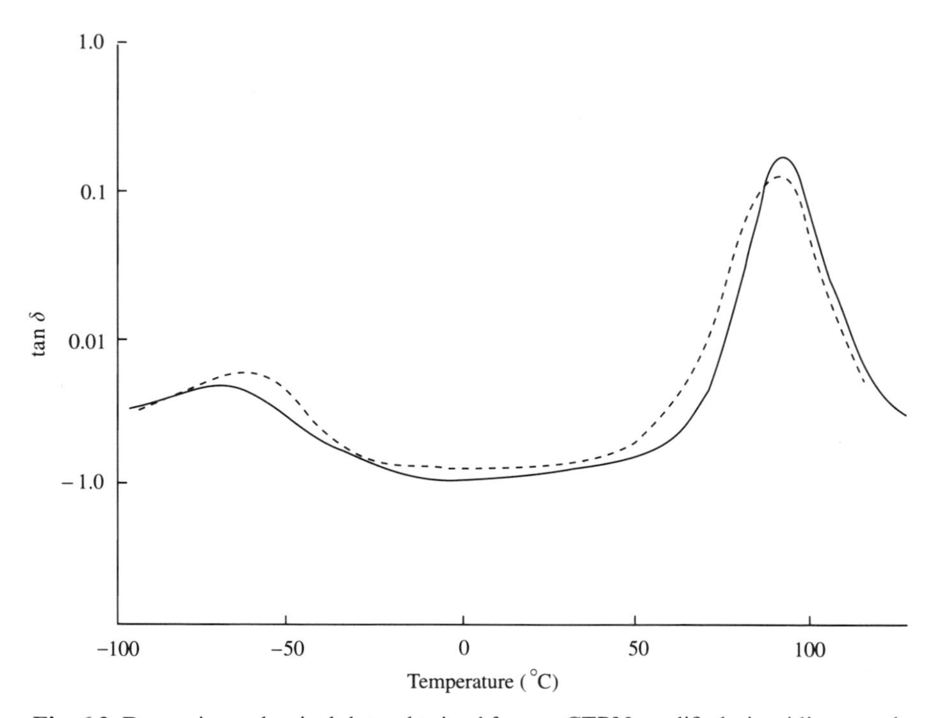

Fig. 6.3 Dynamic mechanical data obtained from a CTBN modified piperidine cured epoxy (——) and its unmodified counterpart (– – –). (Previously appeared in *Journal of Adhesion*, **28**, 231, 1989, S. J. Shaw and D. A. Tod. © Crown Copyright.)

cured unmodified and rubber modified epoxy systems. As indicated, in both cases the mechanical spectra exhibit various peaks or transitions which can indicate a great deal about both the properties and the morphological characteristics of the polymers.

In both cases the most distinctive portion of the mechanical spectrum is the α transition, which is associated with the glass transition temperature of the epoxy. Generally the α transition will be affected by factors such as formulation stoichiometry and degree of cure. With rubber modified polymers, the temperature at which the transition occurs will depend on the amount of elastomer remaining in the matrix following cure and can therefore provide important information about the vitally important phase separation process.

Subzero transitions of the type indicated in Fig. 6.3 have been observed with a wide range of amine cured unmodified epoxies. In particular, the most frequently observed transition, generally occurring in the -50 to $-80\,^{\circ}C$ temperature range, has been referred to as β relaxation [47–52]. Its presence has been attributed to crankshaft rotations of glyceryl units, $—CH_2—CH(OH)—CH_2—O—$, in the epoxy matrix, and has frequently been shown to be dependent on various factors such as formulation stoichiometry, moisture and degree of cure. With rubber modified epoxies, however, complications usually exist with the elastomeric materials often employed generally exhibiting T_g values in the same temperature region as the β relaxation. Consequently, in the rubber modified spectrum shown in Fig. 6.3, the low temperature peak will be composed of both the β relaxation of the epoxy phase and the T_g of the elastomeric phase. Various studies have shown that both the location and the size of the elastomeric T_g peak can provide useful information, particularly with regard to volume fraction of phase separated elastomer [12, 18, 53–55].

Both scanning and transmission electron microscopy (SEM, TEM) have been widely employed to study the morphological characteristics of rubber modified epoxies. Key morphological parameters such as rubber volume fraction, particle size and particle size distribution have been studied successfully using these techniques.

6.5.2 Modification with butadiene–acrylonitrile elastomers

As previously discussed, most of the research conducted into the rubber modification of epoxies has been devoted to the use of butadiene–acrylonitrile elastomers. This is also true of commercial practice with the majority of both toughened adhesive and composite matrix formulations employing these elastomers. It is thus appropriate, in considering mechanical behaviour, to devote most attention to epoxies which have utilized this type of modifier. However, other elastomer types have been considered and will therefore be briefly discussed.

(a) Elastomer type

Various types of butadiene–acrylonitrile elastomers have been studied and considered for use in toughened epoxy formulations. Molecular variations such as type of terminal functionality [19], acrylonitrile concentration [7], and molecular weight [8], have been investigated and shown to be capable of exerting an influence on various important mechanical properties. Generally, carboxyl termination has usually been shown to exert a greater toughening effect than most other functionalities including phenol, epoxy, hydroxyl and mercaptan (Table 6.1) [10]. Consequently, carboxyl-terminated rubbers have been the most extensively employed in commercial formulations. However, more recently attention has focused on amine terminal functionality as a means of introducing rubber modification into a cured epoxy network [55–57]. Formed by simply reacting a carboxyl-terminated rubber with *N*-aminoethylpiperazine, studies have shown that, under certain circumstances, ATBN incorporation can result in toughness enhancement equivalent to that found for CTBN modification. This has been demonstrated in particular by Kunz, Sayre and Assink who, in addition to mechanical property characterization, also carried out a detailed morphological study [55]. Rather surprisingly, although terminal functionality (carboxyl or amine) was shown to have an insignificant effect on toughness as measured in terms of either K_{Ic} or G_{Ic}, one particular parameter, most notably particle shape, was shown to differ dramatically between the two systems. With CTBN modified formulations, rubber particles were essentially spherical in shape, whilst ATBN modification resulted in rubber particles exhibiting highly irregular contours appearing as agglomerated islands of smaller particles. In addition fracture surface studies using scanning electron microscopy revealed major differences in fracture surface appearance. CTBN modified systems resulted in the now well-known 'deep hole' cavity fracture surface which has been demonstrated in many studies. Modification using amine-terminated elastomers, however, resulted in fracture surfaces exhibiting many shallow surface holes, with, in the case of a 15 phr rubber formulation, the surface cavities being virtually on the same level as the primary fracture plane. Particle size differences between

Table 6.1. Effect of terminal functionality on the toughening ability of butadiene–acrylonitrile elastomers

Elastomer	*Functionality*	*Fracture energy ($kJ\,m^{-2}$)*
CTBN	Carboxyl	2.8
PTBN	Phenol	2.6–3.0
ETBN	Epoxy	1.8–2.5
HTBN	Hydroxyl	0.9–2.6
MTBN	Mercaptan	0.2–0.4

From data originally published by C. K. Riew *et al.* in *Advances in Chemistry*, **154**, American Chemical Society, 1976.

the two systems were shown to be negligible, this discounting particle size variation as a possible cause. Instead Kunz, Sayre and Assink proposed that the previously mentioned particle irregularity could account for the fracture surface observations, with the irregularities acting as stress-concentrating defects capable of initiating low strain tear failure giving rise to the relatively flat fracture surfaces. No attempt was, however, made to reconcile this proposition with the insignificant differences in toughness, and indeed modulus, found for the two rubber modified systems.

The absence of terminal reactivity has been shown in several studies to provide a significantly diminished toughening effect thus suggesting that reactivity of the elastomer with the epoxy is a major requirement for substantial toughness enhancement [2, 9]. However, Huang and Kinloch have recently demonstrated that significant toughening can be provided by elastomers lacking in terminal functionality [58]. Although recognizing that for maximum toughness enhancement the use of reactive rubbers was necessary, they suggested that interfacial adhesion forces between particle and matrix arising, for example, from secondary van der Waals forces would appear to be sufficient to provide a significant increase in toughness.

Variation in the level of acrylonitrile contained within the elastomeric modifier has been shown to result in pronounced changes in both properties and polymer morphology. The early work of Rowe, Siebert and Drake demonstrated a maximum in fracture energy, G_{Ic}, with a piperidine cured DGEBA epoxy, when the acrylonitrile content of the CTBN was between 12% and 18% [7]. Additionally, rubber particle size was shown to decrease from approximately 3 μm with a 12% acrylonitrile rubber to 0.2 μm at 25% acrylonitrile. More recently Yee and Pearson have observed similar trends [15].

As previously discussed, the molecular weight of the elastomeric component can play a significant role in the crucial phase separation process. An increase in molecular weight according to equation (6.2) would be expected to increase the Gibbs free energy of mixing, ΔG_m, of the epoxy–curing agent–elastomer system, thus promoting tendencies towards phase separation and two-phase morphology. It is therefore not surprising to recognize that, where this parameter has been studied, increased rubber particle size has been shown to accompany increases in rubber molecular weight [8]. Indeed, such an observation has been applied commercially in certain proprietary formulations where the incorporation of both liquid and solid CTBN rubbers (with the latter exhibiting a significantly higher molecular weight) has been shown to result in a rubber bimodal particle size distribution. Such a morphology, it has been suggested, increases the toughness of a base epoxy to a greater extent than when either rubber modifier is used alone [59, 60].

(b) Influence of rubber concentration

Several investigations have demonstrated the quite substantial influence that rubber concentration can have on toughness [10, 11, 14, 15, 61]. Figure 6.4

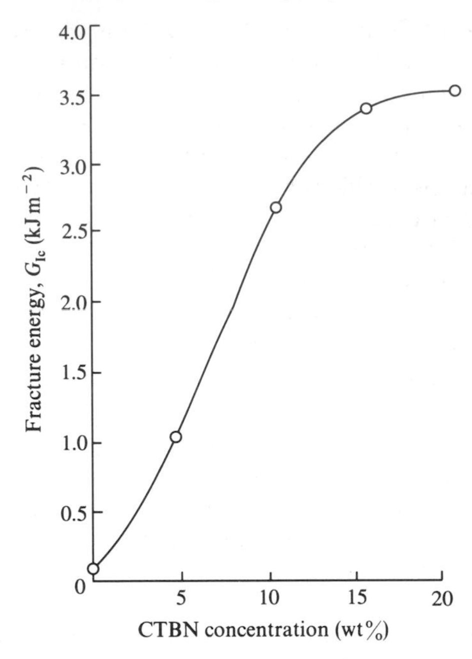

Fig. 6.4 Influence of CTBN concentration on the fracture energy of a bulk rubber modified epoxy. (Constructed from a table published by W. D. Bascom *et al.* in *Naval Engineers Journal*, p. 73, August 1976.)

shows results obtained from the early studies of Bascom *et al.* which were conducted on simple CTBN modified DGEBA epoxies employing piperidine as curing agent [14]. They observed that fracture energy, G_{Ic}, increased dramatically with relatively small increases in rubber concentration, eventually resulting in a maximum toughness value at a CTBN concentration of between 15 and 20 wt%, with further increases in rubber concentration resulting in reductions in G_{Ic}. Bascom *et al.* related this behaviour to changes in morphological character. At rubber concentrations below approximately 20 wt%, electron microscopy revealed particulate morphological characteristics with evidence of small inclusions of elastomer, approximately 2–5 μm in diameter, dispersed within the crosslinked matrix. At higher concentrations, no inclusions were visible under the scanning electron microscope, even at magnifications up to 10 000 ×. A process known as phase inversion has been proposed to account for this transitionary behaviour with, at rubber concentrations in excess of approximately 20 wt%, a reversal in morphological characteristics occurring resulting in epoxy particles embedded in an essentially elastomeric matrix. Within this regime, the mode of failure was found to be dramatically different from that observed with the elastomer particle dispersions. The high CTBN concentration systems exhibited rubber-like behaviour by failing in a tearing fashion in contrast to the more brittle-like behaviour shown by the lower rubber concentration counterparts.

Somewhat similar behaviour has since been observed in more recent studies. For example, Yee and Pearson likewise demonstrated a pronounced increase in G_{Ic} on addition of CTBN elastomer to a DGEBA epoxy cured using piperiidine [15]. Suggestions of a likely maximum in G_{Ic} at CTBN loadings of 20 phr were once again apparent although fracture energy results for rubber modified epoxies having CTBN concentrations in excess of 20 phr were not provided. Kunz, Sayre and Assink, whilst investigating the influence of both CTBN and ATBN incorporation on a polyoxypropyl eneamine cured DGEBA, noted a somewhat atypical correlation between toughness (expressed either as fracture toughness, K_{Ic}, or fracture energy, G_{Ic}) and elastomer concentration [55]. The appropriate data, shown in Table 6.2, indicate that at the lowest rubber concentrations studied, 5 phr, a K_{Ic} increase of approximately 2.5 × and G_{Ic} increases of 400% and 467% result. However, it is noteworthy that, with both elastomer types, further increases in rubber concentration provided little effect on both G_{Ic} and, particularly, K_{Ic} enhancement.

In addition to both G_{Ic} and K_{Ic}, several studies have demonstrated a positive correlation between impact strength and elastomer concentration. Levita, Marchetti and Butta have recently demonstrated the considerable improvements in impact strength which accompany the addition of an ATBN elastomer to a piperidine cured DGEBA [61]. In this work both ATBN concentration and cure temperature were shown to have a substantial effect. A substantial improvement in impact strength (2 ×) was observed at concentrations up to 10 phr, with further concentration increases up to 20 phr providing only a minor impact strength improvement. However, a further substantial increase in impact strength at elastomer concentrations beyond 20 phr, the point at which phase inversion would be expected, was observed.

Recently Sankaran and Chanda have studied the influence of HTBN incorporation on both the mechanical and thermal characteristics of an aromatic amine cured DGEBA epoxy [11]. As with other elastomer types, this

Table 6.2 Influence of elastomer concentration (ATBN or CTBN) on the fracture resistance of an epoxy

Elastomer concentration (phr)	*CTBN*		*ATBN*	
	K_{Ic} ($MN\,m^{-3/2}$)	G_{Ic} ($kJ\,m^{-2}$)	K_{Ic} ($MN\,m^{-3/2}$)	G_{Ic} ($kJ\,m^{-2}$)
0	0.8	0.3	0.8	0.3
5	1.9	1.5	2.0	1.7
10	2.0	2.0	1.9	1.8
15	1.9	2.0	2.0	2.1

Taken from data published by S. C. Kunz *et al.* in *Polymer*, **23**, 1897, 1982.

particular variant also provided a significant enhancement in toughness with, however, an optimum elastomer concentration of just 3 phr. Most interestingly, HTBN incorporation up to 6 phr was actually found to enhance modulus. Such a trend can be regarded unusual, with most studies generally showing a decline in properties such as modulus with increasing levels of elastomer.

(c) Influence of other formulation variables

Proprietary epoxy resin formulations often contain a wide variety of ingredients designed to promote and enhance various processing–mechanical properties as well as to modify cost. As well as modifying properties in their own right they will, in many cases, be capable of exerting an influence on the factors considered important for the successful exploitation of rubber modification as a toughness enhancement technique. No attempt will be made in this account to consider the likely influence of all classes of epoxy additives on the various facets of rubber modification. However, three specific examples of where formulation detail can play a major role in both the rubber modification process and the final properties are worthy of attention.

The first concerns the most obvious formulation variable, namely the resin. Several studies have demonstrated the difficulties which can and invariably do occur when attempts are made to improve the toughness of highly crosslinked epoxies by rubber modification [25, 57, 62, 63]. Figure 6.5, from the work of Pearson and Yee, shows a plot of G_{Ic} against the molecular weight of the epoxy

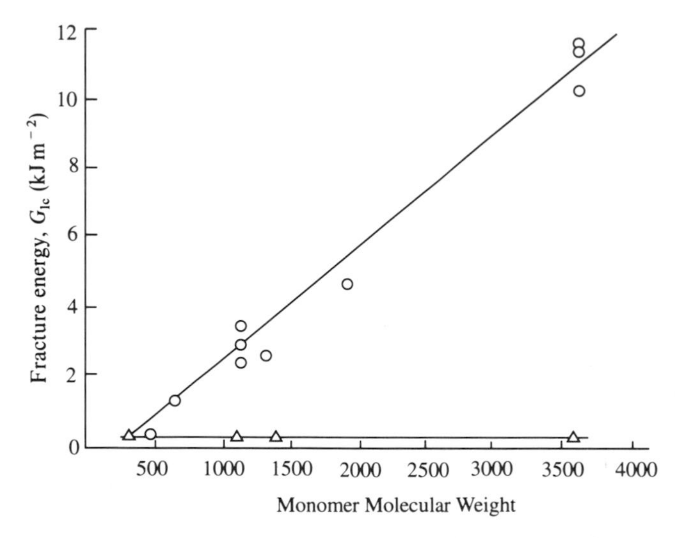

Fig. 6.5 Influence of epoxy monomer molecular weight on the fracture energy of crosslinked unmodified (△) and rubber modified (○) epoxies. (Taken from paper published by Pearson and Yee in *Journal of Materials Science*, **24**, 2571, 1989.)

prior to crosslinking for both unmodified and rubber modified systems [63]. Firstly, the data clearly show that the inherent toughness of the unmodified epoxies is only moderately affected by the initial molecular weight of the resin. However, of particular interest is that the toughness of the rubber modified epoxy is strongly influenced by molecular weight, thus indicating that the toughenability of an epoxy by elastomeric addition will be largely dependent on the crosslink density of the epoxy matrix; the lower the crosslink density, the greater the toughenability. Such difficulties with, for example TGDDM–DDS formulations, the resin system currently favoured for many aerospace applications, have resulted in various studies being conducted aimed at toughness enhancement by other routes, e.g. thermoplastic modification (section 6.4.2).

As described previously, both the type and the concentration of curing agent employed in a rubber modified epoxy formulation will dictate to a large degree the final cured fracture and mechanical properties. Curing agents which contribute to a high crosslink density, in addition to those which are highly reactive and thus not particularly beneficial with regard to phase separation, will generally result in cured formulations exhibiting low fracture toughness values.

Finally, the incorporation of bisphenol A into rubber modified epoxy systems has, as explained previously, been shown to provide substantial toughness benefits. Figure 6.6 shows data obtained from the work of Yee and Pearson which amply demonstrate this point [15].

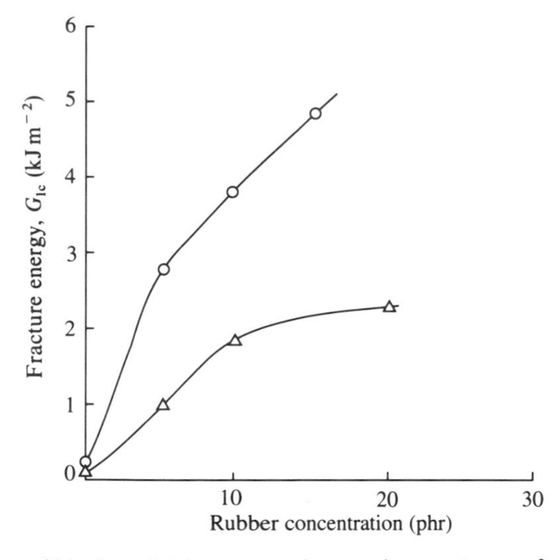

Fig. 6.6 Influence of bisphenol A incorporation on the toughness of a rubber modified epoxy: ○, bisphenol A modified; △, unmodified. (Taken from paper published by Yee and Pearson in *Journal of Materials Science*, **21**, 2462, 1986 (Figure 6).)

(d) Influence of cure conditions

Although it has long been recognized that cure conditions can influence to a large degree the fracture and mechanical properties of unmodified epoxies, it is only relatively recently that detailed attempts have been made to study the effects of cure conditions (principally temperature and time) with rubber modified systems [12, 13, 54].

Working with a CTBN modified piperidine cured DGEBA system, Shaw and Tod concluded that cure conditions, in particular temperature, exerted on toughness as much of an effect as, if not more than a greater effect than, the amount of rubber initially incorporated into the modified formulation [13]. Studying cure temperatures of from 120 °C to 160 °C over cure times of from 2 to 6 h, they obtained the G_{Ic} data indicated in Fig. 6.7. Shown as a contour diagram displaying the cure conditions which would provide G_{Ic} values of from 2 to 12 kJ m^{-2}, the figure shows that, by simply changing cure conditions from 2 h at 120 °C to 6 h at 160 °C, a sevenfold increase in G_{Ic} can result. In addition, the modulus was also found to vary with cure conditions as indicated in Fig. 6.8. Although the influence of cure conditions in this case is rather complex, it is apparent that increasing cure temperature on the whole results in a modulus reduction. Although this may seem intuitively strange, similar behaviour has been observed by other workers and by the current author and colleagues with a similar unmodified variant of the above rubber modified formulation. Free volume related effects have been considered responsible for this anomalous behaviour [64–66].

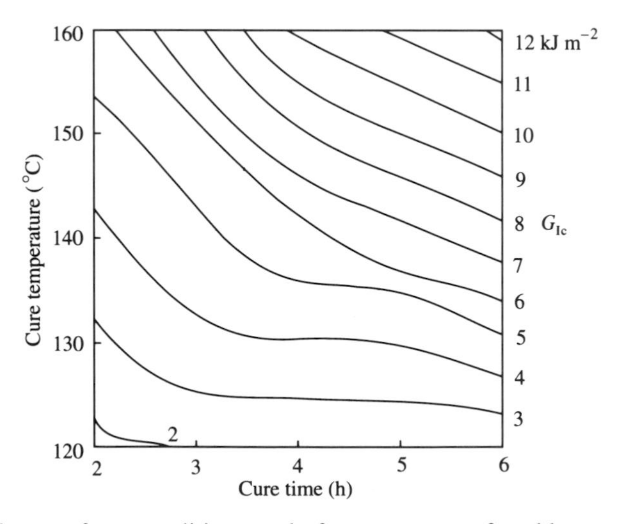

Fig. 6.7 Influence of cure conditions on the fracture energy of a rubber modified epoxy. (Previously appeared in *Journal of Adhesion*, **28**, 231, 1989, S. J. Shaw and D. A. Tod. © Crown Copyright.)

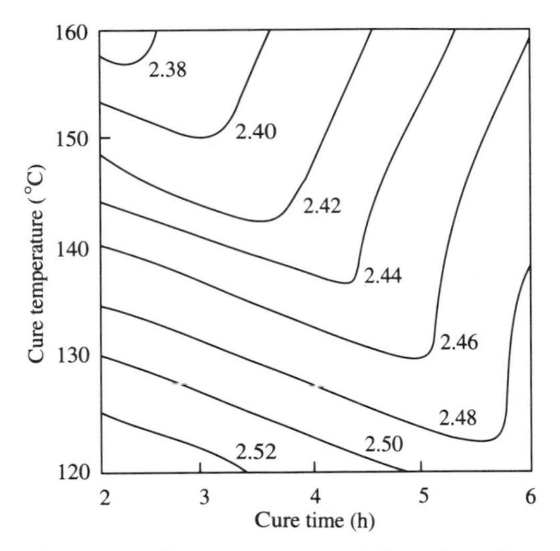

Fig. 6.8 Influence of cure conditions on the modulus of a rubber modified epoxy. (Previously appeared in *Journal of Adhesion*, **28**, 231, 1989, S. J. Shaw and D. A. Tod. © Crown Copyright.)

The work of Shaw and Tod was of particular interest in that the toughness enhancement effects indicated in Fig. 6.7 were obtained with no observable effect on T_g [13]. Evidence for this is provided in Fig. 6.9 which shows results obtained from dynamic mechanical experiments conducted on two identical

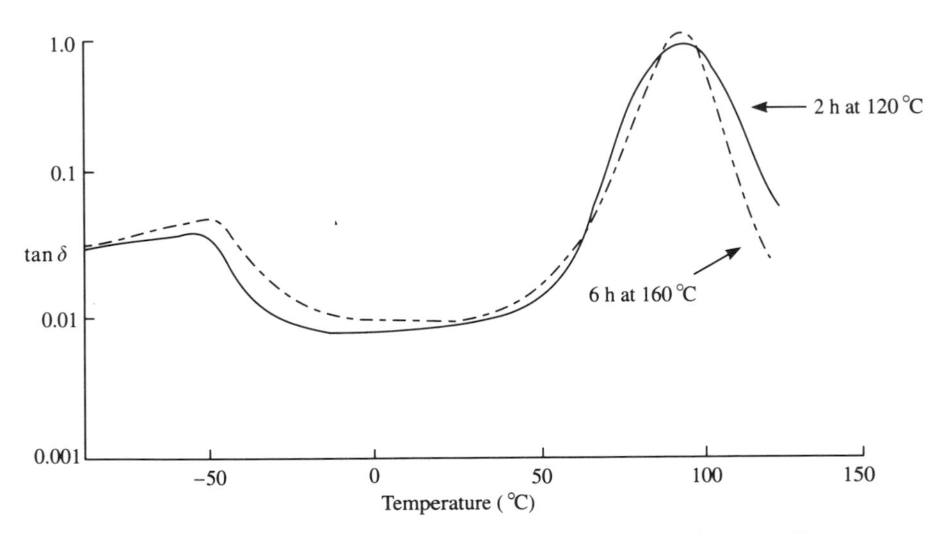

Fig. 6.9 Loss tangent, tan δ, as a function of temperature for a rubber modified epoxy subjected to two cure temperature extremes. (Previously appeared in *Journal of Adhesion*, **28**, 231, 1989, S. J. Shaw and D. A. Tod. © Crown Copyright.)

rubber modified epoxy formulations cured for 2 h at 120 °C and 6 h at 160 °C, i.e. exhibiting the lowest and highest G_{Ic} values respectively (Fig. 6.7). As indicated, although the width of the high temperature transition (α peak) is influenced by cure conditions, the precise location of the peak maximum is not, suggesting in this case that T_g is unaffected by cure conditions (particularly cure temperature). This observation is in conflict with data obtained by Kinloch, Finch and Hashemi who, whilst observing similar fracture energy–cure temperature trends, related an improvement in toughness with a significant reduction in T_g. This they attributed to an increased molecular weight between crosslinks, M_c [67]. Recent work by Truong has, however, agreed with the findings of Shaw and Tod where, once again, substantial toughness improvements with a similar rubber modified formulation were achieved with no sacrifice in T_g [68]. An interesting explanation for these apparent contradictions based on reactivities and extent of reaction has been proposed by Truong [68].

In addition to the properties so far discussed, several studies have highlighted the influence of cure temperature on the morphological characteristics of rubber modified epoxies [12, 13, 54]. In particular, Shaw and Tod [13] have shown that the high cure temperature previously discussed (6 h at 160 °C) leads to the formation of rubber particles larger than those obtained under more moderate cure conditions (2 h at 120 °C). Whether this morphological feature is responsible for, or indeed contributes to, the high levels of toughness obtained with increasing temperature of cure is open to considerable debate. As mentioned previously, rubber particle size and particle size distribution are nowadays viewed by most as having only a minor influence on properties, with rubber phase volume being the dominant morphological feature. This, together with matrix ductility, is believed to be ultimately responsible for the levels of toughness which can be achieved by rubber modification. Thus, in this respect, the findings of Kinloch, Finch and Hashemi would appear to adhere more closely to this generally accepted explanation. However, the ability of the systems studied by Shaw and Tod and by Truong to exhibit increasing levels of toughness whilst maintaining a constant value of T_g is particularly intriguing. Clearly more work is necessary before the complex effect of cure temperature on both morphology and properties can be clearly understood.

(e) Influence of test variables

Both test temperature and testing rate have been shown, in numerous studies, to have significant effects on the fracture behaviour of epoxy resins, both unmodified and rubber toughened [14, 24, 60, 68–77]. This is demonstrated in Figs 6.10 and 6.11, obtained from the extensive studies conducted by Kinloch and coworkers [24, 75]. This work and other studies have demonstrated two major factors. First, an increase in testing rate results in a reduction in fracture toughness, K_{Ic}, with rate dependence being greatest at the highest test tem-

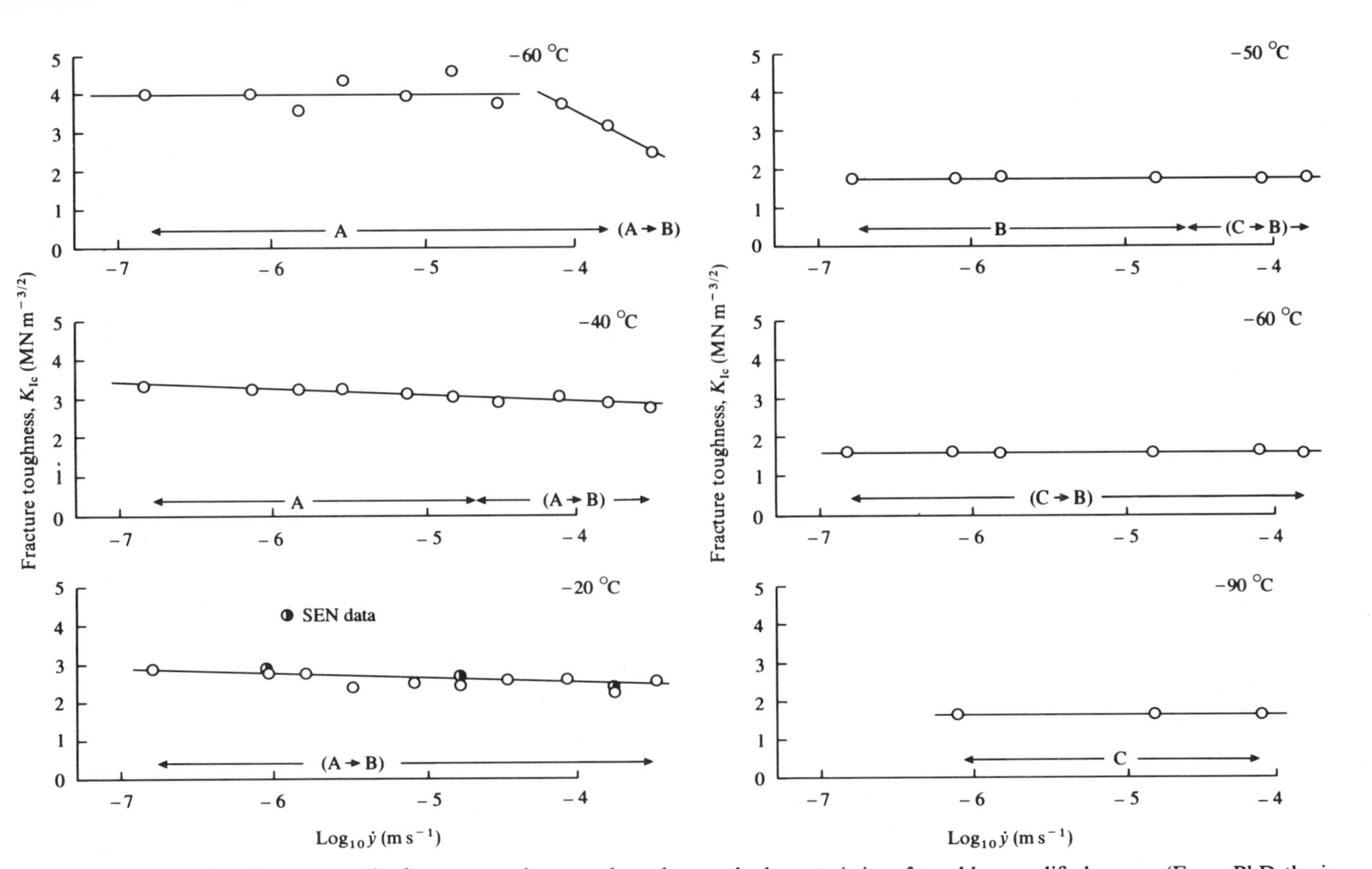

Fig. 6.10 Influence of testing rate on the fracture toughness and crack growth characteristics of a rubber modified epoxy. (From PhD thesis, S. J. Shaw.)

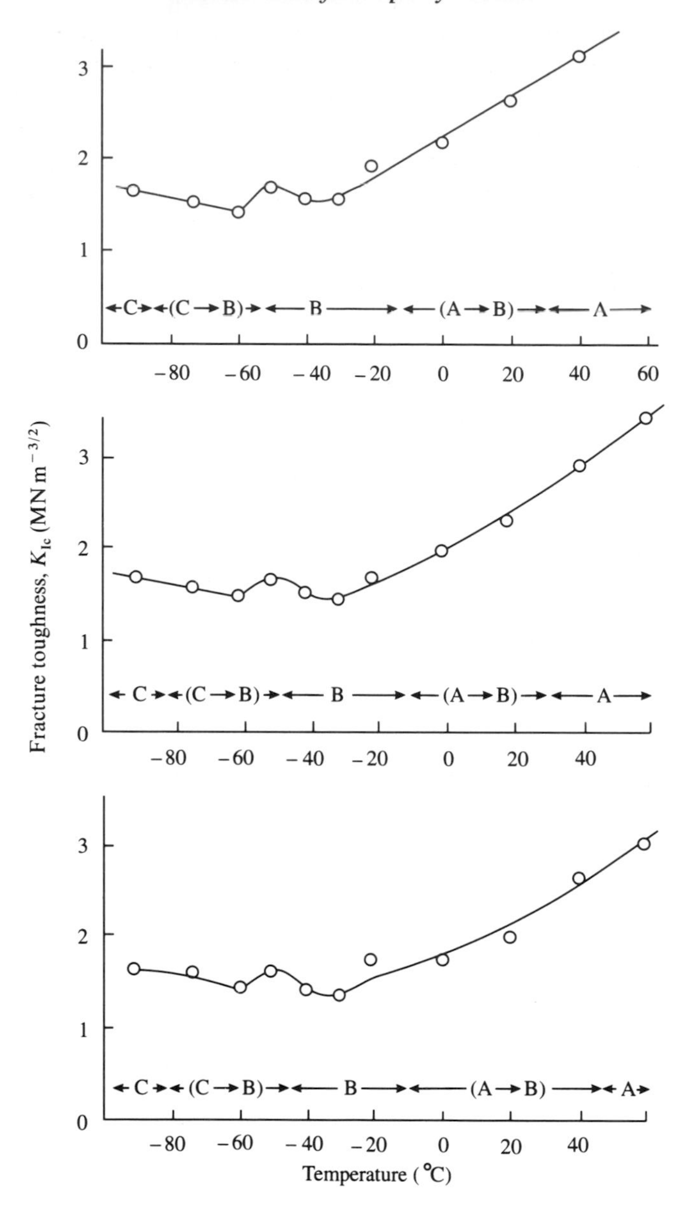

Fig. 6.11 Influence of temperature on the fracture toughness and crack growth characteristics of a rubber modified epoxy. (From PhD thesis, S. J. Shaw.)

peratures. Only at temperatures below about $-40\,°C$ does rate dependence become negligible. Second, as shown in Fig. 6.11, an increase in temperature from approximately $-30\,°C$ produces an increase in K_{Ic}. At temperatures below $-30\,°C$, Kinloch *et al.* further observed two main effects. First, a peak in K_{Ic} at approximately $-50\,^{\circ}C$ was apparent. Second, a decrease in test temperature from $-60\,^{\circ}C$ to $-90\,^{\circ}C$ resulted in a further minor K_{Ic} increase. With regard to the former observation, it was noted that the temperature at which the apparent peak in K_{Ic} occurred was similar to the rubber α peak transition temperature obtained from dynamic mechanical analysis, thus suggesting a correlation between the K_{Ic} peak at $-50\,^{\circ}C$ and the rubber α peak transition. Similar correlations have been observed with other polymeric materials such as polytetrafluoroethylene [78, 79], polysulphones [79] and poly-2, 6-dimethylene-1, 4-phenylene ether [79]. Similarly, Kinloch *et al.* attributed the increase in K_{Ic} which occurs from -60 to $-90\,^{\circ}C$ to the presence of the β relaxation of the epoxy matrix which occurs in this temperature region.

In addition to both fracture toughness, K_{Ic}, and fracture energy, G_{Ic}, crack propagation behaviour has also been shown to be strongly dependent on testing rate and temperature. In their studies, Kinloch, Shaw and Hunston employed an alphabetical notation to describe both crack propagation and fracture surface topography, as indicated in both Figs 6.10 and 6.11 [24]. Type A crack growth, which was generally observed under conditions of relatively high temperatures ($>40\,^{\circ}C$) and low rates, was stable in nature with rate of propagation being governed by the rate of test. Fracture surfaces produced by this crack growth regime were generally rough with pronounced stress whitening. Type B was a notation employed to describe unstable stick–slip crack propagation which occurred at temperatures below approximately $0\,^{\circ}C$, whilst at still lower temperatures a stable continuous form of crack growth, termed type C, was observed.

The observation that increasing test temperature and decreasing testing rate have a similar effect, namely an increase in toughness, clearly suggests that the fracture process is dominated by viscoelastic effects. This interrelationship has in fact been recently expressed in quantitative terms by Hunston *et al.* [80].

6.5.3 Other elastomer types

As briefly mentioned in section 6.2.2, other elastomer types have only received minor attention in comparison with the butadiene–acrylonitrile systems and, where investigations have taken place, these have usually demonstrated the superiority of the butadiene–acrylonitrile rubbers in promoting toughness enhancement.

Notable studies include those conducted by Yorkgitis *et al.* on polysiloxanes [3], Mijovic, Pearce and Foun on fluoroelastomers [4] and Kirshenbaum, Gazit and Bell [5] and Iijima *et al.* [6] on acrylate based elastomers.

Yorkgitis *et al.* employed functionally terminated polydimethylsiloxane, either on its own or in copolymer form with either methyltrifluoropropylsiloxane or diphenylsiloxane, to modify a cycloaliphatic amine cured DGEBA resin [3]. Some of the results obtained are shown in Fig. 6.12. As indicated,

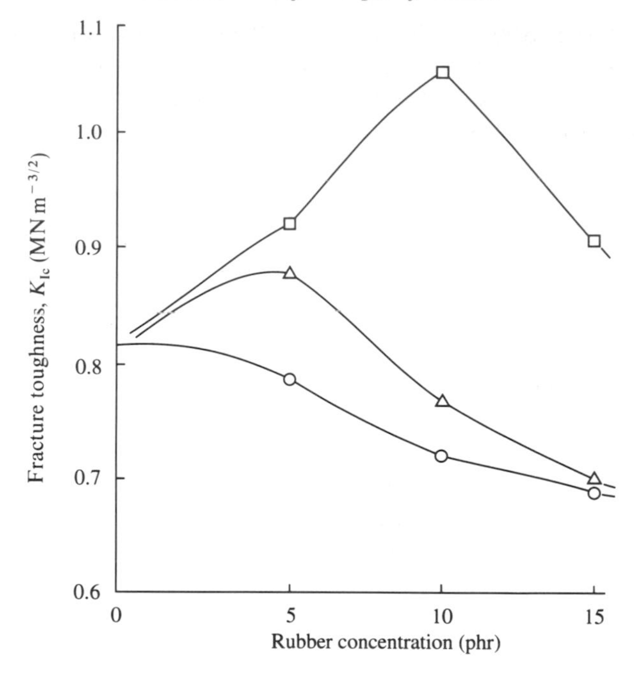

Fig. 6.12 Influence of polysiloxane modification on the fracture toughness, K_{Ic}, of a cycloaliphatic amine cured DGEBA epoxy: ○, pure polydimethylsiloxane; △, 20% methyltrifluoropropylsiloxane; □, 40% methyltrifluoropropylsiloxane. (Taken from paper published by Yorkgitis *et al.* in *Rubber-modified Thermoset Resin* (eds C. K. Riew and J. K. Gillham), *Advances in Chemistry*, **208**, American Chemical Society, 1984.)

only in certain specific cases was elastomer incorporation found to increase K_{Ic}. For example, modification with pure polydimethylsiloxane up to a concentration of 15 wt% resulted in reductions in K_{Ic} with significant improvements only occurring with polysiloxane elastomers containing 40% or more methyltrifluoropropylsiloxane. In this work an interesting correlation between toughness and morphology was observed with low K_{Ic} values being associated with large heterogeneous particles. Higher K_{Ic} values generally related to smaller homogeneous particles that exhibited dilation during the fracture process.

Although incorporation of siloxane elastomers had the expected effect of reducing flexural modulus, Yorkgitis *et al.* concluded that, in certain cases, methyltrifluoropropylsiloxane–polydimethylsiloxane modified resins generally exhibited higher modulus values than their butadiene–acrylonitrile modified counterparts.

Acrylic based elastomers prepared by copolymerizing butyl acrylate, vinylbenzyl glycidyl ether and styrene have been studied with a view to toughness enhancement of epoxies by Iijima *et al.* [6]. They observed that the addition of 20 wt% of a terpolymer based on the above monomers was capable of increasing K_{Ic} of an epoxy system by approximately 80%, and that this was associated with the formation of a two-phase morphology.

6.6 ADHESIVE JOINT PROPERTIES

6.6.1 Fracture mechanics experimental techniques

Before discussing the influence of rubber modification on the properties of epoxy bonded adhesive joints, it is of interest to discuss briefly the experimental techniques which have been employed to characterize this behaviour.

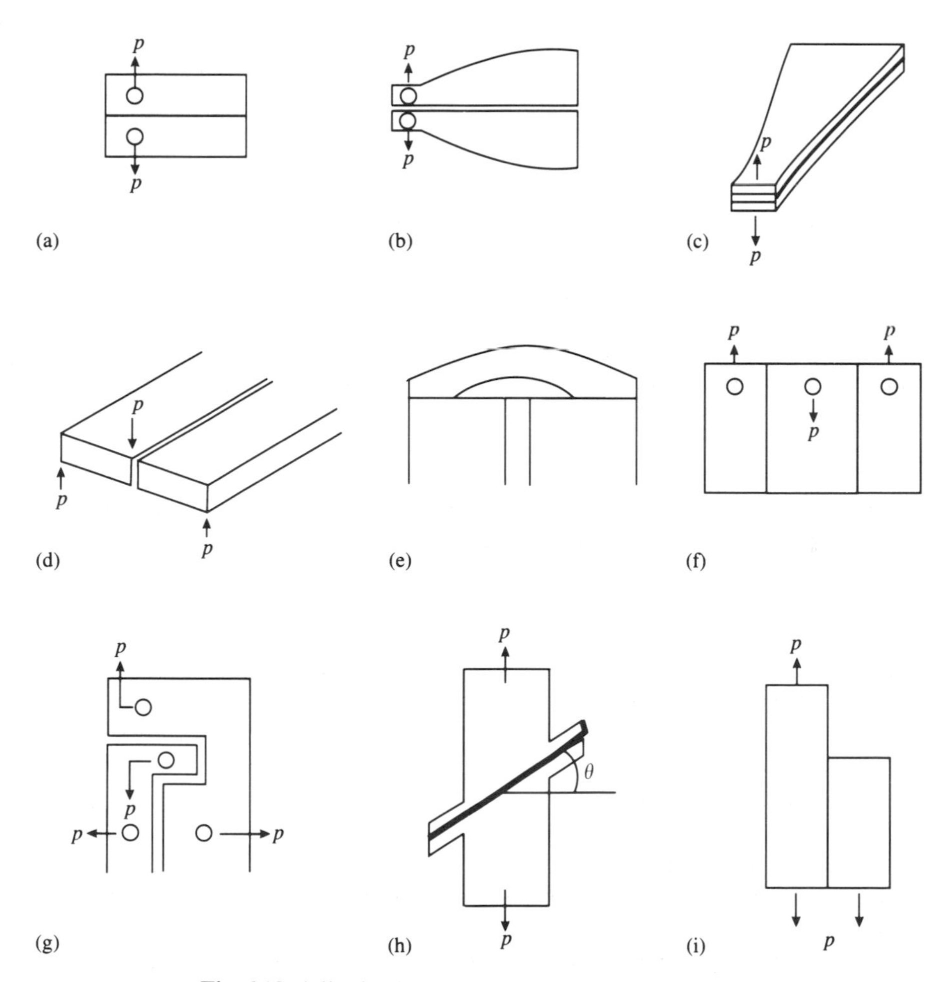

Fig. 6.13 Adhesive joint fracture mechanics specimens.

Although the use of standard adhesive joint test pieces such as lap-shear, wedge and peel joints can be and have been employed for this purpose, most of the more fundamental studies have employed fracture mechanics principles and specimens. It is therefore of interest to consider briefly the various joint types which have been employed to obtain the data discussed later in this chapter.

The number of specimen types which have been employed within this context is quite large and Fig. 6.13 shows some typical examples.

Probably the most popular design which has been employed for determining mode I (tensile opening) fracture energy values is the contoured double cantilever beam specimen (Fig. 6.13(b)). Designed in its initial form by Mostovoy and Ripling, the specimen is contoured so as to provide constant compliance conditions, which allows fracture energy to remain independent of crack length [81–86]. This can be advantageous, particularly in situations where the precise location of a crack tip is difficult to determine. A number of variants of this design have been employed in practice ranging from the parallel sided joint (Fig. 6.13(a)), where constant compliance is not provided [81], to specimens where constant compliance is achieved by an increase in width along the length of the specimen (Fig. 6.13(c)) [87, 88].

In addition to mode I, specimens have been developed and employed for measuring mode II (in-plane shear) and combined mode I–II loads. The so-called scarf joint (Fig. 6.13(h)), developed by Trantina, was designed to impart a combination of both mode I and mode II loads on an adhesive layer, a variation in the angle θ allowing alteration in the mode I–II ratio and thus variability from essentially cleavage to shear [89–92].

The independently loaded mixed-mode specimen (Fig. 6.13(g)) has also been employed to study mixed I–II mode behaviour, this being achievable by independently applying loads perpendicularly as shown [91].

Mode III loading has only received minor attention with regard to both bulk and adhesive joint fracture.

6.6.2 Influence of joint geometry

Numerous studies have shown that various joint geometrical parameters, such as adhesive bond thickness [14, 70, 75, 93], bond width [93] and mode of loading [92], can influence substantially the toughness of a joint comprising a rubber modified epoxy adhesive.

Observed initially by Bascom *et al.* [14, 70] and subsequently by Kinloch and Shaw [75] Fig. 6.14 demonstrates the complex effect bond thickness has on the fracture energy of a rubber modified epoxy, in this case a CTBN modified piperidine cured DGEBA. As indicated, adhesive G_{Ic} passes through a maximum, which we can call G_{Icm}, at a specific bond thickness, t_m. At thicknesses beyond this value, G_{Ic} undergoes a decline until a value is reached which remains essentially constant with increased thickness. This constant G_{Ic}

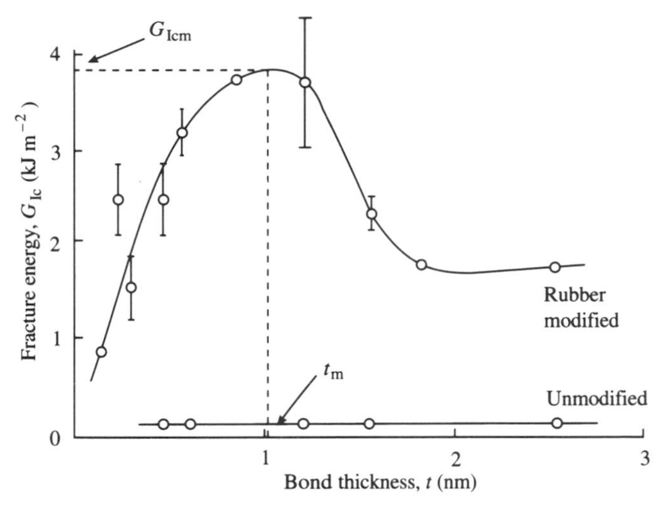

Fig. 6.14 Influence of adhesive bond thickness on the fracture energy of a rubber modified epoxy adhesive and its unmodified counterpart. (From PhD thesis, S. J. Shaw.)

has been shown to be similar to G_{Ic} values obtained from bulk adhesive specimens.

In addition, bond thickness has also been shown to have a significant effect on both crack growth behaviour and fracture surface morphology. Both Bascom *et al.* [14, 70] and Kinloch and Shaw [75] found that at low bond thickness values, i.e. below t_m, crack growth was generally of a stable, continuous nature with the resultant fracture surface being severely stress whitened with the locus of failure at or close to the adhesive–substrate interface. At high bond thicknesses, i.e. greater than t_m, unstable, stick–slip fracture predominated with each crack arrest–initiation point being represented by a curved stress whitened zone on the fracture surface. No evidence of stress whitening or apparent interfacial failure was observed in these fast, unstable growth regions.

The linear relationship shown at the bottom of Fig. 6.14 refers to G_{Ic} values obtained from the corresponding unmodified epoxy adhesive. The trend obtained is similar to that found by previous workers such as Mostovoy, Ripling and Bersch who, using both amine and anhydride cured epoxies, observed no significant effect of bond thickness on adhesive G_{Ic} over a thickness range of approximately 0.05–0.5 mm [83, 85]. Clearly the presence of the rubber particles within the modified epoxy is imparting characteristics which lead to the complex relationship shown and it is of interest to consider the mechanisms responsible for this behaviour.

The bond thickness effect shown in Fig. 6.14 has been discussed by Kinloch and Shaw in terms of a crack tip plastic zone restriction–constraint mechan-

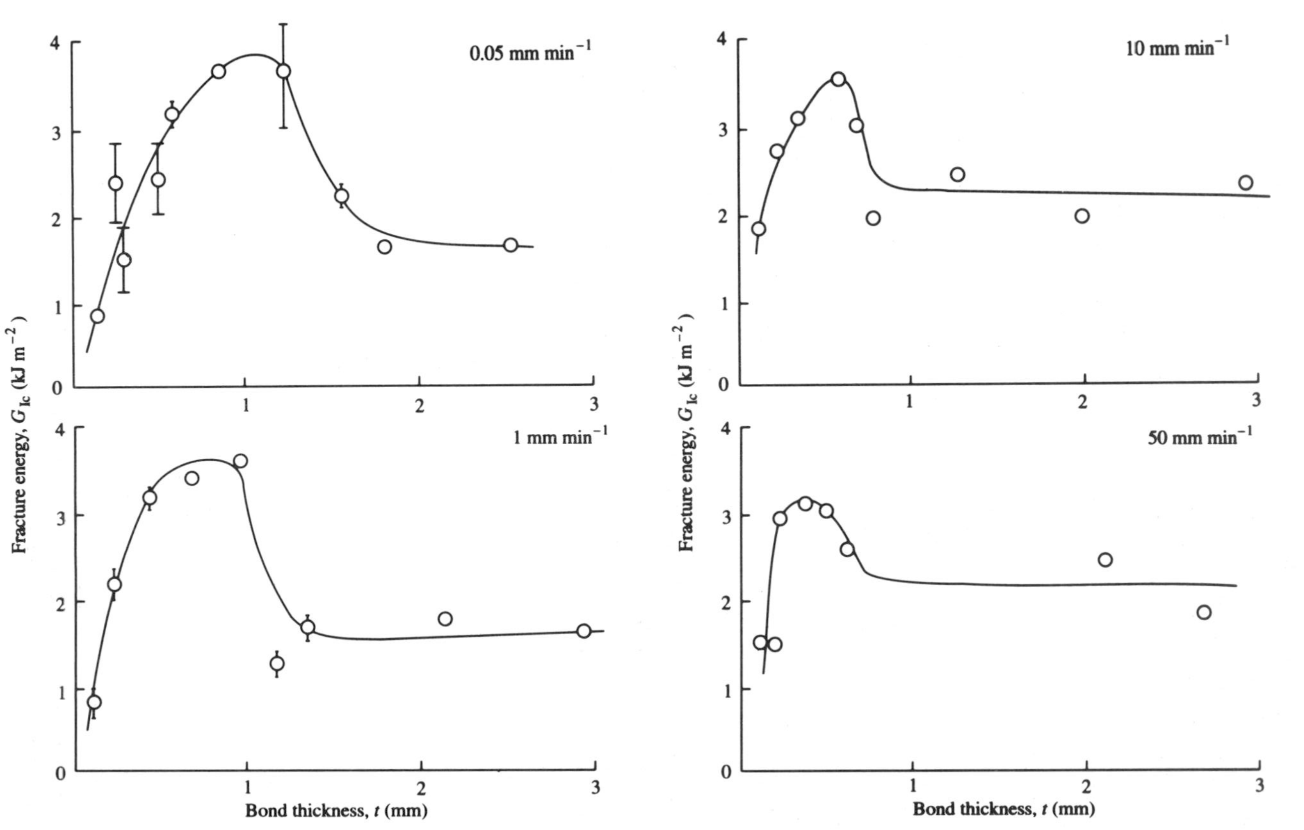

Fig. 6.15 Effect of testing rate on the fracture energy–bond thickness relationship of a rubber modified epoxy adhesive. (Previously appeared in *Journal of Adhesion*, **12**, 59, 1981, A. J. Kinloch and S. J. Shaw. © Crown Copyright.)

ism, the basis of which is outlined in Fig. 6.15 [75]. As indicated, crack tip plasticity and in particular the size and shape of the plastic zone form the basis of this mechanism. Because of the relatively large crack tip plastic zones which can develop in rubber modified epoxies, in reality plastic zone dimensions can approximately equate with the bond thickness dimensions shown in Fig. 6.14. Thus at bond thickness values less than the plastic zone diameter at the tip of the crack, one can perceive a restriction in the development of this zone in the bond thickness direction thus reducing plastic zone volume and hence toughness. Additionally, studies conducted by Wang, Mandel and McGarry [94], using finite element analytical techniques, together with recent experimental observations [93] have shown that constraints imposed on an adhesive layer can dictate to a large degree the distance over which the principal stress responsible for the plastic zone exists. Within this context, increased constraint brought about by, for example, a reduction in bond thickness can produce an elongation of the crack tip plastic zone which can thus, all other factors being equal, increase plastic zone volume and hence toughness. We can, therefore, view the fracture energy–bond thickness relationship as being due to two essentially competing effects, plastic zone restriction and constraint, which between them can influence both the shape and the volume of the crack tip plastic zone and hence the toughness of the adhesive layer. Kinloch and Shaw have suggested that the maximum volume of plastic zone, and hence the maximum fracture energy, G_{Icm}, occurs when the maximum degree of constraint exists, at a given bond thickness, commensurate with the condition that no restriction on the development of the plastic zone from the substrates exists [75]. Studies conducted with a rubber modified epoxy over a range of rates and temperatures have shown that this condition occurs, as shown in Table 6.3, when the plane stress plastic zone roughly equates with t_m, i.e. the

Table 6.3 Comparison of adhesive bond thickness, t_m, at maximum fracture energy, G_{Icm}, and plane stress plastic zone diameter

Temperature (°C)	*$\log_{10}$ (rate) ($m\,s^{-1}$)*	*t_m (mm)*	*Plastic zone diameter (mm)*
20	−6.08	1.0	0.85
20	−4.78	0.8	0.70
20	−3.78	0.55	0.49
20	−3.08	0.4	0.43
50	−4.66	1.1	1.6
37	−4.66	0.9	1.16
25	−4.66	0.6	0.57
0	−4.66	0.5	0.39
−20	−4.66	0.25	0.15
−40	−4.66	0.1	0.05

Data originally published by A. J. Kinloch and S. J. Shaw in *Journal of Adhesion*, **12**, 59, 1981. Crown Copyright.

bond thickness at the fracture energy maximum. The decline in adhesive fracture energy at bond thicknesses less than t_m can therefore be attributed to the presence of the high modulus substrates restricting development of the plastic zone in the bond thickness direction. Although extension of the plastic zone along the bond line will increase, plastic zone restriction will be dominant, resulting in a G_{Ic} decline. At t_m, G_{Ic} of the adhesive layer will be greater than that obtained for bulk polymer owing simply to the enhanced plastic zone volume brought about by the constraint effect. As bond thickness is increased beyond t_m, restriction in the bond thickness direction will clearly no longer apply and, with constraint on the adhesive layer reducing as bond thickness is increased, a reduction in plastic zone extension, and therefore volume, will result. This will cause G_{Ic} to fall until a bond thickness is reached whereupon the influence of the substrates will be negligible, resulting in an approximate equivalence between adhesive and bulk G_{Ic}.

In addition to fracture investigations under mode I loading conditions, studies under mode II and combined mode I–II combinations have also been carried out. Table 6.4 shows fracture energy data obtained for two types of epoxy adhesive subjected to loading under mode I, mode II and combined mode I–II conditions. For the unmodified system (a DGEBA epoxy cured with tetraethylenepentamine), the mode I value as expected is the lowest. Although introducing a mode II component increases fracture energy by a factor of approximately 2, a high fracture energy is only achieved when virtually all of the mode I component is removed. The toughened adhesive, however, as shown, exhibits far more complex behaviour. The mixed mode I–II values are substantially lower than either mode I or II values with the two extreme loading conditions providing approximately equal fracture energy values. A maximum in the I–II fracture energy at $\phi = 45°$ is also apparent. These interesting features have been addressed by Bascom and Oroshnik [92].

Table 6.4 Effect of loading mode on the fracture energy of epoxy resin adhesives

	Fracture energy ($kJ\ m^{-2}$)				
	G_{Ic}		$G_{I.IIc}$*		G_{IIc}†
Adhesive		30°	45°	60°	
DGEBA–TEPA (unmodified epoxy)	0.07	0.15	0.13	0.15	1.45
DGEBA–CTBN–piperidine (toughened epoxy)	3.40	0.31	0.50	0.31	3.55

* Obtained from scarf joint specimen.
† Obtained from independently loaded mixed-mode joint.
Data originally published by A. J. Kinloch and S. J. Shaw in *Developments in Adhesives 2* (ed. A. J. Kinloch), Applied Science Publishers, 1981. © Crown Copyright.

6.6.3 Influence of test variables

Both testing rate and test temperature have been shown to have pronounced effects on both the toughness and the crack propagation behaviour of rubber modified epoxy bonded joints [73, 75, 93]. Dealing firstly with rate, the influence of this parameter on the fracture energy–bond thickness relationship

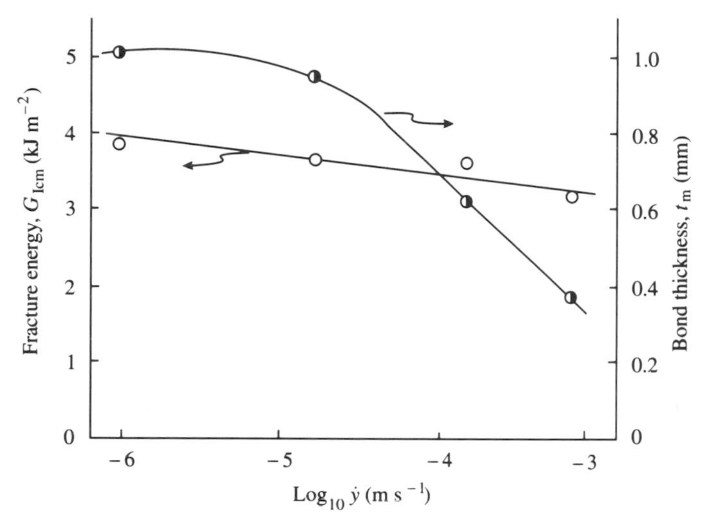

Fig. 6.16 Influence of testing rate on the rubber modified adhesive fracture energy maximum, G_{Icm}, and bond thickness, t_m. (From PhD thesis, S. J. Shaw.)

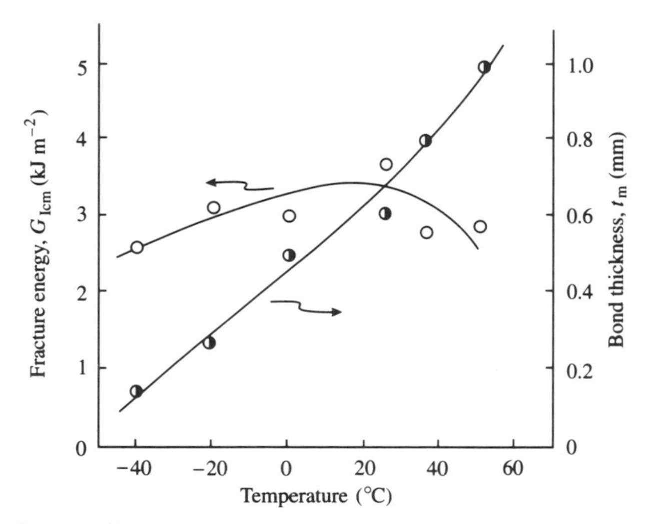

Fig. 6.17 Influence of temperature on the rubber modified adhesive fracture energy maximum, G_{Icm}, and bond thickness, t_m. (From PhD thesis, S. J. Shaw.)

discussed above is demonstrated in Fig. 6.15 for a series of contoured double cantilever beam joints bonded using a rubber modified epoxy and tested at 20 °C [75, 93]. This shows that the general trend previously discussed for one particular test rate is maintained over a fairly wide range of rates. Furthermore, it can be seen that an increase in rate produces two main effects, first a reduction in the value of the maximum fracture energy, G_{Icm}, and second a reduction in the bond thickness, t_m, corresponding to G_{Icm}. In addition, crack propagation was also affected with increasing rate generally resulting in a reduction in the degree of stable crack growth. Somewhat similar effects have been observed with test temperature [73], with reductions in temperature having similar effects on the fracture energy–bond thickness relationship to those of increasing rate. As with the bulk fracture data previously discussed, this interrelation between rate and temperature once again emphasizes the

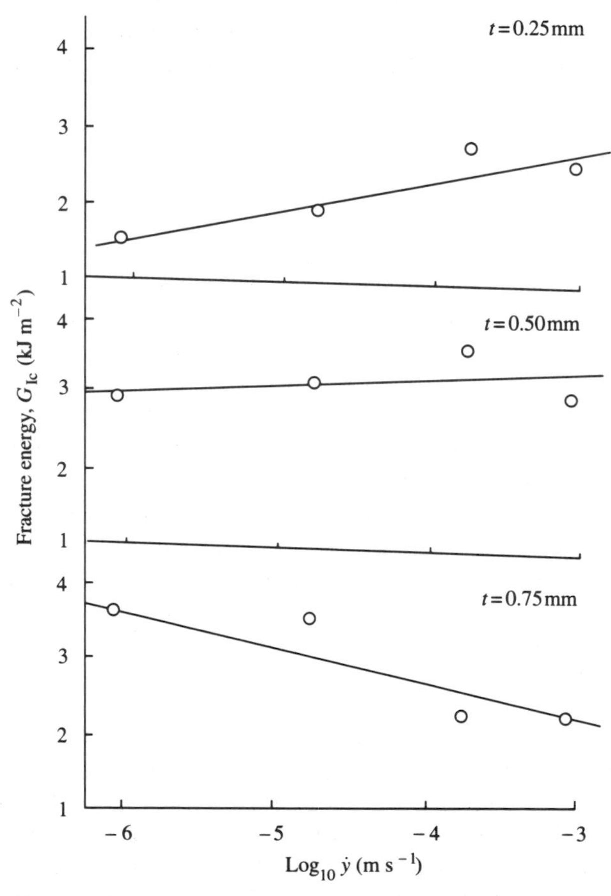

Fig. 6.18 Rubber modified epoxy adhesive fracture energy, G_{Ic}, as a function of testing rate for three bond thicknesses. (From PhD thesis, S. J. Shaw.)

importance of the viscoelastic response of the material in influencing fracture behaviour.

The practical consequences of the effects of rate and temperature on both G_{Icm} and t_m are demonstrated in Figs 6.16 and 6.17 [16]. As indicated, the selection of the optimum value of bond thickness, t_m, to achieve maximum toughness will depend on the service conditions the joint is likely to experience, which would obviously impose considerable design difficulties in attempting to obtain maximum potential from these materials. It is also of particular interest to recognize that the existence of the bond thickness effect exerts a complicating influence on the dependence of adhesive G_{Ic} on rate. As discussed previously, bulk rubber modified epoxy specimens generally exhibit a reduction in toughness with increasing rate. For adhesive joints, however, considerably more complex behaviour exists as indicated in Fig. 6.18. As shown, increasing rate can produce either an increase or a decrease in G_{Ic}. Consequently the trend can be the same as, or opposite to, that seen in bulk samples. The reasons for this complex behaviour have been explained by Kinloch and Shaw in terms of the plastic zone restriction mechanism discussed previously [75].

6.7 COMPOSITE FRACTURE

Fibre reinforced composite materials, particularly those based on thermoset matrix resins such as epoxies, have been shown to be vulnerable to impact damage with delamination being the principal mode of failure. Since the presence of delamination in a composite can substantially reduce composite performance, particularly under compression loading, considerable effort has been devoted to the development of resins having improved resistance to delamination. Since rubber modification has been shown to be capable of promoting dramatic improvements in the toughness of epoxies, the use of this modification approach has been proposed as a potential means of promoting delamination resistance and damage tolerance in composites. Unfortunately much of the work conducted has not provided optimistic results. For example, early work conducted by McGarry and coworkers found that rubber modification provided no improvement in the interlaminar fracture energy of an epoxy composite even though the toughness of the bulk epoxy was increased tenfold [95]. Similarly Scott and Phillips found only a twofold increase in interlaminar fracture energy with a unidirectional CFRP composite using a CTBN modified piperidine cured epoxy as matrix [96]. Scott and Phillips rather tentatively attributed this unsubstantial difference in behaviour to a suppression of the toughening mechanism responsible for toughness enhancement. Later work conducted by Bascom *et al.* produced more optimistic results with a twentyfold increase in bulk toughness resulting from rubber modification, this being translated into an eightfold increase in interlaminar toughness [97].

In an attempt to establish a detailed relationship between bulk resin toughness and composite interlaminar fracture behaviour, Hunston in 1984 conducted a detailed analysis of both bulk and composite fracture data [98]. In this study a wide range of both polymer toughness values and material types ranging from brittle and rubber modified epoxies to thermoplastic polymers such as polysulphones and polyetherimide were investigated. The relationship he observed is shown in Fig. 6.19. This shows a clear correlation between bulk resin toughness and composite interlaminar fracture energy. For resins with low fracture energy (less than approximately 300 J m^{-2}) the composites are tougher than the resin. However, above about 300 J m^{-2} there exists an approximate three-to-one reduction in the composite interlaminar fracture energy compared with that for the bulk resin. Making certain assumptions regarding fibre–fibre and interply spacing, Hunston demonstrated a qualitative similarity between crack tip plastic zone size for polymer with a bulk fracture energy at or close to the intersection shown in Fig. 6.19 and interply distance, thus indicating similarity in behaviour between composites and adhesive joints. Hunston therefore proposed that, with particularly brittle polymers, bulk toughness will be directly transferred to the composite and thus interlaminar fracture energy will be at least as high as the bulk G_{Ic}. In addition, since crack propagation in a composite will invariably involve other energy absorption mechanisms, interlaminar G_{Ic} will usually be greater than

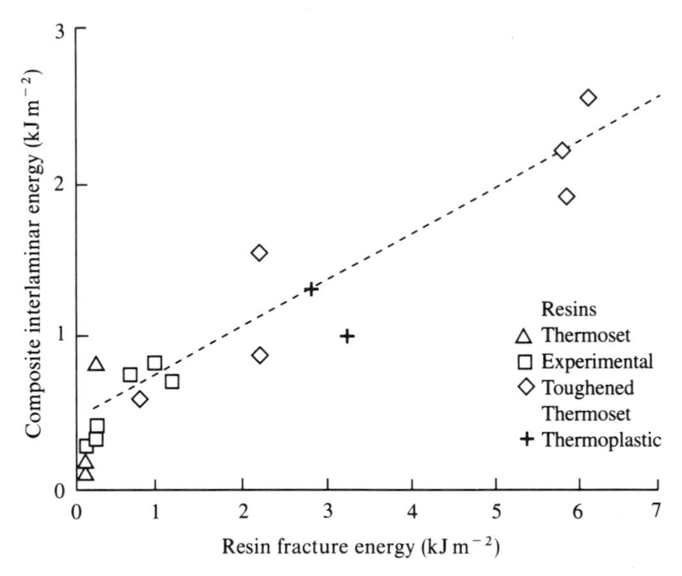

Fig. 6.19 Relationship between bulk resin fracture energy and composite interlaminar fracture energy: △, thermoset resin; □, experimental resin; ◇, rubber modified thermoset resin; +, thermoplastic. (Taken from paper published by D. L. Hunston in *Composites Technology Review*, **6**, 176, 1984.)

bulk G_{Ic}. With tougher matrix resins such as rubber modified epoxies, a similar situation to that previously described with adhesive joints will prevail, whereby the restrictive influence of the fibres will prevent full development of the crack tip plastic zone, and thus have a harmful effect on toughness. Thus although the rubber modification of epoxy matrix resins can be and indeed has been shown to translate into improved composite toughness characteristics, full utilization of resin toughness is generally not realized. This is discussed fully in Chapter 3.

REFERENCES

1. Potter, W. G. (1970) *Epoxide Resins*, Iliffe, London.
2. May, C. A. (1988) *Epoxy Resins, Chemistry and Technology*, 2nd edn, Dekker, New York.
3. Yorkgitis, E. M., Trau, C., Eiss, N. S., Hut, T. Y., Yilgor, I., Wilkes, G. L. and McGrath, J. E. (1984) *Adv. Chem. Ser.*, **208**, 137.
4. Mijovic, J., Pearce, E. M. and Foun, C. C. (1984) *Adv. Chem. Ser.*, **208**, 293.
5. Kirshenbaum, S. L., Gazit, S. and Bell, J. P. (1984) *Adv. Chem. Ser.*, **208**, 163.
6. Iijima, T., Tomoi, M., Yamasaki, J. and Kakiuchi, H. (1990) *Eur. Polym. J.*, **26**(2), 145.
7. Rowe, E. H., Siebert, A. R. and Drake, R. S. (1970) *Mod. Plast.*, **47**, 110.
8. McGarry, F. J. and Willner, A. M. (1968) *Report R68–8*, MIT, Cambridge, MA.
9. Siebert, A. R. and Riew, C. K. (1971) *Org. Coat. Plast. Chem.*, **31**, 552.
10. Riew, C. K., Rowe, E. H. and Siebert, A. R. (1976) *Adv. Chem. Ser.*, **154**, 326.
11. Sankaran, S. and Chanda, M. (1990) *J. Appl. Polym. Sci.*, **39**, 1459.
12. Chan, L. C., Gillham, J. K., Kinloch, A. J. and Shaw, S. J. (1984) *Adv. Chem. Ser.*, **208**, 235.
13. Shaw, S. J. and Tod, D. A. (1989) *J. Adhes.*, **28**, 231.
14. Bascom, W. D., Cottington, R. L., Jones, R. L. and Peyser, P. (1975) *J. Appl. Polym. Sci.*, **19**, 2545.
15. Yee, A. F. and Pearson, R. A. (1986) *J. Mater. Sci.*, **21**, 2462.
16. Sultan, J. N. and McGarry, F. (1973) *Polym. Eng. Sci.*, **13**, 29.
17. Meeks, A. C. (1974) *Polymer*, **15**, 675.
18. Bucknall, C. B. and Yoshii, T. (1978) *Br. Polym. J.*, **10**, 53.
19. Mertz, E. H., Claver, G. C. and Baer, M. (1956) *J. Polym. Sci.*, **22**, 325.
20. Kunz-Douglas, S., Beaumont, P. W. R. and Ashby, M. F. (1980) *J. Mater. Sci.*, **15**, 1109.
21. Bucknall, C. B. and Smith, R. R. (1965) *Polymer*, **6**, 437.
22. Donald, A. M. and Kramer, E. J. (1982) *J. Mater. Sci.*, **17**, 1871.
23. Newman, S. and Strella, S. (1965) *J. Appl. Polym. Sci.*, **9**, 2297.
24. Kinloch, A. J., Shaw, S. J. and Hunston, D. L. (1983) *Polymer*, **24**, 1355.
25. Pearson, R. A. and Yee, A. F. (1983) *Polym. Mater. Sci. Eng.*, **49**, 316.
26. Goodier, J. N. (1933) *J. Appl. Mech.*, **55**, 39.
27. Lange, F. F. and Radford, K. C. (1971) *J. Mater. Sci.*, **6**, 1197.
28. Broutman, L. J. and Sahu, S. (1971) *Mater. Sci. Eng.*, **8**, 98.
29. Mallick, P. K. and Broutman, L. J. (1975) *Mater. Sci. Eng.*, **18**, 63.
30. Young, R. J. and Beaumont, P. W. R. (1977) *J. Mater. Sci.*, **12**, 684.
31. Moloney, A. C., Kausch, H. H. and Stieger, H. R. (1983) *J. Mater. Sci.*, **18**, 208.
32. Moloney, A. C., Kausch, H. H. and Stieger, H. R. (1984) *J. Mater. Sci.*, **19**, 1125.
33. Spanoudakis, J. and Young, R. J. (1984) *J. Mater. Sci.*, **19**, 473.

34. Moloney, A. C., Kausch, H. H., Kaiser, T. and Beer, H. R. (1987) *J. Mater. Sci.*, **22**, 381.
35. Lange, F. F. (1970) *Philos. Mag.*, **22**, 983.
36. Evans, A. G. (1972) *Philos. Mag.*, **26**, 1327.
37. Green, D. J., Nicholson, P. S. and Emberg, J. D. (1979) *J. Mater. Sci.*, **14**, 1657.
38. Maxwell, D., Young, R. J. and Kinloch, A. J. (1984) *J. Mater. Sci. Lett.*, **3**, 9.
39. Kinloch, A. J., Maxwell, D. and Young, R. J. (1985) *J. Mater. Sci.*, **20**, 4169.
40. Bucknall, C. B. and Partridge, I. V. (1983) *Polymer*, **24**, 639.
41. Diamont, J. and Moulton, R. S. (1984) 29th Natl SAMPE Symp., Vol. 29, p. 422.
42. Hedrick, J. L., Yilgor, I., Wilkes, G. L. and McGrath, J. E. (1985) *Polym. Bull.*, **13**, 201.
43. Cerere, J. A., Hedrick, J. L. and McGrath, J. E. (1986) 31st SAMPE Symp., p. 583.
44. Knott, J. F. (1973) *Fundamentals of Fracture Mechanics*, Butterworths, London.
45. Kinloch, A. J. and Young, R. J. (1983) *Fracture Behaviour of Polymers*, Applied Science, London.
46. Williams, J. G. (1984) *Fracture Mechanics of Polymers*, Ellis Horwood, Chichester.
47. Dammont, F. R. and Kwei, T. K. (1967) *J. Polym. Sci.*, **5**, 761.
48. Delatycki, O., Shaw, J. C. and Williams, J. G. (1969) *J. Polym. Sci.*, **7**, 753.
49. Arridge, R. G. C. and Speake, J. H. (1972) *Polymer*, **13**, 44.
50. Hirai, T. and Kline, D. E. (1973) *J. Appl. Polym. Sci.*, **17**, 31.
51. Williams, J. G. (1979) *J. Appl. Polym. Sci.*, **23**, 3433.
52. Shaw, S. J., Tod, D. A. and Griffith, J. R. (1988) In *Adhesives, Sealants and Coatings for Space and Harsh Environments* (ed. L. H. Lee), Plenum, New York, p. 45.
53. Kalfoglou, N. K. and Williams, H. L. (1973) *J. Appl. Polym. Sci.*, **17**, 1377.
54. Manzione, L. T., Gillham, J. K. and McPherson, C. A. (1981) *J. Appl. Polym. Sci.*, **26**, 889.
55. Kunz, S. C., Sayre, J. A. and Assink, R. A. (1982) *Polymer*, **23**, 1897.
56. Pulliam, L., Siebert, A. R. and Drake, R. (1989) *Adhes. Age*, (July), 18.
57. Hwang, J.-F., Manson, J. A., Hertzberg, R. W., Miller, G. A. and Sperling, L. H. (1989) *Polym. Eng. Sci.*, **29**, 1466.
58. Huang, Y. and Kinloch, A. J. (1992) *Polym. Commun.*, **33**, 1330.
59. Ting, R. Y. and Moulton, R. J. (1980) 12th Natl SAMPE Tech. Conf., p. 265.
60. Bascom, W. D., Ting, R. Y., Moulton, R. J., Riew, C. K. and Siebert, A. R. (1981) *J. Mater. Sci.*, **16**, 2657.
61. Levita, G., Marchetti, A. and Butta, E. (1985) *Polymer*, **26**, 1110.
62. Lee, W. H., Hodd, K. A. and Wright, W. W. (1985) *Adhesives, Sealants and Encapsulants Conf.*, ASE '85, Day 1, November 1985, p. 145.
63. Pearson, R. A. and Yee, A. F. (1989) *J. Mater. Sci.*, **24**, 2571.
64. Enns, J. B. and Gillham, J. K. (1983) *J. Appl. Polym. Sci.*, **28**, 2567.
65. Gupta, V. B., Orzal, L. T., Lee, C. Y.-C. and Rich, M. J. (1985) *Polym. Eng. Sci.*, **25**, 812.
66. Sancakter, E., Jozari, H. and Klein, R. M. (1983) *J. Adhes.*, **15**, 241.
67. Kinloch, A. J., Finch, C. A. and Hashemi, S. (1987) *Polym. Commun.*, **28**, 322.
68. Truong, V.-T. (1990) *Polymer*, **31**, 1669.
69. Young, R. J. and Beaumont, P. W. R. (1976) *J. Mater. Sci.*, **11**, 779.
70. Bascom, W. D. and Cottington, R. L. (1976) *J. Adhes.*, **7**, 333.
71. Yamini, S. and Young, R. J. (1977) *Polymer*, **18**, 1075.
72. Gledhill, R. A., Kinloch, A. J., Yamini, S. and Young, R. J. (1978) *Polymer*, **19**, 574.
73. Yamini, S. and Young, R. J. (1979) *J. Mater. Sci.*, **14**, 1609.
74. Kunz, S. and Beaumont, P. W. R. (1981) *J. Mater. Sci.*, **16**, 3141.
75. Kinloch, A. J. and Shaw, S. J. (1981) *J. Adhes.*, **12**, 59.
76. Bitner, J. R., Rushford, J. L., Rose, W. S., Hunston, D. L. and Riew, C. K. (1982) *J. Adhes.*, **13**, 3.

77. Low, I.-M. and Mai, Y.-W. (1989) *J. Mater. Sci.*, **24**, 1634.
78. Vincent, P. I. (1974) *Polymer*, **15**, 111.
79. Kisbenyi, M., Birch, M. W., Hodgkinson, J. M. and Williams, J. G. (1979) *Polymer*, **20**, 1289.
80. Hunston, D. L., Kinloch, A. J., Shaw, S. J. and Wang, S. S. (1984) In *Adhesive Joints* (ed. K. L. Mittal), Plenum, New York, p. 789.
81. Mostovoy, S. and Ripling, E. J. (1966) *J. Appl. Polym. Sci.*, **10**, 1351.
82. Mostovoy, S. and Ripling, E. J. (1969) *J. Appl. Polym. Sci.*, **13**, 1083.
83. Mostovoy, S., Ripling, E. J. and Bersch, C. F. (1971) *J. Adhes.*, **3**, 125.
84. Mostovoy, S. and Ripling, E. J. (1971) *J. Appl. Polym. Sci.*, **15**, 641.
85. Mostovoy, S. and Ripling, E. J. (1971) *J. Appl. Polym. Sci.*, **15**, 611.
86. Ripling, E. J., Mostovoy, S. and Bersch, C. F. (1971) *J. Adhes.*, **3**, 145.
87. Mostovoy, S. and Ripling, E. J. (1975) Fracture characteristics of adhesive joints. *Final Report, Contract N00019-75-C-0271*, Naval Air Systems Command, Washington, DC.
88. Brussat, T. R., Chin, S. T. and Mostovoy, S. (1977) Fracture mechanics for structural adhesive bonds. *TR-77-163*, Air Force Materials Laboratory, Dayton, OH.
89. Trantina, G. G. (1972) *J. Compos. Mater.*, **6**, 192.
90. Trantina, G. G. (1972) *J. Compos. Mater.*, **6**, 371.
91. Bascom, W. D., Cottington, R. L. and Timmons, C. O. (1977) *Appl. Polym. Symp.*, **32**, 165.
92. Bascom, W. D. and Oroshnik, J. (1978) *J. Mater. Sci.*, **13**, 1411.
93. Shaw, S. J. (1984) PhD Thesis, City University, London.
94. Wang, S. S., Mandel, J. F. and McGarry, F. J. (1978) *Int. J. Fract.*, **14**, 39.
95. McKenna, G. B., Mandel, J. F. and McGarry, F. J. (1974) 29th Annu. Tech. Conf., Reinforced Plastics/Composites Division, Society of Plastics Industry, Paper 13-C.
96. Scott, J. M. and Phillips, D. C. (1975) *J. Mater. Sci.*, **10**, 551.
97. Bascom, W. D., Bitner, J. L., Moulton, R. J. and Siebert, A. R. (1980) *Composites*, **11**, 9.
98. Hunston, D. L. (1984) *Compos. Tech. Rev.*, **6**, 176.

7

Toughened polyamides

R. J. Gaymans

7.1 INTRODUCTION

Dry polyamides (PAs) are semiductile materials at room temperature and tough above their glass transition temperature (T_g). The deformation mechanism in the neat PA is shear yielding. Above T_g excessive shear yielding can take place owing to the lowering of the yield strength at T_g. Unfortunately, at T_g the modulus decreases strongly too. Below T_g the energy absorption is mainly in the notch region and therefore mainly in the crack initiation step. Polyamides can be made tough by plasticizing with water and by (block) copolymerization. With these methods the impact improvement is obtained at a considerable loss of modulus [1]. An effective method of impact modification is by blending in rubber. In this way the toughness can be increased by many factors while the tensile strength and modulus decrease approximately proportionally to the rubber concentration [2, 3]. The observed impact level of these blends, as measured on notched Izod samples, is 50–100 kJ m^{-2}. Because of their very high impact strength these blends are often referred to as 'super tough'. The development originated from DuPont where they managed to obtain very fine dispersions of elastomers in PA [2, 4]. As there is a big difference in polarity between the PA and most rubbers, only fine dispersions of the (EP) rubber could be obtained after modification of the interphase.

The toughness is mainly studied by the notched Charpy and Izod methods. The samples are tested dry, i.e. dry as molded or dried before testing. On rubber modification two new transitions appear in the temperature–toughness graph (Fig. 7.1), one near T_g of the rubber and one at the brittle-to-tough transition (T_{bt}) between T_g of the rubber and T_g of the PA. At T_g of the rubber the impact modifier becomes active (region B). On the impacted sample in the notch region stress whitening is apparent (Fig. 7.2(a)), while the fracture surface over the rest of the sample is smooth. The energy absorption is mainly due to deformation during the crack initiation and little due to crack propagation. The crack propagation is unstable in region B.

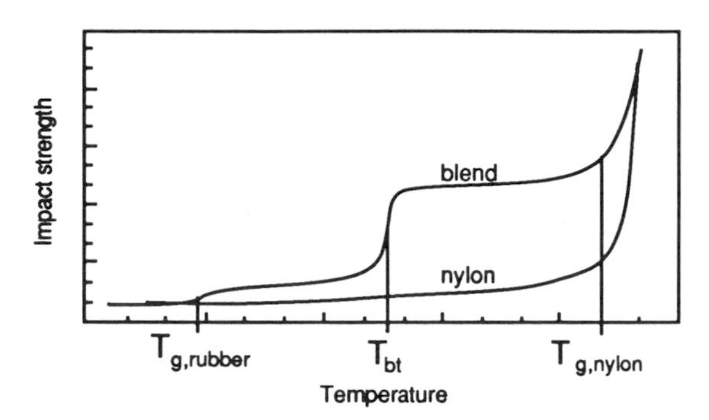

Fig. 7.2 Stress whitening (s.w.) on fractured samples: (a) brittle region B; (b) tough region C. (Reproduced from R. J. M. Borggreve, PhD Thesis, 1988.)

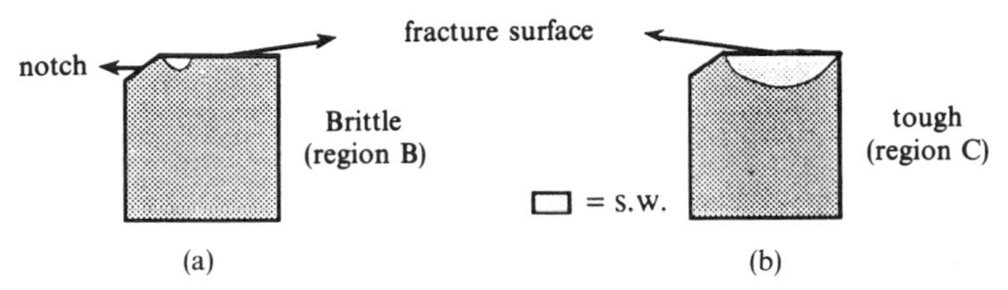

Fig. 7.2 Stress whitening (▫) on fractured samples: (a) brittle region B; (b) tough region C. (Reproduced from R. J. M. Borggreve, PhD Thesis, 1988.)

At T_{bt} (region C) not only the notch area but the whole fracture area of the impacted sample becomes stress whitened (Fig. 7.2(b)), and the crack propagation is stable. The strain energy stored in the sample up to the point of fracture becomes equal to the energy dissipated in the creation of the fracture surfaces [6].

Some properties of rubber modified PA are given in Table 7.1.

Important parameters in the rubber toughening of PA are

- the rubber concentration,
- the particle size,
- the type of rubber, and
- distribution of the rubber particles.

For PA–rubber blends with a particle dispersion the modulus of the blends was found to decrease linearly with the rubber volume fraction [8]. The yield strength decreased with the volume fraction (Φ_R) according to the Ishai and Cohen equation (7.1) at both low (5% min^{-1}) [8,9] and very high (60 000% min^{-1}) deformation rates (Fig. 7.3) [9]. This behavior is irrespective of the particle size [8]. At high rates the temperature effect is bigger than at low deformation rates.

$$\sigma_{\text{blend}} = \sigma_{\text{PA}}(1 - \Phi_R^{2/3}). \tag{7.1}$$

Table 7.1 Properties of PA 6,6 and PA 6,6–rubber blend [7]

	*Zytel E 101L**		*Zytel ST 801†*	
	DAM‡	*50% relative humidity*	*DAM*	*50% relative humidity*
Tensile strength at break (MPa)	83	77	52	41
Elongation at break (%)	60	>300	60	210
Flexural modulus (GPa)	2.8	1.2	1.7	0.9
Impact strength, notched Izod ($J\,m^{-1}$)				
23 °C	53	112	910	1170
−40 °C	32	27	160	140
Water absorption saturation (%)	8.5	–	6.7	–

* General-purpose PA 6,6.
† Rubber toughened PA 6,6.
‡ Dry as molded.

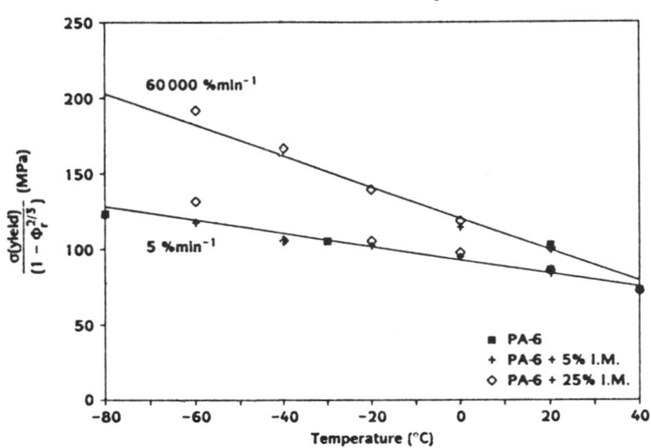

Fig. 7.3 Yield strength of PA–rubber blends as a function of temperature and impact modifier (IM) content: ■, PA 6; +, PA 6–5% IM; ◇, PA 6–25% IM. The yield strength data are corrected for the rubber concentration with equation (7.1). (Reproduced from M. Bosma, IUPAC Symposium, Mechanisms of Polymer Strength and Toughness, Prague, 1990.)

To study the impact toughness of these materials few researchers have applied the LEFM approach [6, 10–13]. For this there are several reasons. For studying the plane strain fracture toughness K_{Ic}, the plane strain conditions must be fulfilled. This means that for these super tough materials the thickness must be at least 70 mm [10]. The J integral method that is more applicable (sample thickness 15 mm) for these systems is, however, difficult to perform at high test speeds [10]. Sunderland *et al.* developed a special test rig for testing at high strain rates ($50\,s^{-1}$) [12]. Sue and Yee found that by increasing the test speed by two decades J_c decreased from 40 to $30\,kJ\,m^{-2}$ [14]. They found too that the size of the deformation zone changed with strain rate. The reported J values are therefore dependent on the test speed, which means that these blends cannot be characterized by a single parameter [10, 13].

Next to the excellent impact properties the PA–rubber blends also have excellent flexing behavior [3, 15, 16]. Important in this respect as well is that fracture is not spontaneous but preceded by stress whitening [3].

7.2 BLEND FORMATION

7.2.1 Extrusion process

PA–rubber blends are mostly made by a reaction blending process. PA can be toughened with a rubber if the particle size of the rubber is of the order of 0.1–2.0 μm [17–19]. The interfacial tension between PA and olefinic rubbers is too high for a stable fine dispersion to be obtained [20]. The interfacial tension can be lowered by increasing the polarity of the rubber, by adding an interfacial agent or by chemical bonding at the interphase. Fine dispersions can be obtained from thermoplastic elastomers (segmented block copolymers) with polyester and polyamide containing segments [20–22]. A chemical bonding can be obtained by functionalization of the PA with unsaturated groups that can react with the rubber or by modifying the rubber with groups that are reactive towards the PA. Mostly the rubbers are modified with the reactive maleic anhydride (MA) [8, 19, 21–35] (Chapter 5).

(a) Dispersion parameters

The blending of a rubber in PA has two aspects, the breaking up of the particles and the distribution of the particles. The distribution of the particles is governed by reorientations of the melt [36]. The amount of reorientation induced depends very much on the type of blending apparatus used. Suitable machines are internal mixers, twin-screw extruders and single-screw extruders with special mixing elements.

The breaking up of the particles is governed by the viscous forces and the interfacial forces [37, 38]. The Weber number (We) gives the ratio between the viscous and interfacial forces:

$$\mathrm{We} = \frac{\text{viscous forces}}{\text{interfacial forces}} = \frac{\eta_m \dot{\gamma}}{\sigma / R} \tag{7.2}$$

where η_m is the matrix viscosity, $\dot{\gamma}$ is the shear rate, σ is the interfacial tension and R is the particle radius. The Weber number is also dependent on the viscosity ratio of the phases, on whether it is a shear or longitudinal flow, on the viscoelastic nature [39, 40] and on the concentration of the second phase [41]. In shear flow the optimal conditions are at a viscosity ratio, of the dispersed phase over the matrix phase (η_d/η_m), of unity [20, 21, 37, 38].

Wu [20] studied (PA 6,6 and EP rubbers) the influence of interfacial tension (σ), viscosity ratio (η_d/η_m) and shear rate ($\dot{\gamma}$) on the dispersion process in a corotating twin-screw extruder. The observed effects were in line with results obtained with Newtonian fluids. The particle size changed linearly with the shear rate, matrix viscosity (η_m) and interfacial tension (σ). The Weber number had a minimum for $\eta_d/\eta_m = 1$ and increased with the concentration of the dispersed phase (Fig. 7.4) [20, 41].

In polymer blends the breakup of droplets is not limited to the region $\eta_d/\eta_m < 4$ as in Newtonian fluids. The relationship for the critical Weber number as measured is

$$\mathrm{We} = \frac{\eta_m \gamma R}{\sigma} = \begin{cases} 4(\eta_d/\eta_m)^{0.84} & \text{for} \quad \eta_d/\eta_m > 1 & (7.3) \\ -4(\eta_d/\eta_m)^{0.84} & \text{for} \quad \eta_d/\eta_m < 1. & (7.4) \end{cases}$$

For higher concentrations this equation must be corrected [41]. A fine dispersion can be obtained if the matrix viscosity is high, the shear rate is high, the interfacial tension is low and the viscosity ratio is near unity.

If the interface is modified with MA, the interfacial tension at 280 °C in the melt can be lowered from 10 to 0.25 mJ m^{-1} [20]. This is a decrease in

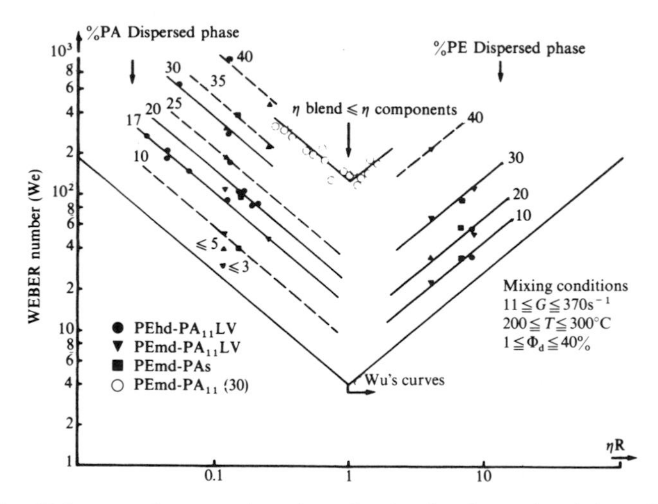

Fig. 7.4 The Weber number as a function of η_d/η_m for the melt mixing of PA 6–PE using a Brabender mixer (mixing conditions: 11 s^{-1} ⩽ G ⩽ 370 s^{-1}; 200 °C ⩽ T ⩽ 300 °C; 1% ⩽ Φ_d ⩽ 40%): ●, PEHD–PA 11LV; ▼, PEMD–PA 11LV; ■, PEMD–PA 6; ○, PEMD–PA 11. (Reproduced from G. Serpe, J. Jarrin and F. Dawans, *Polymer Engineering and Science*, 1990.)

interfacial tension by a factor of 40 and the particle size of the rubber in the blend is decreased by approximately the same factor [19, 20].

(b) Extrusion blending process

PA blends are usually made in a twin-screw extruder. In a twin-screw extruder high shear rates can be obtained and a considerable amount of reorientation of the melt is also taking place. The rate of coalescence seems to depend on the Weber number compared with the critical We and the local concentration of the dispersed phase [41]. The dispersion rate in an extruder is high. Oostenbrink and Gaymans [42] studied the particle size of the system PA 6–EPM-g-MA at different positions in the twin-screw extruder (Fig. 7.5). After the first set of kneading blocks the rubber in the blend had already nearly reached its final size.

Blends can also be made by injection molding a dry blend. In this way the blending step in the extruder is omitted. The results are as a whole not so good. Better dispersions can be obtained if a mixing head is mounted at the tip of an injection molding screw. With a Twente Mixing Ring, which is very suitable for injection molding screws, blends with a fair-to-good dispersion could be obtained [43].

(c) Reactions at the interface

For the modification of the rubber mainly maleic anhydride (MA) [32, 33, 44–47], maleic esters [24], fumaric acid and other unsaturated acids

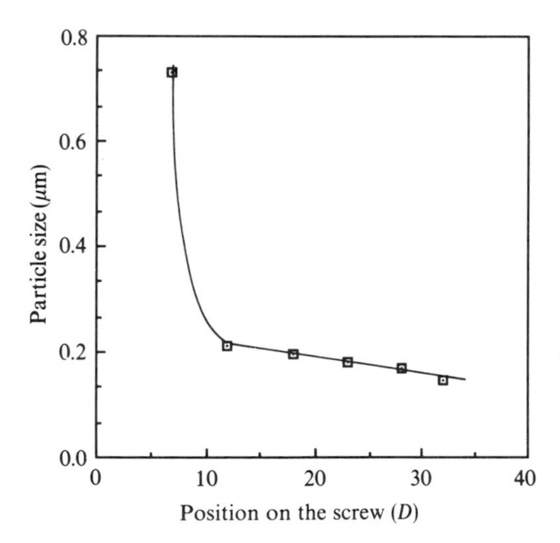

Fig. 7.5 Blending PA 6–EPM-g-MA (90:10) with a 25 mm, twin-screw extruder (extruder length:screw diameter ratio, 33:1): particle size as a function of position (D) on the screw [42].

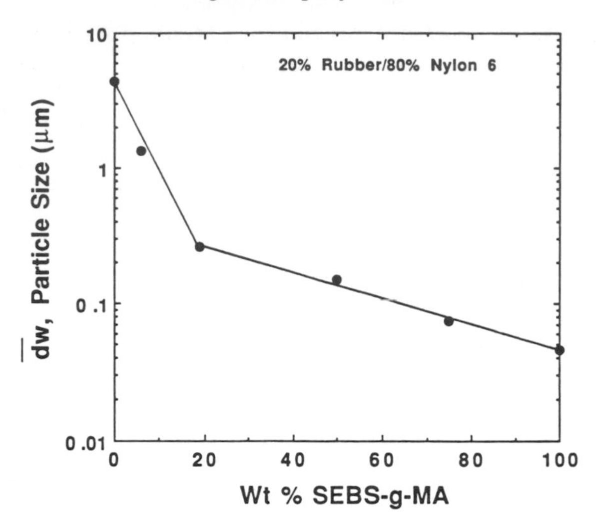

Fig. 7.6 Particle size in PA 6–SEBS (80:20) blends as a function of the amount of SEBS-g-MA. (Reproduced from A. J. Oshinski, H. Keskkula and R. D. Paul, *Polymer*, 1992.)

[12, 47] are used. The modification of the rubber is usually a post-polymer reaction whereby the unsaturated acid is grafted onto the rubber [44, 45]. During the reaction blending process the modified rubber reacts at the interphase with the PA.

It is expected that the grafted MA (-g-MA) is preferential at the interface. If the -g-MA concentration in the rubber is reduced, particularly if this is achieved by mixing a modified with an unmodified rubber, the particle size increases (Fig. 7.6) [18, 19]. This suggests that the interfacial tension increases with decreasing -g-MA concentration. This increase is, however, not linear, which reflects something of a preference of -g-MA groups for the interface. The time for this reaction is only a few minutes. The reactions that can take place between the PA and the -g-MA are the reaction of MA with a free amine or of the hydrolyzed MA with an amide group forming a strong amide or imide linkage (Fig. 7.7) [32, 33, 35, 45]. If the amine concentration is increased the impact strength appears to be higher, presumably reflecting the efficiency of the grafting [32, 35]. As the grafting efficiency is increased the particle size is expected to decrease. However, the particle size is thought to have little effect on the impact strength in the tough region [18, 48]. Inhomogeneous distributions in the blend give a lowering of the impact toughness.

An analysis of the graft PA layer on the rubber particles showed that the amount of PA grafted is a function of the -g-MA and the particle size [33] and can be as high as 40% of the rubber concentration [32, 33]. Borggreve and Gaymans found that with increased -g-MA the amount of grafted PA per unit surface area (weight per cent) and the molecular weight of the grafted PA chains

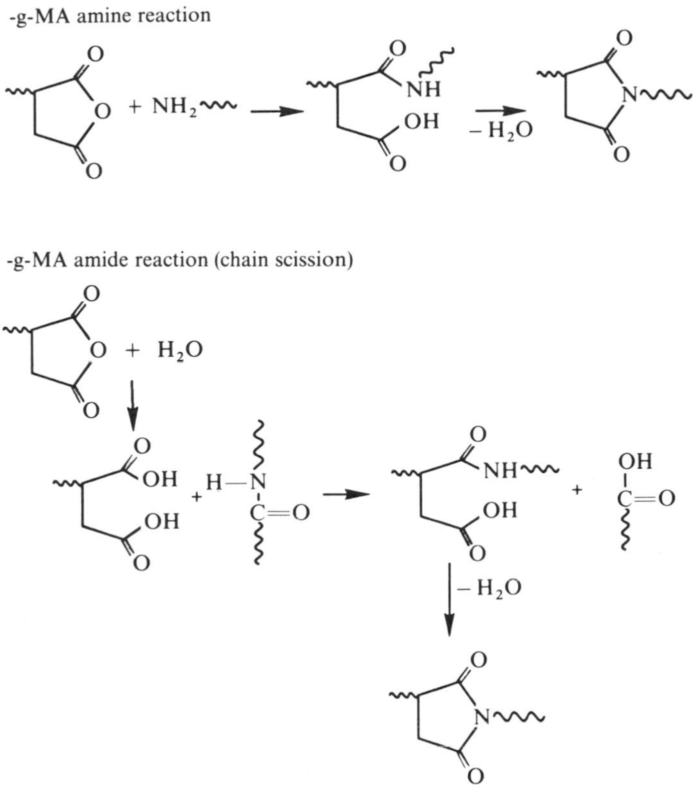

-g-MA amine reaction

-g-MA amide reaction (chain scission)

Fig. 7.7 Reaction scheme for PA with maleic anhydride groups.

decreased [33]. This suggests that because of the presence of -g-MA severe chain scission has taken place, which means that the reaction is mainly by the amide route [33]. Lawson, Hergenrother and Matlock concluded from the molecular weight of the non-grafted PA that the grafting was dominated by the reaction with the amine end groups and little with the amide groups [32].

As we have seen above (Fig. 7.5) a small particle size is rapidly established. Increasing the reaction temperature gives only bigger particle sizes [43]. This suggests that the reaction rate at the interphase is not the rate-controlling step of the dispersion process.

(d) Rheological properties of the blend

The melt viscosity of blends changes linearly with composition if no special interactions at the interphase are present. The viscosity of blends deviates positively if at the interface there is chemical bonding [19]. Therefore, rheological measurements give insight into this interaction behavior. At the same time, these data give information on the processability of these materials.

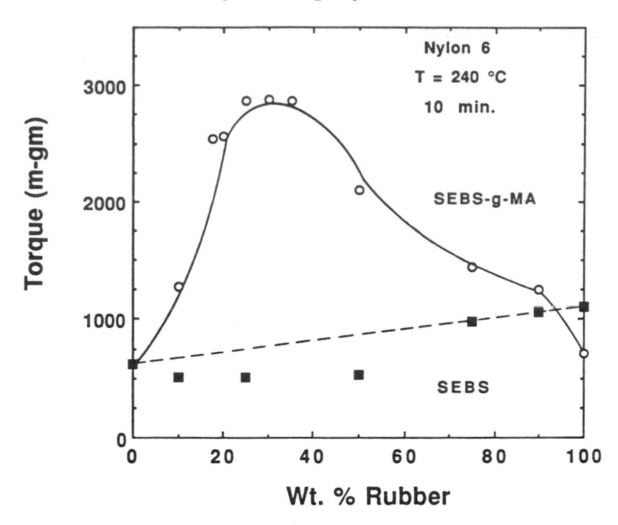

Fig. 7.8 Brabender torque for blends of PA 6 with SEBS (■) and with SEBS-g-MA (○) after 10 min at 240 °C and 60 rev min^{-1}. (Reproduced from A. J. Oshinski, H. Keskkula and R. D. Paul, *Polymer*, 1992.)

The positive deviation depends both on the amount of interaction and on the particle size of the dispersed phase [41]. By changing the interaction these two effects take place simultaneously. Oshinski, Keskkula and Paul [19] measured the torque of a Brabender mixer during blending of PA with styrene–hydrogenated butadiene–styrene triblock copolymers (SEBS). They observed a strong increase in torque with rubber concentration with an MA modified rubber (SEBS-MA) and hardly any effect with an unmodified rubber (Fig. 7.8). Results of Serpe, Jarrin and Dawans [41] indicate that for big particles (more than 10 μm) this positive deviation effect is small.

7.2.2 Reactor process

Rubber modified PSs are usually made in a reactor process, for which the formation of a block copolymer is important [49]. Only a few papers on PA have appeared describing the formation of toughened systems by a reactor process. As the optimum particle size is critical, the polarity difference being big and the polymerization conditions being harsh, as yet few systems have been reported that combine a high toughness with good tensile properties. The systems that have been described concern mainly anionically polymerized RIM PA [50–52]. The synthesis of PA 6 is usually hydrolytic, starting from caprolactam with a few per cent of water. If a VK type reactor is used, no mechanical stirring is applied and the reaction times at 240–260 °C are of the order of 15 h. If an autoclave process is used stirring is possible and the reaction times are much shorter. PA 6,6 is usually synthesized in an autoclave process with a stirred reactor. The starting mass is here PA 6,6 salt with at least 15% water. The reaction takes 30–60 min at 275–300 °C. The rubber has to withstand these reaction conditions without degradation. Another method of synthesizing PA from lactams is by anionic polymerization. This anionic polymerization is applied in cast and RIM systems. With anionic polymerization the reaction is much faster and can be carried out at lower temperatures and in shorter times. The rubber modification of these systems is the same as with the hydrolytic method but at less severe conditions. For RIM systems it is also important that the viscosities of starting liquids are low.

(a) Segmented block copolymers

One method of rubber modification is synthesizing segmented block copolymers with soft blocks. The types of rubber segments which are used are polyethers, aliphatic polyesters and functionalized polybutadiene. If the block length of the rubber phase is 1000–2000 g mol^{-1}, the phase separation is in a laminar form. Owing to this, both the modulus and the yield strength decrease strongly. The elongation at break increases and the materials are tough [53]. If the longer blocks are used as in RRIM [52] micro-phase separation takes place and more complex structures develop. The high toughness is still obtained with a considerable loss of tensile properties (Fig. 7.9) [50–52].

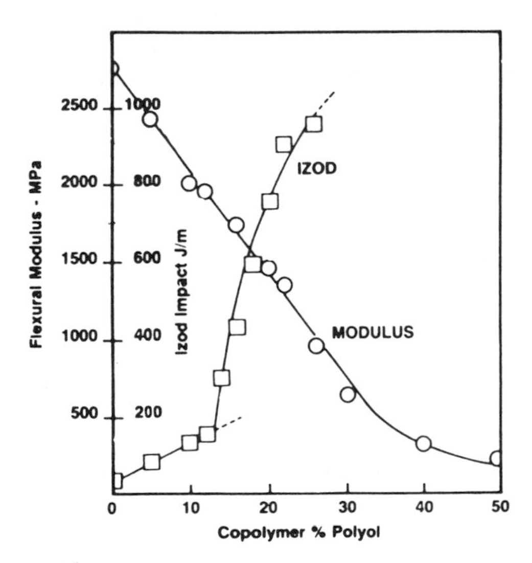

Fig. 7.9 Modulus–impact balance for RIM PA block copolymers: ○, modulus; □, Izod impact strength. (Reproduced from J. D. Gabbert and R. M. Hendrick, *Polymer Processing and Engineering*, 1986.)

(b) Dissolved rubbers

As the usual rubbers are highly viscous, it will be very difficult to disperse them in a reactor under low shear. To reduce the viscosity the rubber can be dissolved in a solvent. It is also possible to dissolve the rubber or the rubber–solvent mixture in the starting reaction mass. In particular, caprolactam is a strong solvent.

Cimmino and coworkers studied the reaction blending of PA 6 with EPM and EPM-g-MA [29, 54]. The rubbers were dissolved in xylene before being added to the caprolactam. This mixture reacted for 4 h at 260–270 °C under vigorous mechanical stirring. If only EPM was used the particle sizes obtained were of the order of 2.5–17.5 μm [29, 54]. The particle size is about the same as when no solvent was used or if the blends were made by melt mixing [55]. If an EPM–EPM-g-MA (18:2, weight per cent) rubber mixture was used the particle sizes were about in the same range as if only EPM was used. With EPM–EPM-g-MA (15:5) and xylene the particle size was 1–5 μm. This is better than without solvent (5–35 μm) and comparable with that of a melt blended sample (1.25–8.75 μm).

Some rubbers dissolve in caprolactam (e.g. nitrile rubber, PA-block copolymers) and the blends so obtained have improved impact properties; however, the tensile properties decrease strongly [56, 57].

(c) Core–shell particles

Another way of obtaining the required phase structure is by starting from core–shell particle impact modifiers. Several core–shell systems have been developed consisting of a rubbery core and a hard shell, e.g. acrylate rubber–PMMA and polybutadiene–SAN. Extrusion blends with core–shell impact modifiers give good impact behavior [58, 59]. For the reactor process the agglomerates have only to be dispersed. The core–shell material is supplied as agglomerates and these have to be broken up. Without any specific interactions this breaking up in the reactor is only partial. On subsequent injection molding the dispersion is better. The dispersion could be improved by modifying the shell [59–61]. A side effect of compatible shells is that caprolactam (partly) dissolves the shell. This compatibilization of the shell might ease the dispersion of the particles, but it increases the viscosity of the lactam mixture and changes the structure of the interfacial layer of the blend. By starting from a latex instead of agglomerate particles an aggregate-free system can easily be obtained [60].

7.3 PARAMETERS AFFECTING IMPACT TOUGHNESS

7.3.1 Influence of structure

For rubber-induced toughening the rubber is preferably present as a dispersed phase. If the rubber phase contains inclusions [34, 62], the dispersed phase concentration increases and the modulus and yield strength of the blend decrease.

(a) Rubber concentration

As the rubber concentration increases, so does the impact strength (Fig. 7.10) [8, 17, 19, 58, 63]. This increase in impact strength is up to 30 vol.% approximately linear with concentration both at $-40\,°C$ and above the brittle-to-tough transition temperature [8, 17, 19]. At the same time T_{bt} of the blend

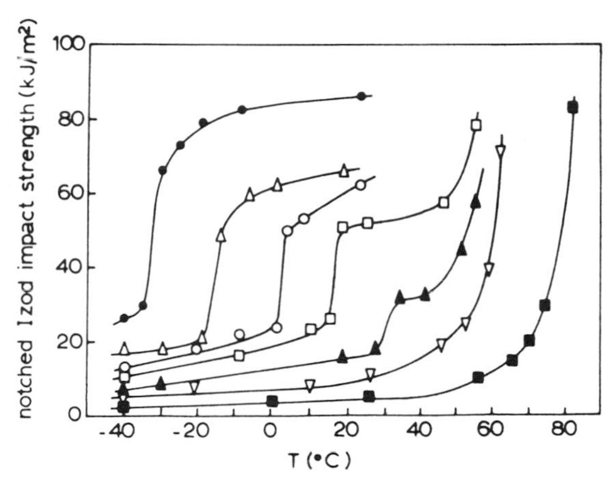

Fig. 7.10 Impact strength of PA–EPDM blends as a function of temperature (particle size, 0.3 μm), showing the influence of rubber concentration (volume per cent): ■, 0; ▽, 2.6; ▲, 6.4; □, 10.5; ○, 13.0; △, 19.6; ●, 26.1. (Reproduced from R. J. M. Borggreve *et al.*, *Polymer*, 1987.)

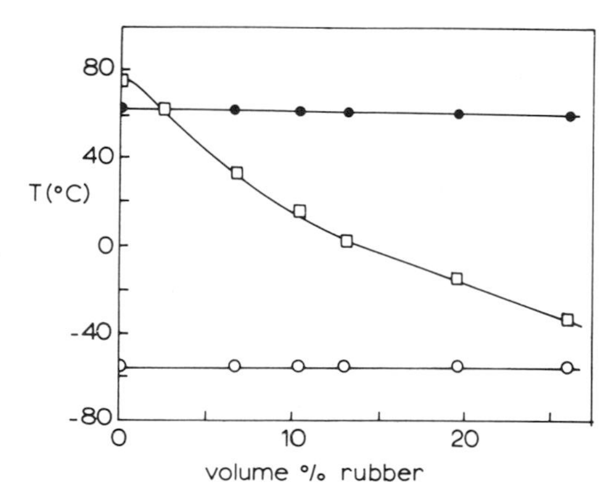

Fig. 7.11 Influence of rubber concentration on transition temperatures: □, T_{bt}; ●, $T_{g,\ PA}$; ○, $T_{g,\ EPDM}$. (Reproduced from R. J. M. Borggreve *et al.*, *Polymer*, 1987.)

decreases although the T_g of the PA and EPDM phases remain constant (Fig. 7.11) [8, 17, 19]. Blends with high rubber concentrations (more than 30 vol.%) show a marked drop in impact strength [63, 64].

(b) *Particle size*

With decreasing particle size T_{bt} decreases to lower temperatures [8, 18, 48]. For the rubber to be effective the particle size has to be small (0.1–2 μm). Very small particles are not effective. The impact strength values as a function of particle size (Fig. 7.12) for a 20 wt% blend at room temperature (RT) show that there is a lower limit at about 0.1 μm [18, 19, 48, 65, 66] and an upper limit at about 1 μm [18, 19, 48]. The lower limit has not yet been explained. The reason for this might be that the material with very fine particles mainly consists of an interphase layer and no neat nylon or rubber is present any longer. Wu [67] found the interfacial thickness to be of the order of 50 nm. Oostenbrink, Molenaar and Gaymans [18] observed no shift in the glass transition temperatures for these fine dispersions. Oshinski, Keskkula and Paul [19] found that the tan δ peak of the rubber was smaller and the tan δ peak of the PA bigger, both by about 15%. The crystallinity of PA was also lowered by a few per cent. These effects are not enough to conclude that in fine blends the interfacial layer concentration is so large that it has changed the properties of the system drastically. Another hypothesis is that fine particles are more difficult to cavitate [18, 48, 66, 68]. Gent [69] showed that the cavitation starts from a defect and that a decreasing defect size increases the cavitational stress.

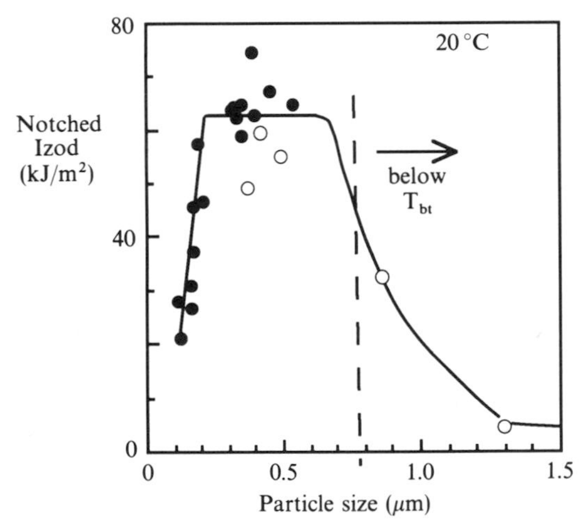

Fig. 7.12 Impact strength of PA 6–EP blends as a function of particle size (26 vol. % EP rubber; 20 °C): the different symbols refer to different manufacturing methods. (Reproduced from A. J. Oostenbrink, L. J. Molenaar and R. J. Gaymans, PRI Conference on Polymer Blends, 1990.)

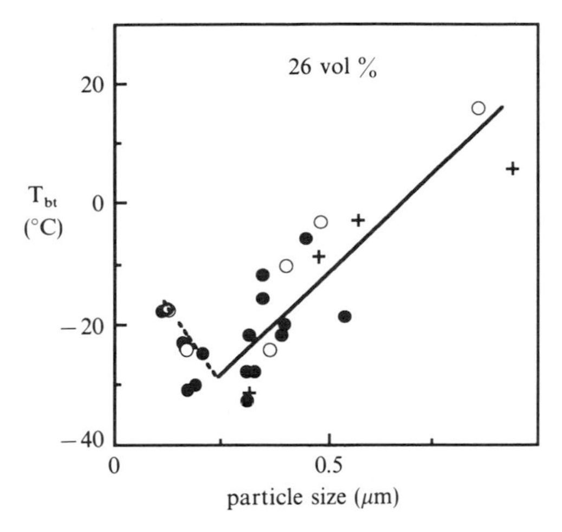

Fig. 7.13 Brittle-to-tough transition temperature of PA 6–EP blends as a function of particle size showing lower limit (26 vol. % EP rubber): the different symbols refer to different manufacturing methods. (Reproduced from A. J. Oostenbrink, L. J. Molenaar and R. J. Gaymans, PRI Conference on Polymer Blends, 1990.)

Decreasing the particle size might decrease the defect size or decrease the number of particles which have a defect.

The observed upper limit in particle size in Fig. 7.12 is the critical particle size for this 26% blend at T_{bt}, 20 °C. T_{bt} decreases with decreasing particle size (Fig. 7.13). Here too a lower limit can be observed at about 0.1 μm, although here the lower limit is less convincing [18, 48]. It is surprising that the level of impact strength in the range between the upper and lower limits seems little dependent on the particle size (Fig. 7.12) [14, 18, 48]. However, with very small particle sizes the level of impact strength in the tough region falls off with temperature. This suggests that with increasing temperature the very finely dispersed rubber in the blend cavitates with more difficulty [48].

(c) Particle size distribution

The particle size distribution in PA–rubber blends is usually small and in the range 1.4–2.1 μm [70]. The particle size distribution increases if two rubbers are used, particularly if they were not carefully mixed before they were blended in [43, 70]. The effect of particle size distribution on the impact level is thought to be small [71].

(d) Particle distribution

With a good mixing device, such as a twin-screw extruder, not only can the particle size be controlled but also a homogeneous distribution of particles can be obtained. If the system has a poor distribution with flocculated particles, then a lower impact strength is observed [71]. The particle distribution should therefore be as homogeneous as possible.

(e) *Single structural parameter*

As seen above, the Izod impact strength increases approximately linearly with rubber concentration both in the brittle and in the tough regions. The Izod impact strength is independent of the type of rubber [72] and the particle size as long as it is between the upper and lower limits. T_{bt}, however, depends on the type of rubber [22, 70], rubber concentration and particle size. For this transition from unstable to stable crack propagation the rubber concentration and particle size effect might be combined into a single structural parameter. Wu [17] evaluated several models, of which the interparticle distance model seems to fit his results best, in particular the data for high rubber concentrations. The other model that gives a reasonable fit is the interfacial area model. In the interparticle distance model the ligament thickness is the controlling factor (Fig. 7.14). The ligament thickness can be calculated by assuming a cubic lattice:

$$\mathrm{ID} = d[(\pi/6\Phi)^{1/3} - 1] \tag{7.5}$$

where ID is the interparticle distance (ligament thickness), d the particle diameter and Φ the rubber volume fraction. At RT the critical interparticle distance was found for the EP rubber to be 0.3 μm (Fig. 7.14) [17]. With decreasing interparticle distance T_{bt} also decreased (Fig. 7.15) [8, 73]. For

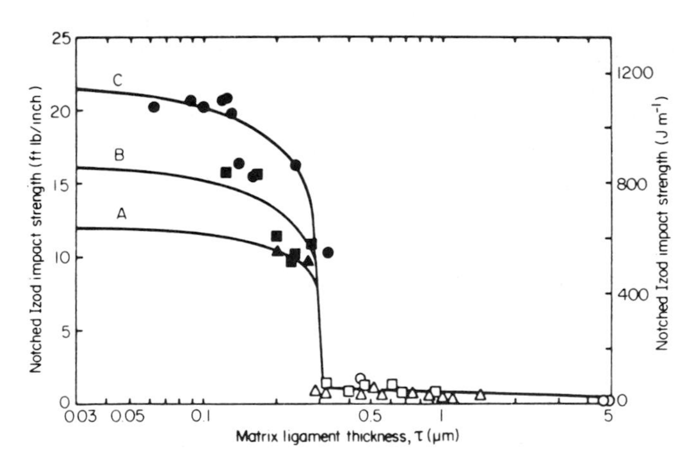

Fig. 7.14 Notched Izod impact strength of PA 6, 6–EP rubber blends vs interparticle distance: curve A, $\Phi_r = 0.128$; curve B, $\Phi_r = 0.189$; curve C, $\Phi_r = 0.306$. (Reproduced from S. Wu, *Polymer*, 1985.)

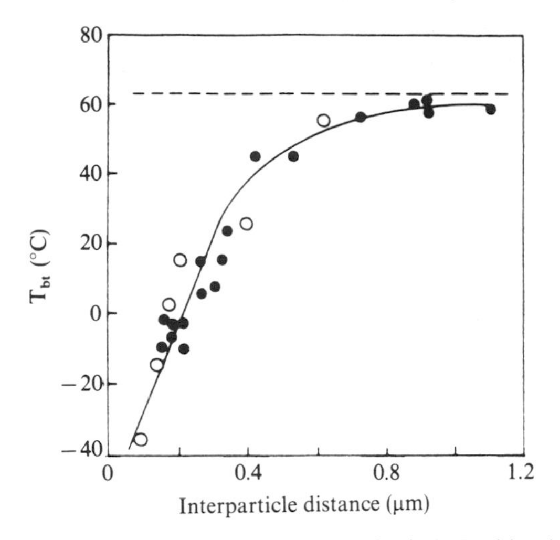

Fig. 7.15 Brittle-to-tough transition temperature of PA 6–EP blends as a function of interparticle distance: ○, rubber concentration; ●, particle size. (Reproduced from R. J. M. Borggreve, R. J. Gaymans, and A. R. Luttmer, *Makromolekulare Chemie, Macromolecular Symposia*, 1988.)

$T_{bt} = -25\,°C$ ID is approximately 0.1 μm. ID is typically a matrix property. This model does not describe the effect of the type of rubber on T_{bt}.

(f) Particle–matrix interface

Blends with -g-MA on the rubber in the range 0.1–0.7 wt% have an identical particle size–T_{bt} relationship [33]. Thus the interfacial tension influences the dispersion process but not the impact behavior at constant particle size. Wu [67] suggests that the interfacial strength for adhesion in blends has only to be 1000 J m^{-2}, which is the tearing strength of the rubber. This level of interfacial strength can already be obtained with van der Waals bonding. In this way the rubber fails by cavitation before the particle delaminates. In fracture surfaces of blends delamination was only observed with unmodified EP rubbers [67]. The question that is still unanswered is whether delamination at the PA–rubber interface gives poorer impact behavior than cavitation of the rubber.

7.3.2 Type of rubber

The function of the rubber is to induce a toughening mechanism. Therefore it is expected that the type of rubber matters. The type of rubber has little influence on the notched Izod impact strength in the tough region [72] but a strong effect on T_{bt} (Fig. 7.16) [22, 72]. As with the influence of particle size, this

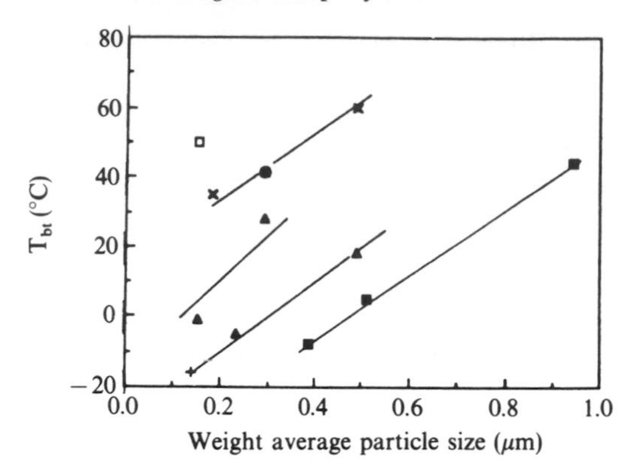

Fig. 7.16 T_{bt} as a function of particle size for PA 6–13 vol.% rubber blends with different types of rubbers: ■, EPDM; +, EPR; ×, LDPE; ▲, Keltaflex; □, ●, polyester TPE. (Reproduced from R. J. M. Borggreve, R. J. Gaymans and J. Schuijer, *Polymer*, 1989.)

indicates that the deformation which triggers the stable crack propagation is a different process from the deformation which determines the toughness level. The effect of the type of rubber on T_{bt} could not be correlated with the tensile strength or elongation at break of the elastomers. A good correlation was found with the tensile modulus (Fig. 7.17) [74]: the lower the modulus of the rubber the lower was T_{bt} (constant rubber concentration and particle size). The polyetherester thermoplastic elastomers do not follow the line of the olefinic rubbers. According to Gent and coworkers, the cavitational stress depends on the tensile modulus [68, 69, 75]. The cavitation process must depend on the

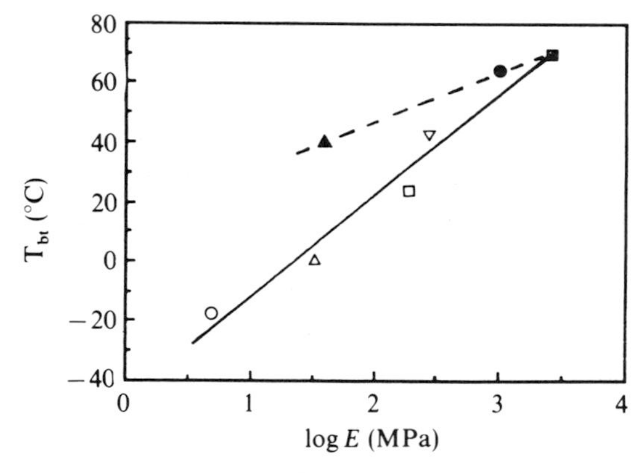

Fig. 7.17 T_{bt} as a function of rubber modulus for PA 6–13 vol.% rubber blends ($d_w = 0.3\,\mu m$): ■, PA; ▽, LDPE; □, Keltaflex; △, EPR; ○, EPDM; ▲, ●, polyester TPE. (Reproduced from R. J. Gaymans, R. J. M. Borggreve and A. J. Oostenbrink, *Makromolekulare Chemie, Macromolecular Symposia*, 1990.)

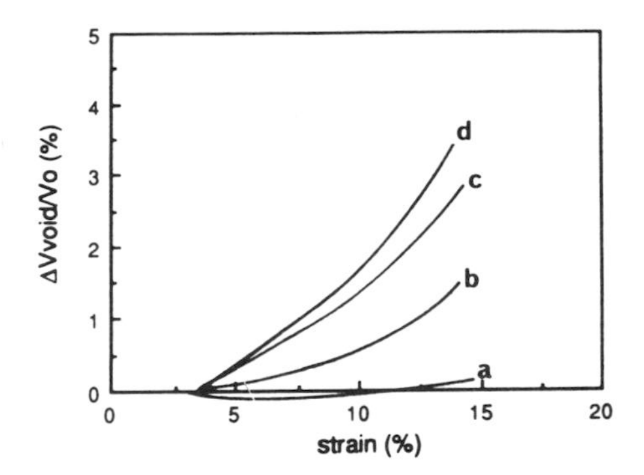

Fig. 7.18 Volume change as a function of applied strain for PA 6–13 vol.% rubber blends ($d_w = 0.3$ μm): curve a, EPDM; curve b, EPR; curve c, Keltaflex; curve d, LDPE. (Reproduced from R. J. M. Borggreve, R. J. Gaymans and H. M. Eichenwald, *Polymer*, 1989.)

bulk properties. Possibly both the tensile modulus and the cavitational stress are related for example by the cohesive energy density of the rubber. Another explanation is given by Wu [76]. He suggested that different rubbers may generate different amounts of internal stresses. To study the cavitational behavior, volume strain experiments were carried out [13, 77, 78]. The blend with the best impact behavior cavitated most easily (Fig. 7.18). The amount of cavitation as a function of the strain seems little dependent on the particle size in the range 0.5–1.6 μm. With increasing rubber concentration the amount of cavitation increased. The volume strain effect is proportional to the rubber concentration. This suggests that the cavitation of particles is not influenced by neighboring particles. The onset of cavitation is at low strains (2–4%) [13, 77, 78], which is still in the elastic strain region of the PA.

7.3.3 Matrix material

PA is a semiductile material that has (when dry) a low notched impact strength at high deformation rates. The crack propagation is unstable. At high temperatures or when plasticized, so that the test temperature is above T_g, the impact strength is very high. The reported impact strength in the literature is mostly on dry as molded or dried samples as that is the area where most is to be gained. The dry as molded method gives higher impact values [42].

(a) *PA type*

The PA rubber toughened systems studied most are based on PA 6 or 6,6. If the molecular weight of the PA is higher the melt viscosity increases and a fine dispersion is obtained more easily. In particular, for dispersing viscous rubbers a high melt viscosity of the PA phase is an advantage. A higher molecular weight has also the advantage that the number of entanglements per chain is higher. In this way the material is more resistant to craze formation and a somewhat higher impact strength can be expected. In blends the effect of the molecular weight seems small. A high molecular weight extrusion grade PA 6 ($M_n = 35\,000$) as compared with an injection molding grade ($M_n = 13\,000$) with similar blend morphology does not seem to give a lower T_{bt} or higher impact strength in the tough region [42]. Rubber modification destroys the spherulitic structure but the material is still crystalline. The crystallinity as measured by differential scanning calorimetry (DSC) is slightly lower than in the neat material [45]. The crystallization rate is, however, dramatically lower [79]. This decrease is related to the increase in melt viscosity observed in PA–EPR-g-MA blends. Blends with low crystalline material (random copolymer PA 6 or 6,6) gave a similar impact behavior to that of the neat polymers [26, 80]. Partially aromatic amorphous PA with T_g well above 100 °C can also be toughened with rubber very well [26, 63]. The tensile properties of PA well below T_g are little dependent on the crystallinity [26]. Thus the small effect of crystallinity on impact properties seems logical.

PA 6 and 6,6 are in many respects very similar and if they have the same blend morphology their impact behaviors are similar too. For PA 6 the blend structure is mostly sperical with few inclusions [19]. For PA 6,6 non-spherical dispersions and inclusions in the rubber particles have been reported [34]. Oshinski, Keskkula and Paul [34] explained this difference as being due to the chemical structure (diamine–diacid compared with the amino acid). This suggests that PA 6, which has only one amine end group in each chain, is monofunctional towards the polyfunctional EP-g-MA. At the interface comb-like structures might be formed. PA 6,6 can have two amine end groups in one chain and these chains can form with the polyfunctional EP-g-MA network structures. These networks can increase the viscosity of the rubber phase and in that way change the dispersion process.

(b) Influence of plasticizer

Another way of obtaining tough PA is by plasticization [1, 81]. There are several plasticizers for PA, in particular methanol and water. By adding a plasticizer the glass transition temperature can be lowered by 90 °C to that of a non-hydrogen-bonded system. In a PA blend the water take-up is reduced by as much as the rubber fraction. The water concentration in the PA phase is thus not affected by the presence of the rubber. The T_g shift due to the water is independent of the rubber concentration. The mechanism here is that, as T_g of the blend is lowered, the yield strength is lowered and excessive deformation can take place. This mechanism is different from that in blends where, below T_g of the nylon, the yield strength is lowered by relieving the triaxial tension ahead of the crack tip. These two mechanisms of lowering the yield strength are not synergistic [81].

7.4 FRACTURE GRAPHICS

By the deformation to fracture, stress whitening is apparent in the fracture zone (Fig. 7.2). In this stress whitened region, with scanning electron microscopy (SEM), large cavities can be observed [47, 48, 72]. In the cavities (fragments of) rubber particles are present but no crazes [5, 13, 74, 82]. The often-suggested craze deformation mechanism in these systems can be doubted. Ramsteiner [82] examined the stress whitened zone with transmission electron microscopy (TEM) and observed that extensive cavitation had taken place inside the rubber particle. SEM of a slowly deformed notched sample in three-point bending revealed a cavitation zone (Fig. 7.19). The cavities lie in a semicircle around the crack and are elongated. This suggests that the matrix material between the cavities is deformed, probably by shear banding. The radius of the crack tip is of the order of 0.5–1 μm. In the region 2.5–1000 μm, the effect of notch radius is small [2]. The stress whitened zone of impacted samples has a thickness of approximately 0.35–1.5 mm. In this stress whitened zone three regions can be observed (Fig. 7.20). Approaching the fracture plane, first a thick region of round cavities can be observed (Fig. 7.20, region III). The cavities increase in size as they lie nearer the fracture plane [47, 48, 72]. In the second region, which lies closer to the fracture plane (50–200 μm), the cavities

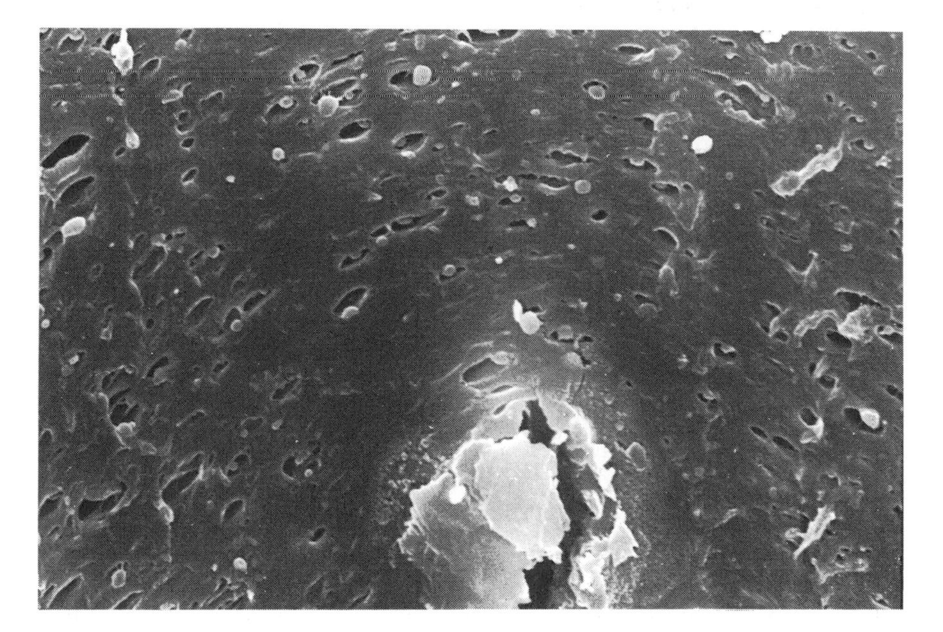

Fig. 7.19 Deformation zone ahead of the crack tip after three-point bending at low speed of a PA 6–EP blend (13 vol. %). (Reproduced from R. J. M. Borggreve, PhD Thesis, 1988, and R. J. Gaymans, R. J. M. Borggreve and A. J. Oostenbrink, *Makromolekulare Chemie, Macromolecular Symposia*, 1990.)

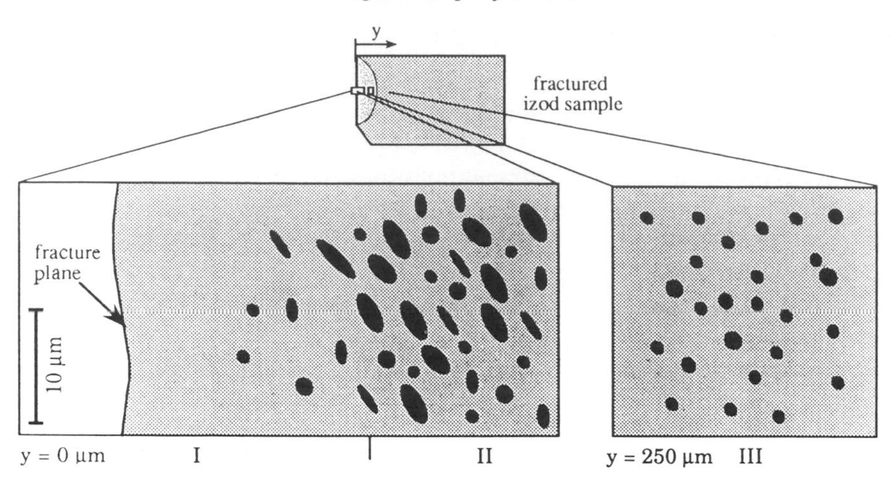

Fig. 7.20 A sketch of a deformation zone of an impacted sample perpendicular to the fracture surface: zone I, near the fractured surface; zone II, 5–100 μm; zone III, 0.1–1.5 mm. (Reproduced from A. J. Oostenbrink *et al.*, PRI Conference on Deformation Yield and Fracture of Polymers, 1991.)

are highly deformed (Fig. 7.20, region II). The width–length ratios are here of the order of 3–10. The matrix material between the cavities has been fibrillated (strained up to 200–600%) [12, 13, 47, 72]. A small region exists next to the fractured surface where no cavitation is apparent, 3–10 μm thick (Fig. 7.20, region I) [48, 72]. The fact that here no cavities are visible, while ahead of the crack tip cavitation has taken place, suggests that the highly deformed material has relaxed. Relaxation of orientation in PA can only take place in the melt. This suggests that the matrix in the small region next to the fractured surface has been molten while fractured. The temperature increase in the fracture surface must have been at least 200 °C. This temperature effect is much bigger than the 9 °C measured with a thermocouple [67]. The earlier reported observation that, in the fractured surface of tough samples, no cavities could be seen [67] might be due to this fracture surface melting effect.

The energy absorption is partly due to cavitation and partly due to shear banding. The heat of fusion of the thin layer around the fracture surface can at most be $2\,kJ\,m^{-2}$. Cavitation is thought to be an important dissipative mechanism [47]. Others think that shear banding is the major energy absorbing mechanism [13, 73].

In the brittle fractured samples a very fine craze layer could be seen [2, 83].

7.5 TOUGHENING MECHANISMS

Evidently the PA can be made tough by rubber modification if the rubber particles are small (0.1–2 μm). Since 1976, the year DuPont brought these systems onto the market, the explanation for the toughening behavior has changed considerably and is still a matter of discussion. Important parameters which affect the toughening behavior are

- matrix material,
- type of rubber (the cavitational behavior of the rubber),
- rubber concentration,
- rubber particle size, and
- rubber particle distribution.

7.5.1 Crazing model

PA–rubber blends show considerable stress whitening in the deformation zone. In line with the deformation behavior of HIPS crazing was thought to be the cause for this [17, 83]. This crazing of the matrix was triggered by the stress field overlap of neighboring particles [83]. The function of the rubber particles was to create these stress fields in the matrix. To do this the modulus of the dispersed phase had only to be lower than that of the matrix. If this difference in modulus is more than a factor of 10, then no extra effect is expected [22]. However, the rubber modulus was found to be an important parameter (Fig. 7.17) [74] and a very low modulus of the rubber is an advantage. Another area where the crazing model is difficult to apply is at the low rubber concentration (6.25%) [8]. At low rubber concentrations the stress field overlap is minimal but still a stable crack propagation can be obtained. Wu suggested in 1985 [17] that crazing was one of the energy absorbing mechanisms and calculated that the amount of energy absorbed in this way was 25%. Others could not observe any crazes in these systems [8, 13, 47, 82].

In the more complex system PA–(PPO–rubber), the PPO–rubber phase consists of PPO in which the rubber is dispersed. In this PA–(PPO–rubber) system both crazing and shear yielding take place [14, 64, 78, 84]. The deformation is here initiated by the cavitation of the dispersed rubber in the PPO. The deformation stress field overlap is thought to be a critical feature [64, 84]. In the range from 0.1 to 10 μm the PPO–rubber particle sizes do not seem to influence the impact strength [14]. The material just ahead of the crack tip has shear yielded and at the crack tip crazes are formed (Fig. 7.21) [84]. As the crack propagates the craze zone is transformed into a shear zone. The crazes are initiated in the PPO–rubber phase and are extended into the PA phase. In this way the hydrostatic tension is relieved and allows the shear yielding mechanism to be activated and bulk shear yielding can take place. [84].

7.5.2 Cavitation model

The observed stress whitening in the deformation zone is due to cavitation of the rubber [5, 62, 74, 82]. With volume strain experiments it is apparent that

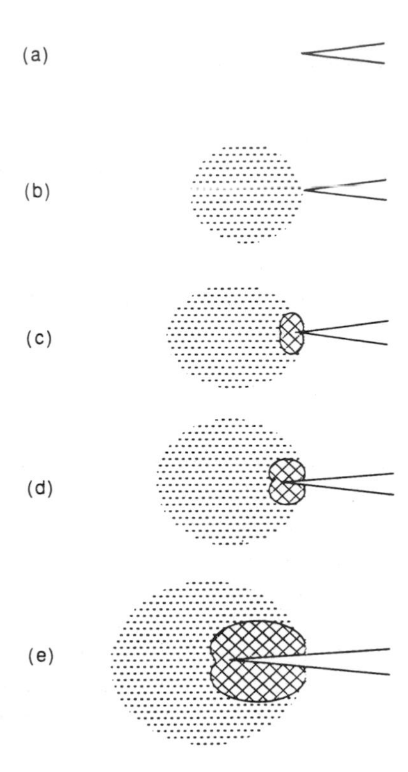

Fig. 7.21 A sketched sequence of the toughening mechanisms in PA–(PPO–rubber): (a) the initial starter crack; (b) formation of a craze zone in front of the crack tip when the specimen is initially loaded; (c) formation of initial shear yielded plastic zone around the crack tip when the hydrostatic tension is relieved by the formation of crazes; (d) once the buildup of shear strain energy reaches a critical value, the material begins to undergo shear yielding and the crack propagates; (e) a damage zone surrounds the propagating crack before the crack becomes unstable. (Reproduced from H. J. Sue and A. F. Yee, Churchill Conference Papers, 1988.)

the onset of cavitation is at low strains (2–6%) [13, 77], which means that under triaxial stress the rubber cavitates while the matrix material is still in the elastic region. After the cavitation the triaxial stress state is locally relieved and the yield strength is lowered. The deformation process is initiated in the rubber and not in the matrix or at the interface. The cavitation depends on the cavitational stress, which must be a function of the cohesive energy density (CED), the entanglement density and inhomogeneities in the material [85, 86]. The entanglement density is dependent on the molecular weight of the material. The lower the molecular weight and the smaller the molecular weight distribution are, the easier is the cavitation [87]. The cavitation is initiated by heterogeneities in the material [75]. As heterogeneities one can think of precavities or foreign particles. In each rubber particle an inhomogeneity must be present, otherwise the cavitation is not initiated. If the particle size is decreased, the number of particles is considerably increased. At some stage one must reach the point where not all the particles have inhomogeneities. The observed lower limit in particle size might be where a significant proportion of the particles do not contain inhomogeneities. Another explanation for the lower limit in rubber particle size is that it is found where the interface thickness is a significant part of the particle volume [88]. From volume strain experiments it was observed that the volume increase was linear with the rubber concentration and independent of the particle size in the range 1.6–0.5 μm [77]. This means that the cavitation of particles is not influenced by neighboring particles.

The main effect of void formation at all test speeds is an acceleration of the shear yielding mechanism in the PA matrix [13, 89, 90]. Cavitation of the rubber particles will result in a local decrease in the hydrostatic component of stress and a correspondent increase in the deviatoric (shear) component, and with that a higher stress concentration factor [13]. The rubber in a blend can be precavitated slightly by tensile straining at a low strain rate. In this way the cavitation is not optimal as the triaxial stress state is not fully achieved. On precavitated samples notched Izod studies were made. It was found that the impact behavior of the precavitated samples is certainly as good as that of the unstrained samples. This suggests that the cavitation process is not the major energy absorbing mechanism and that precavitation even helps to improve the impact behavior [72].

7.5.3 Interparticle distance model

The brittle-to-tough transition temperature is a function of rubber concentration, particle size and type of rubber. For describing T_{bt} the structural parameters can be combined. The best fit is the interparticle distance (ligament thickness) model (Fig. 7.14). This means that the ligament thickness is important for T_{bt}, the transition from unstable to stable crack propagation. The level of impact strength in the tough region is mainly dependent on the rubber concentration and not on the particle size and the type of rubber. Thus the impact strength values are not dependent on the ligament thickness. At first it was thought that the ligament thickness could be explained as a stress field overlap effect [17]. However, the stress field overlap is independent of the particle size. Therefore the ligament thickness effect cannot be explained by the stress field overlap model [70].

Another explanation that was given for this ligament thickness was the transition from plane strain to plane stress [5, 71, 73]. This ligament thickness was just the thickness at which the transition would take place. PA being an easily shearable material, the plane strain to plane stress transition is for flat plates of the order of millimeters. As a result of the spherical cavities the plane strain to plane stress transition must be at much smaller thicknesses, otherwise cavitation in the neighborhood of cavities would not take place. The triaxial stress state is only locally relieved by cavitation of particles so that neighboring particles can still cavitate. The amount of material that is relieved of the triaxial stress depends on the cavity radius. The ligament thickness is independent of the particle size. Thus the ligament thickness parameter can only be explained by the transition from plane strain to plane stress if the radius of the cavity is independent of the radius of the rubber particle. Certainly, there seems to be a delicate balance between the cavitation in a triaxial stress state and the relief of the triaxial stress state due to this cavitation. A third explanation is given by Bucknall [90] who points out that the major energy absorption is through the extensive formation of fibrils in the matrix. The fibrillation is a consequence of cavitation and the diameters of the fibrils are controlled by the interparticle spacings. The extent of fibrillation seems to depend on the fibril diameter [13, 90, 91]. As the ligament thickness becomes thinner the chance of inhomogeneities in the ligament becomes smaller and the tensile stress of the fibrils is increased.

7.5.4 Percolation models

Lately two percolation models have been proposed to explain the brittle-to-tough transition [92–94].

The percolation model described by Margolina and Wu [92] is based on interconnecting ligaments. They tried to verify the model with the impact levels of brittle and tough samples. The verification with the impact levels has been questioned, as the impact strength is not correlated with the ligament thickness [95]. Another discussion point is the effect of broadening of particle sizes and particle distributions. Inhomogeneous distributions always ease the percolation. A poorly distributed blend, however, gives poorer impact properties [70]. A modified version of the model was proposed in which the impact data were corrected for the rubber concentration effect on the macroscopic yield strength [93]. As the impact levels are little dependent on the particle size and are now also corrected for the rubber concentration, it is not clear what this modified model means.

The percolation in the model of Sjoerdsma is the percolation of overlapping stress fields [94]. He assumed the width of the deformation zone to be constant. It is known for nylon–rubber blends that the width of the stress whitened zone is not constant [67]. Therefore it is questionable whether it is allowed to take the width of the shear deformation zone as constant. Secondly, it is not yet clear whether the fracture process can be described with an overlapping stress field model.

7.6 RUBBER TOUGHENED COMPOSITES

With rubber modification the impact strength of PA can be considerably increased at some cost to the tensile properties [8]. The addition of fillers to PA increases the modulus and yield strength and decreases the elongation at break. For optimal properties one would like to have a material with a high modulus, high yield strength, high impact toughness and good flexural properties. A way of achieving this goal is by long fiber reinforced systems [96–98], but these good tensile properties are obtained at the cost of ease of production and ease of processing. Another way of obtaining high strength, high modulus and high impact resistant systems is by adding both filler and rubber to the PA. The reinforcement can be with particulate, plate-like or fibrous fillers. Some properties of rubber toughened composites are given in Table 7.2 [7].

7.6.1 Compounding

The compounds of nylons are made on a mixing extruder. As the rubber particle size is a critical parameter and as the blending is more critical than the compounding step, one might ask whether it is better first to make the blends and then to add the filler or whether another order of mixing should be followed. Smith *et al.* [99] found that for the system PA 6,6–EPM-g-MA–glass fiber it was advantageous firstly to make a blend before adding the fiber, compared with adding all ingredients as a dry blend to the extruder. For the system PA 6–EPM-g-MA–glass fiber five different mixing orders were evaluated [100]. It was found that if there is rubber at the interface of the fiber and the PA then the tensile and impact properties are poor. The best results

Table 7.2 Properties of glass reinforced PA 6,6 and PA 6,6–rubber blend [7]

	*Zytel 70 G35HSL**		*Zytel 80 G33HS11*†	
	DAM	*50% relative humidity*	*DAM*	*50% relative humidity*
Tensile strength at break (MPa)	205	138	145	110
Elongation at break (%)	3	4	4	5
Flexural modulus (GPa)	10.2	7.6	6.9	5.1
Impact strength, notched Izod ($J\,m^{-1}$)				
23 °C	139	160	219	235
−40 °C	107	104	136	152
Water absorption saturation (%)	5.5	–	4.0	–

* 35% short glass filled PA 6,6.
† 33% short glass filled Zytel ST801 (rubber modified PA 6,6).

were obtained by first forming the interface of the glass fiber and the PA before the rubber was added. If the coating on the glass fiber was well chosen, good results could also be obtained by mixing in first the rubber and then the glass fiber. Compounding in two steps seems to give slightly better mechanical properties than compounding in one extrusion step [100]. If compounding was carried out in one step with a twin-screw extruder, then the best results were obtained by dosing the PA at the first port, the glass fiber at the second port and the rubber at the fifth port, and by applying vacuum at the fourth port. A similar preference of mixing order was observed with glass beads [42].

7.6.2 Particle reinforced PA blend

The influence of glass spheres (20 μm coated with γ-amylpropylsilane) on the impact toughness is that, as the sphere concentration is increased, T_{bt} shifts to higher temperatures and the impact toughness at RT is gradually decreased (Fig. 7.22) [42]. In the case with only low interfacial bonding the compounds had much lower impact values and T_{bt} shifted to higher temperatures.

7.6.3 Glass fiber reinforced PA blends

The influence of glass fibers on the toughening behavior of blends is that at already very low fiber concentrations (0.2% and 1%) T_{bt} shifts to much higher temperatures [100]. From 5 vol.% onwards the 'super tough' behavior is completely absent (Fig. 7.23). At 15 vol.% glass fiber the toughness is higher again. Compared with the blend, the matrix deformation next to the fracture plane is strongly suppressed by adding glass fibers. The notched Izod impact values of the blend compounds are a factor of 2 higher than without the rubber

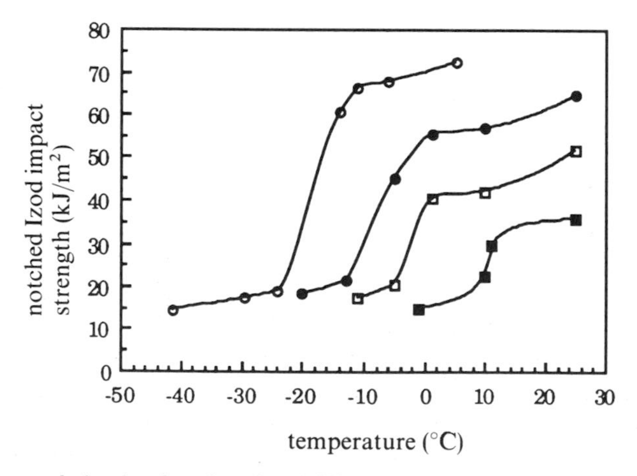

Fig. 7.22 Impact behavior for glass bead filled PA 6–EP blends (26 vol.% EP) as a function of bead content (volume per cent): ○, 0; ●, 2; □, 5; ■, 15 [42].

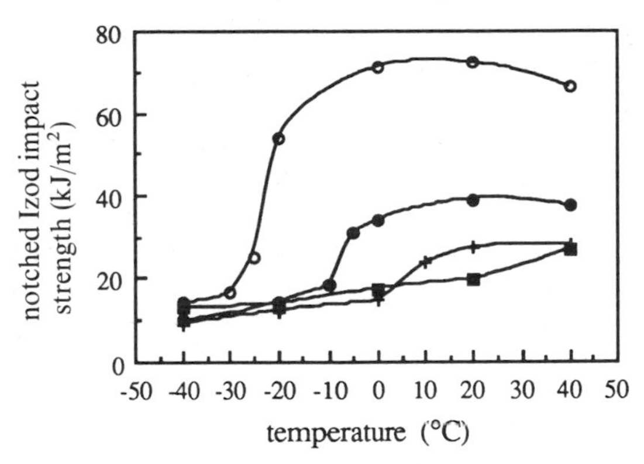

Fig. 7.23 Impact behavior for glass fiber filled PA 6–EP blends (26 vol.% EP) as a function of fiber content (volume per cent): ○, 0; ●, 0.2; +, 1; ■, 5. (Reproduced from R. J. Gaymans *et al.*, PRI Conference on Deformation and Fracture of Composites, 1991.)

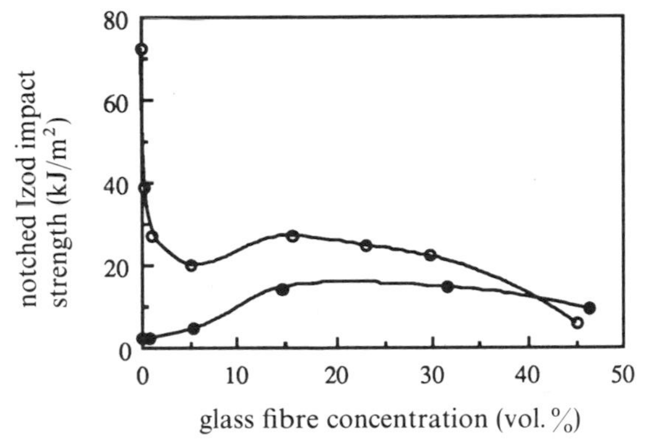

Fig. 7.24 Impact behavior of PA 6 (●) and PA 6–EP blends (26 vol.% EP) (○) as a function of fiber content. (Reproduced from R. J. Gaymans *et al.*, PRI Conference on Deformation and Fracture of Composites, 1991.)

present (Fig. 7.24) [94, 100–103]. In the presence of the rubber the modulus and yield strength are lower. The increase in toughness is due to higher energy absorptions in the crack propagation step [101]. The fiber pull-out mechanism changes from a relatively clean pull-out for PA–glass fiber (Fig. 7.25(a)) to a sheathed pull-out for the blend with glass fiber (Fig. 7.25(b)) [59, 100, 101]. In sheathed pull-out the fibers are coated with a thick layer of matrix material. The thickness of this sheathed layer increases with rubber concentration (Fig. 7.26) [100, 101], temperature [100] and deformation rate [101]. This sheathed pull-out process only takes place at high deformation rates [101]. This means that LEFM, measured at low deformation rates, cannot be used for describing the fracture toughness under impact conditions. Bailey and

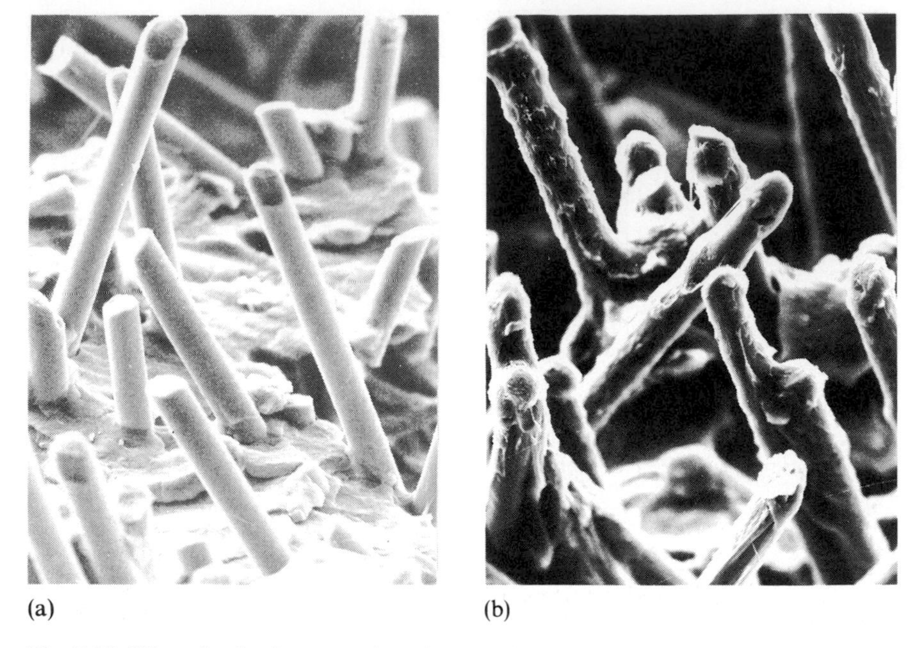

Fig. 7.25 Fibers in the fracture plane in (a) PA 6 and (b) PA 6–EP blend. (Reproduced from R. J. Gaymans *et al.*, PRI Conference on Deformation and Fracture of Composites, 1991.)

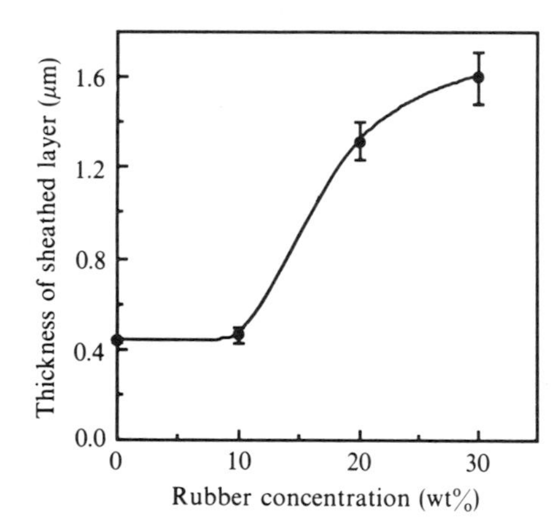

Fig. 7.26 Thickness of the matrix layer around a pulled-out fiber as a function of the rubber concentration in the blend (15 vol.% glass fibers). (Reproduced from R. J. Gaymans *et al.*, PRI Conference on Deformation and Fracture of Composites, 1991.)

Bader [101] think that the sheathed pull-out is due to the fact that the rubber lowers the shear strength of the matrix to a value which is lower than the interfacial strength. However, this does not explain the effect of deformation rate. On micrographs taken perpendicular to the fractured surface for PA–

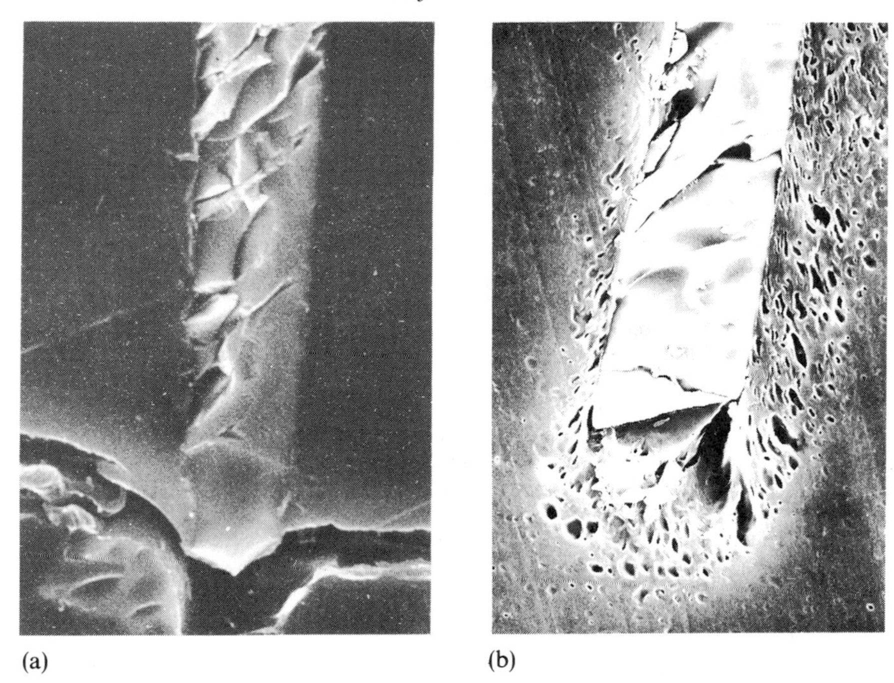

(a) (b)

Fig. 7.27 Fiber in the fracture zone of an impacted sample perpendicular to the fracture surface in (a) PA 6 and (b) PA 6–EP blend. (Reproduced from R. J. Gaymans *et al.*, PRI Conference on Deformation and Fracture of Composites, 1991.)

glass fiber the pictures show only cracks (Fig. 7.27(a)). In the blend the strained fibers were found to have a layer of cavitated material around them (Fig. 7.27(b)) [100]. This cavitated layer has elongated cavities, which means that the layer is strongly deformed. Also, some cavitation zones were observed between fibers. Thus the blend cavitates around the strained fibers and between the fibers and subsequently shear yielding takes place. The cavitated layer seems to correspond to the sheathed layer. The cavitation might be enhanced by increasing the deformation rate. If the bonding between the blend and the fiber is poor then the mechanical properties are also poor. After impact the fibers in the fracture surface appear to be clean. Cavitation around the fiber has then not taken place [100].

REFERENCES

1. Kohan, M. L. (1973) *Nylon Plastics*, SPE Monograph, Wiley, New York.
2. Flexman, E. A. von, (1979) *Kunststoffe*, **69**, 172.
3. Epstein, B. N. and Adams, G. C. A. (1985) PRI Conf. on Toughening of Plastics, London, July 1985, Paper 10.
4. Epstein, B. N. (1979) US Patent 4,174,385 (to DuPont).
5. Borggreve, R. J. M. (1988) PhD Thesis, University of Twente.

6. Vu-Khanh, T. (1988) *Polymer*, **29**, 1979.
7. DuPont (1990) Technical Bulletin on Zytel Nylon Resins, E-96368, January 1990.
8. Borggreve, R. J. M., Gaymans, R. J., Schuijer, J. and Ingen Housz, A. J. (1987) *Polymer*, **28**, 1489.
9. Bosma, M. (1990) IUPAC Symp., Mechanisms of Polymer Strength and Toughness, Prague, July 1990, p. 22.
10. Hadhemi, S. and Williams, J. G. (1987) *Polymer*, **27**, 384.
11. Saleemi, A. S. and Nairn, J. A. (1988) *ACS Polym. Prepr.*, **29**(2) 155.
12. Sunderland, P., Kausch, H. H., Schmid, E. and Arber, W. (1988) *Makromol. Chem., Macromol. Symp.*, **16**, 365.
13. Bucknall, C. B., Heather, P. S. and Lazzeri, A. (1989) *J. Mater. Sci.*, **16**, 2255.
14. Sue, H. J. and Yee, A. F. (1989) *J. Mater. Sci.*, **24**, 1447.
15. Hahn, M. T., Hertzberg, R. W. and Manson, J. A. (1983) *J. Mater. Sci.*, **18**, 3551.
16. Hahn, M. T., Hertzberg, R. W. and Manson, J. A. (1986) *J. Mater. Sci.*, **21**, 39.
17. Wu, S. (1985) *Polymer*, **26**, 1855.
18. Oostenbrink, A. J., Molenaar, L. J. and Gaymans, R. J. (1990) PRI Conf. on Polymer Blends, Cambridge, July 1990, Paper E3.
19. Oshinski, A. J., Keskkula, H. and Paul, R. D. (1992) *Polymer*, **33**, 268.
20. Wu, S. (1987) *Polym. Eng. Sci.*, **27**, 335.
21. Xanthos, M. (1988) *Polym. Eng. Sci.*, **28**, 1392.
22. Borggreve, R. J. M., Gaymans, R. J. and Schuijer, J. (1989) *Polymer*, **30**, 71.
23. Illing, G. (1981) *Angew. Makromol. Chem.*, **95**, 83.
24. Greco, R., Lanzetta, G., Maglio, G., Malinconico, M., Martuscelli, E., Palumbo, R., Ragosta, G. and Scarinzi, G. (1986) *Polymer*, **27**, 299.
25. Hert, M., Guetdoux, L. and Lebez, J. (1987) *Angew. Makromol. Chem.*, **154**, 111.
26. Epstein, B. N., Latham, R. A., Dunphy, J. F. and Pailagan, R. U. (1987) *Elastomerics*, **119**, 10.
27. Braun, D. and Illing, W. (1987) *Angew. Makromol. Chem.*, **154**, 179.
28. Immirizi, B., Lanzetta, N., Laurienzo, P., Maglio, G., Malinconico, M., Martuscelli, E. and Palumbo, R. (1987) *Makromol. Chem.*, **188**, 951.
29. D'Orazio, L., Mancarella, C. and Martuscelli, E. (1988) *J. Mater. Sci.*, **23**, 161.
30. Chen, C. C., Fontan, E., Min, K. and White, J. L. (1988) *Polym. Eng. Sci.*, **28**, 69.
31. Willis, J. M. and Favis, B. D. (1988) *Polym. Eng. Sci.*, **28**, 1416.
32. Lawson, D. F., Hergenrother, W. L. and Matlock, M. G. (1988) *ACS Polym. Prepr.*, **29**(2), 193.
33. Borggreve, R. J. M. and Gaymans, R. J. (1989) *Polymer*, **30**, 63.
34. Oshinski, A. J., Keskkula, H. and Paul, R. D. (1992) *Polymer*, **33**, 284.
35. Hu, G.-H., Holl, Y. and Lambla, M. (1989) PPS Conference, Rolduc, 1989.
36. Gale, G. M. (1982) *Plast. Rubber Proc. Appl.*, **2**, 347.
37. (a) Taylor, G. I. (1932) *Proc. R. Soc. London, Ser. A*, **138**, 41.
    (b) Taylor, G. I. (1934) *Proc. R. Soc. London, Ser. A*, **146**, 50.
38. Karam, H. J. and Bellinger, J. C. (1986) *Ind. Eng. Chem. Fundam.*, **7**, 576.
39. Elmendorp, J. J. (1986) *Polym. Eng. Sci.*, **26**, 418.
40. Elmendorp, J. J. and Vergt, A. K. van der (1986) *Polym. Eng. Sci.*, **26**, 1332.
41. Serpe, G., Jarrin, J. and Dawans, F. (1990) *Polym. Eng. Sci.*, **30**, 553.
42. Oostenbrink, A. J. and Gaymans, R. J. Unpublished results.
43. Oostenbrink, A. J., Molenaar, L. J. and Gaymans, R. J. (1990) PRI Conf. on Polymer Blends, Cambridge, July 1990, Paper A21.
44. Hoght, A. H. (1988) Antec 1988, Atlanta, GA, Paper 553.
45. Oostenbrink, A. J. and Gaymans, R. J. (1992) *Polymer*, **33**, 3086.
46. Speroni, F., Castoldi, E., Fabbri, P. and Casiraghi, T. (1989) *J. Mater. Sci.*, **24**, 2165.

47. Cimmino, S., D'Orazio, I., Greco, R., Maglio, G., Malinconico, M., Mancarella, M., Martuscelli, E., Palumbo, R. and Ragosta, G. (1984) *Polym. Eng. Sci.*, **24**, 48.
48. Oostenbrink, A. J., Dijkstra, K., Van Der Wal, A. and Gaymans, R. J. (1991) PRI Conf. on Deformation Yield and Fracture of Polymers, Cambridge, April 1991, Preprints Paper 50.
49. Bucknall, C. B. (1977) *Toughened Plastics*, Applied Science, London.
50. Gabbert, J. D. and Hendrick, R. M. (1986) *Polym. Process. Eng.*, **4**, 359.
51. Bengemann, M. and Menges, G. (1988) *Kem. Kemi*, **15**, 122.
52. Loos, J. L. M. van der and Geenen, A. A. van (1985) *ACS Symp. Ser.*, **270**, 181.
53. Borggreve, R. J. M. and Gaymans, R. J. (1988) *Polymer*, **29**, 1441.
54. Cimmino, S., D'Orazio, L., Greco, R., Maglio, G., Malinconico, M., Mancarella, C., Martuscelli, E., Musto, P., Palumbo, R. and Ragosta, G. (1985) *Polym. Eng. Sci.*, **25**, 193.
55. Greco, R., Malinconico, M., Martuscelli, E., Ragosta, R. and Scarinzi, G. (1988) *Polymer*, **29**, 1418.
56. Russo, S., Alfonso, G. C., Tuturro, A. and Pedemonte, E. (1984) Eur. Patent 0,146,983 (to Enichimica).
57. Alfonso, G. C., Dondero, G., Russo, S. and Tuturro, A. (1986) Proc. 17th Europhysics Conf., Macromolecular Physics, Prague, 1986, p. 426.
58. Sederel, L. C., Mooney, J. and Weese, R. H. (1987) Polymer Blends and Alloys Conf., Strasbourg, June 1987.
59. Leblanc, D., Reginster, L. and Sederel, L. C. (1990) Compolloy 90, London, January 1990.
60. Udipi, K. (1988) *J. Appl. Polym. Sci.*, **36**, 117.
61. Lefelar, J. A. and Udipi, K. (1989) *Polym. Commun.*, **30**, 38.
62. Ban, L. L., Doyle, M. J., Disko, M. M. and Smith, G. R. (1988) *Polym. Commun.*, **29**, 163.
63. Neuray, D. and Ott, K. H. (1981) *Angew. Makromol. Chem.*, **98**, 213.
64. Hobbs, S. Y., Dekkers, M. E. J. and Watkins, V. H. (1989) *J. Mater. Sci.*, **24**, 2025.
65. Gelles, E., Modic, M. and Kirkpatrick, J. (1988) Antec 1988, p. 513.
66. Bowman, C. (1990) PRI Conf. on Polymer Blends, Cambridge, July 1990, Paper C3.
67. Wu, S. (1983) *J. Polym. Sci.*, **21**, 699.
68. Gent, A. N. and Tompkins, D. A. (1969) *J. Polym. Sci. A-2*, **7**, 1483.
69. Gent, A. N. (1990) *Rubber Chem. Technol.*, **63**, 949.
70. Cimmino, S., Coppola, F., D'Orazio, I., Greco, R., Maglio, G., Malinconico, M., Mancarella, M., Martuscelli, E. and Ragosta, R. (1986) *Polymer*, **27**, 1874.
71. Wu, S. (1988) *J. Appl. Polym. Sci.*, **35**, 549.
72. Dijkstra, K., Oostenbrink, A. J. and Gaymans, R. J. (1991) PRI Conf. on Deformation Yield and Fracture of Polymers, Cambridge, April 1991, Preprints Paper 39.
73. Borggreve, R. J. M., Gaymans, R. J. and Luttmer, A. R. (1988) *Makromol. Chem., Macromol. Symp.*, **16**, 195.
74. Gaymans, R. J., Borggreve, R. J. M. and Oostenbrink, A. J. (1990) *Makromol. Chem., Macromol. Symp.*, **38**, 125.
75. Gent, A. N. and Lindley, P. B. (1969) *Proc. R. Soc. London, Ser. A*, **249**, 2520.
76. Wu, S. (1990) *Polym. Eng. Sci.*, **30**, 753.
77. Borggreve, R. J. M., Gaymans, R. J. and Eichenwald, H. M. (1989) *Polymer*, **30**, 78.
78. Hobbs, S. Y. and Dekkers, M. E. J. (1989) *J. Mater. Sci.*, **24**, 1316.
79. Martuscelli, E., Riva, F., Sellitti, C. and Silvestre, C. (1985) *Polymer*, **26**, 270.
80. Roura, M. J. (1981) Eur. Patent 0,034,704 (to DuPont).

81. Gaymans, R. J., Borggreve, R. J. M. and Spoelstra, A. B. (1989) *J. Appl. Polym. Sci.*, **37**, 479.
82. Ramsteiner, F. (1983) *Kunststoffe*, **73**, 148.
83. Hobbs, S. Y., Bopp, R. C. and Watkins, V. H. (1983) *Polym. Eng. Sci.*, **23**, 380.
84. Sue, H. J. and Yee, A. F. (1988) Churchill Conf. Papers, Cambridge, 1988, Paper 38.
85. Wu, S. (1989) *J. Polym. Sci. B*, **27**, 723.
86. Kramer, E. J. (1984) *Polym. Eng. Sci.*, **24**, 219.
87. Brown, N. and Ward, I. M. (1983) *J. Mater. Sci.*, **18**, 1405.
88. Michler, G. H. (1990) *Makromol. Chem., Macromol. Symp.*, **38**, 195.
89. Bucknall, C. B. (1988) *Makromol. Chem., Macromol. Symp.*, **16**, 209.
90. Bucknall, C. B. (1990) *Makromol. Chem., Macromol. Symp.*, **38**, 1.
91. Hinrichsen, G. To be published.
92. Margolina, A. and Wu, S. (1988) *Polymer*, **29**, 2170.
93. Wu, S. and Margolina, A. (1990) *Polymer*, **31**, 972.
94. Sjoerdsma, S. D. (1989) *Polym. Commun.*, **30**, 106.
95. Gaymans, R. J. and Dijkstra, K. (1990) *Polymer*, **31**, 971.
96. Crosby, J. M. and Drye, T. R. (1987) *J. Reinf. Plast. Compos.*, **6**, 162.
97. Gore, C. R., Cuff, G. and Ciabelli, D. A. (1986) *Mater. Eng.*, **103**, 47.
98. Bijsterbosch, H., Gaymans, R. J. and Groot, N. H. M. (1991) PRI Conf. on Deformation and Fracture of Composites, Manchester, March 1991, Preprints Paper 43.
99. Smith, G. R., Cartasegna, S., Stuyver, J. and Wong, W. K. (1989) Personal communications.
100. Gaymans, R. J., Oostenbrink, A. J., Bennekom, A. C. M. van and Klaren, J. E. (1991) PRI Conf. on Deformation and Fracture of Composites, Manchester, March 1991, Paper 23.
101. Bailey, R. S. and Bader, M. G. (1985) ICCM V, p. 947.
102. Leach, D. C. and Moore, D. R. (1985) *Composites*, **16**, 113.
103. Fahnler, F. and Merten, D. (1985) *Kunststoffe*, **75**, 157.

8

Toughened polyesters and polycarbonates

D. J. Hourston and S. Lane

8.1 INTRODUCTION

Although many polyesters are known, this chapter is concerned only with those which have been developed for engineering applications. Polyethylene terephthalate and polybutylene terephthalate are both semicrystalline thermoplastics which have found increasing use within the last decade, both in pure form and as toughened blends. Their physicochemical properties are very similar to those of polyamides, with which they compete for many applications. Few papers concerned with the properties and toughening mechanisms of rubber modified polybutylene terephthalate and the neat polymer have been published in the open literature, in contrast to the relatively large number which have appeared on toughened polyamides in recent years. The two terephthalates will be considered in separate sections of this chapter.

The only other polyester which has found widespread use as an engineering thermoplastic is the polycarbonate of bisphenol A. In contrast to the terephthalates, this polymer is an amorphous material, which is relatively tough without modification, and can be produced in transparent form. The fracture mechanisms of polycarbonate and its toughened blends have received considerable attention for many years. The mechanisms involved in the toughening of polyesters are not well understood, and a consensus of opinion has yet to be established.

8.2 POLYBUTYLENE TEREPHTHALATE

Polybutylene terephthalate (PBT) has been used as an engineering thermoplastic since the early 1970s. The polymer is semicrystalline with a melting point of 222–232 °C, depending on thermal history, and a glass transition temperature of approximately 50 °C. It undergoes a unique, reversible crystalline transformation when subjected to low levels of stress [1–3]. Crystallization from the melt occurs rapidly, even in the absence of promoters, and the polymer has excellent mould flow characteristics and resistance to solvents. The main uses of PBT are in the manufacture of injection moulded articles for domestic, electrical and automotive applications, and in the production of bristles for paint and tooth brushes. Although PBT has relatively high unnotched impact strength and gives no break in the unnotched Izod test at 23 °C and − 40 °C, the notched Izod strength is surprisingly low (typically 2.5–3.0 kJ m^{-2}). Attempts to enhance the fracture properties of PBT by the addition of elastomeric impact modifiers have resulted in the commercial development of several rubber toughened materials such as Pocan (Bayer), Ultradur (BASF) and Gaftuf (Gaf).

Although many industrial research groups have obtained patents for the toughening of PBT, very few papers have appeared in academic journals. In the first part of this section, a general review of the literature is presented according to the type of impact modifier and this is followed by short synopses of the more important recent papers.

8.2.1 Acrylate rubbers

Horlbeck and Bittscheidt [4] prepared a blend containing 15% of a butyl acrylate–methyl methacrylate copolymer which had a notched Izod impact strength of 20.4 kJ m^{-2}. This increased to 53.9 kJ m^{-2} after heating the material at 190 °C for 5 h under nitrogen.

A thirtyfold increase in impact strength compared with unmodified PBT was reported by Kasuga, Takahashi and Tajima [5] for a blend containing 20% of an acrylate rubber and 10% of Hytrel 4050 thermoplastic elastomer. In a subsequent patent by Takahashi and Tajima [6], a silane coupling agent (2%) was used in combination with a polyacrylate-based impact modifier.

Inoue [7] obtained a tough material by blending 4% of an ethyl acrylate–ethylene copolymer (1:4) with PBT, although only a very modest improvement in impact strength was claimed [8] for a blend containing a glycidyl methacrylate–ethylene copolymer.

8.2.2 SBR and related latex rubbers

Wang and Chung [9] obtained good Izod impact strength at − 40 °C for a blend containing 20% of an acrylonitrile–butadiene–styrene (ABS) core–shell impact modifier with a particle size as low as 90 nm. In a similar patent, Binsack *et al.* [10] reported an Izod impact strength of 28 kJ m^{-2} at room temperature for a material containing 20% of an ABS copolymer. Their impact modifier was 80% polybutadiene and had an average particle size of approximately 0.5 μm. In another patent from the same group [11] a similar latex rubber was used together with 0.8% of polytetrafluoroethylene (PTFE) to produce a blend with an enhanced Izod impact strength of 52 kJ m^{-2} compared with 39 kJ m^{-2} without PTFE. The Bayer group [12] also obtained a patent for the use of butadiene–styrene copolymer impact modifier with an average particle size of 0.4 μm.

Triglycidyl isocyanurate and phosphite stabilizers were used by Blaschke, Klinkenberg and Vollkomer [13] to increase the Izod impact strength of a blend containing 15% of a methyl methacrylate–butadiene–styrene copolymer (20:60:20) from 12 kJ m^{-2} to 44 kJ m^{-2}. Chiolle and Andreoli [14] claim a 4.5-fold increase in impact strength at −40 °C with the use of a methyl methacrylate–butadiene–styrene copolymer (20%) in combination with a terephthalic acid–1,4-butanediol–polyoxytetramethylene glycol block copolymer (5%).

The Mitsubishi company [15] also used a graft copolymer prepared from butadiene, styrene and acrylates to produce a twofold increase in the impact strength of PBT at −30 °C. In a recent patent, Bronstert, Schwaben and Echte [16] demonstrated that a butadiene–styrene (91:9) block copolymer with 1,3-diaminopropane end groups gave a notched impact strength of 20.1 kJ m^{-2} compared with 13.1 kJ m^{-2} without the amine end groups and 12.0 kJ m^{-2} with no rubber.

8.2.3 EPDM and EPR rubbers

Lee, Ju and Chang [17] polymerized butanediol and dimethylterephthalate in the presence of suspended EPR rubber particles to give a toughened blend. However, better impact properties were obtained by preparing a material of the same composition by conventional melt blending of the polyester and rubber. The incorporation of an elastomeric impact modifier during the polymerization of PBT has also been reported by Martuscelli and coworkers [18, 19] and their papers are discussed in more detail later in this chapter.

Sheer [20] obtained a notched Izod impact strength of 10.9 kJ m^{-2} by blending PBT with 28% of an ethylene–fumaric acid–1,4-hexadiene–norbornadiene–propylene graft copolymer together with 7% of polybutylene glycol. The impact strength was only 16.5 kJ m^{-2} in the absence of the polyether.

The addition of a small amount of bismaleimide to PBT–EPDM blends gave enhanced impact properties after solid-state post-condensation of the blend [21] and 1% of an oligomeric diisocyanate was found to enhance the properties of a blend containing a maleic anhydride–dibutyl maleate grafted rubber [22].

Reactive impact modifiers have also been prepared by grafting EPDM rubbers with anhydrides [18, 19, 23, 24], acrylates [25], esters [26], imides [18, 19] and a mixture of styrene and acrylonitrile [27]. Inoue and coworkers [28] have grafted an EPR rubber with glycidyl methacrylate groups.

Interfacial reaction has also been achieved in PBT–EPDM blends by extruding unmodified rubbers with a modified PBT containing a few per cent of maleate groups [29–31]. The carbon–carbon double bonds in the two components react via an ene and/or a radical addition mechanism.

8.2.4 EVA rubbers

Pilati [32] and Pilati and Pezzin [33] have polymerized PBT in the presence of an ethylene–vinyl acetate (EVA) copolymer and obtained rubber particles in the range 1.6–3.0 μm. The particle size decreased as the vinyl acetate content and the reaction time increased.

A glass-filled PBT–EVA blend [34] gave an Izod impact strength of 24 kJ m^{-2} at 23 °C and 17 kJ m^{-2} at − 40 °C when stabilized with phosphite and isocyanurate additives. The impact strengths were reduced to 14 kJ m^{-2} and 10 kJ m^{-2} at 23 °C and − 40 °C respectively, in the absence of stabilizers.

8.2.5 Polycarbonate with acrylate rubbers

A modest improvement in Izod impact strength was achieved [35] by compounding PBT with 15% polycarbonate and 20% of an allyl methacrylate–butyl acrylate–methyl acrylate–methyl methacrylate (0.4:79.6:2:18) copolymer. Neuray *et al.* [36] also report good impact properties for a blend containing 50% polycarbonate and 7.5% butadiene–methyl methacrylate copolymer.

The use of a butadiene–methyl methacrylate–styrene copolymer impact modifier in combination with polycarbonate is reported by Verhoeven and Pearson [37]. The material had a brittle–ductile transition temperature of − 30 °C.

A series of papers from the General Electric Company [38, 39] in which the addition of polycarbonate enhanced the impact strength of blends containing a latex rubber with a core–shell type morphology is discussed in section 8.3.

8.2.6 Miscellaneous rubbers

Horlbeck and Mumcu [40] report an Izod impact strength of 8.1 kJ m^{-2} at 23 °C for a blend containing 15% of a polyamide–polyester–polyether copolymer (35 parts dodecandioic acid, 30 parts laurolactam and 35 parts polybutylene glycol). This increased to 28.8 kJ m^{-2} when the material was heated at 200 °C for 3 h in an atmosphere of nitrogen before moulding. A twofold increase in impact strength by adding 20% of a polycaprolactone–PBT copolymer to pure PBT is claimed in a patent by the Toyobo company [41]. A blend containing 50% of a polyurethane prepared from adipic acid, 1,4-butanediol, ethylene glycol and methyldi-*p*-phenylene isocyanate had a notched Izod impact strength of 83 kJ m^{-2} compared with

$2.7\,kJ\,m^{-2}$ for pure PBT [42]. Enhanced impact properties have been reported [43] for blends containing acrylonitrile–butadiene copolymers.

8.2.7 Review of important recent papers

Ragosta and coworkers [26] prepared a series of blends using ethylene–propylene copolymers functionalized with dibutylsuccinate (DBS) groups. By maintaining a constant PBT:EPR ratio (80:20 by weight) and varying the degree of grafting from 0 to 7.2 wt% of rubber, they were able to demonstrate the beneficial effects of interfacial reaction. The Charpy impact strength of the blends was measured as a function of temperature in the range from $-30\,°C$ to $20\,°C$. Although the addition of unmodified EPR produced a modest toughening effect at $20\,°C$, the Charpy impact strength increased significantly with the degree of grafting over the whole temperature range. The blend with the highest degree of grafting gave a fivefold increase in the Charpy impact strength compared with pure PBT at $20\,°C$.

The tensile properties of the blends were shown to be inferior to those of pure PBT and although the values of modulus and stress at break were largely unaffected by the degree of grafting, the strain at break increased significantly with DBS content. SEM analysis of the blends revealed a reduction in particle size as the degree of grafting increased, from approximately 5–15 μm with the unmodified EPR rubber to 2–5 μm in the blend with the highest DBS content. It was also observed that detachment of rubber particles from the polyester matrix in the absence of grafting was not seen in any of the compatibilized blends. The authors suggest that the rubber particles act as sites of stress concentration and that a 'multicraze' mechanism is activated when the stress around a particle exceeds the yield stress of the matrix. The degree of toughening is increased by a reduction in the particle size of the dispersed phase and/or by increased particle to matrix adhesion. Intermolecular interactions between the DBS and PBT ester groups are also postulated.

In a subsequent paper from the group at Arco Felice, Greco and coworkers [24] have used an EPR rubber modified with 2.3 wt% of maleic anhydride to investigate the effect of processing conditions on blend morphology and impact properties. Melt blending was performed in a Haake Rheocord apparatus, and a series of blends was produced by varying the roller speed from 4 to $64\,rev\,min^{-1}$. It was shown that the impact resistance of pure PBT was significantly reduced when the roller speed exceeded $48\,rev\,min^{-1}$, and this effect was attributed to mechanical degradation of the PBT since the intrinsic viscosity of the polymer showed an analogous trend. The roller speed also affected the morphology of the reactive blends, with a reduction in the average particle size, and a narrowing of the particle size distribution as the mixing speed was increased to $32\,rev\,min^{-1}$. Although no further changes in the blend morphology were observed at higher roller speeds, the Charpy impact strength continued to increase, reaching a maximum at $48\,rev\,min^{-1}$, above which the impact properties deteriorated presumably as a result of polyester degradation. Thus, a relatively narrow processing window exists in which efficient dispersion of the rubber is achieved without damage to the matrix. A tough morphology is only possible when the interfacial tension between the incompatible polymers is reduced by chemical reaction. Evidence for the formation of a PBT–EPR copolymer was obtained from dynamic mechanical analysis of the blends. The T_g values of the polyester and rubber shifted to lower and higher temperatures, respectively, with increasing roller speed.

A different mechanism for optimizing the morphology of PBT–EPR blends has also been reported by the Italian group [18]. The procedure was initially developed using polyamide–EPR blends [44] and involves the addition of a reactive elastomer during the polymerization of the thermoplastic matrix. In these systems, the EPR rubber was modified with dibutylsuccinate or *N*-hydroxyethyl succinimide using a solution method for the grafting reaction. Both groups react with the hydroxyl end groups of the PBT molecules by transesterification reactions, and small amounts of PBT–EPR copolymers were extracted from the blends with trifluoroacetic acid.

The morphologies of the blends were investigated by SEM and found to be influenced by the reactivity and number of grafts, and by the length of the polymer chains and the melt viscosity when the rubber was added. The best results were achieved by adding a rubber containing a low content (0.5 wt%) of hydroxyethyl succinimide groups when the number-average molecular weight of the PBT had reached approximately 10 000. DSC analysis of the materials showed that although the melting point of the PBT remained constant, T_g and the degree of crystallinity both decrease in the blends.

The impact properties and fractographic analysis of the materials were presented in a later publication [19], and significant differences were observed between the two types of grafts. As predicted, the blend containing hydroxyethyl succinimide functionalized rubber gave the best impact properties and was found to contain a bimodal distribution (*ca.* 0.5 μm and 5.0 μm) of well-adhering particles. The dibutylsuccinate grafts gave a relatively narrow particle size distribution and significantly lower impact strength, and examination of the fracture surface indicated that the matrix–particle interface was weaker. The authors suggest that the rubber particles act as initiators and terminators of crazes. The high impact strength of the succinimide-grafted material is attributed to the presence of relatively large rubber particles which provide favourable sites for craze initiation, together with much smaller particles which act as effective craze terminators.

A series of papers have been published by a research group at the General Electric Company [38, 39] in which blends of PBT and polycarbonate were toughened with a core–shell impact modifier. They demonstrated a synergism between the toughening effects of the polycarbonate and the impact modifier, especially below room temperature. Examination of the blends revealed that the two polyesters were essentially incompatible, and that the impact modifier was located only in the polycarbonate domains. These blends are discussed in more detail in section 8.3.

Another procedure for reducing interfacial tension during the melt blending of PBT with rubbers was developed in a collaborative research programme between Lancaster University and Akzo Plastics [29–31]. It was shown, using model compound reactions, that a modified PBT, containing a few mole per cent of maleate groups (M–PBT) could react with unsaturated groups in EPDM and polydiene elastomers via ene and/or radical mechanisms. The thermomechanical properties of the M–PBT were not significantly different from those of pure PBT and the reactivity of the polyester matrix could be altered by mixing M–PBT and PBT in different proportions.

Using a combination of instrumented impact testing and transmission electron microscopy, it was shown that the impact properties improved, and the average particle size of the rubber decreased, as the mixing time was increased. In the blends which contained no M–PBT or an unreactive rubber, the impact properties and morphology were unaffected by the residence time in the mixing apparatus. By using this procedure, chemical modification of rubbers, which often results in instability and premature crosslinking, is not necessary and the degree of reaction between the two components of the blend is not dependent on the concentration of polyester end groups.

The group at Lancaster University have also used transmission electron microscopy to investigate the failure mechanisms of rubber modified PBT samples [45]. When notched samples are fractured a stress whitened, plastically deformed region is formed as the crack propagates. This is often referred to as the process zone, and it is the mechanism by which energy is dissipated in this region which determines the toughness of the sample. A series of transmission micrographs were obtained in the three orthogonal planes (Fig. 8.1)

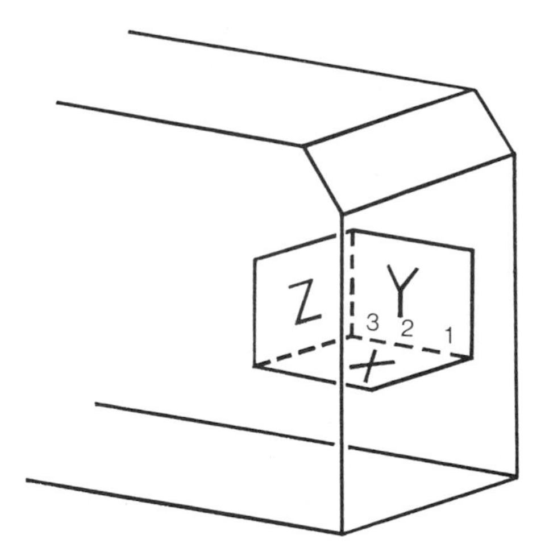

Fig. 8.1 Definition of the planes referred to in Figs 8.2–8.8.

through the process zone of a nitrile rubber–PBT blend which had undergone ductile fracture (Figs 8.2–8.8). It can be seen from the micrographs in the *x* plane that the stress whitening is a result of cavitation within the rubber particles, rather than debonding at the particle–matrix interface or voiding within the matrix (e.g. crazing or shear banding). The proportion of cavitated particles increases through the process zone and appears to be independent of the size of the particle. It is also apparent that extensive yielding of the matrix material has occurred behind the fracture surface, although the region of cavitation extends well beyond that of plastic deformation. When the micrographs of the *y* and *z* planes are examined, it can be seen that the plastic deformation has occurred predominantly in the shear plane, causing the rubber particles to adopt a flattened, ellipsoidal shape. The same effect was observed in a commercial grade of toughened PBT which is thought to contain an ABS impact modifier. However, it is not known whether cavitation is a prerequisite for, or incidental to, the shear yielding process, which is likely to absorb most of the impact energy.

In a more recent paper [21], the influence of particle size and interfacial adhesion in rubber modified PBT has been investigated. The two factors are closely related and the morphology of incompatible blends is usually controlled by altering interfacial tension during the blending procedure as discussed earlier. It is, therefore, very difficult to vary the level of adhesion between

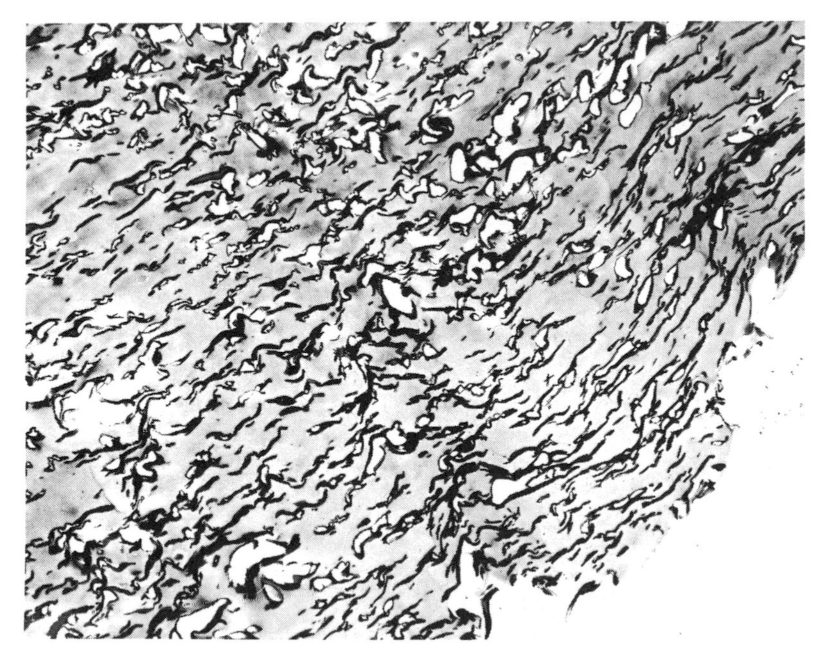

Fig. 8.2 Transmission electron micrograph taken in a plane through the process zone of a nitrile rubber–PBT blend: *x* plane, region 1.

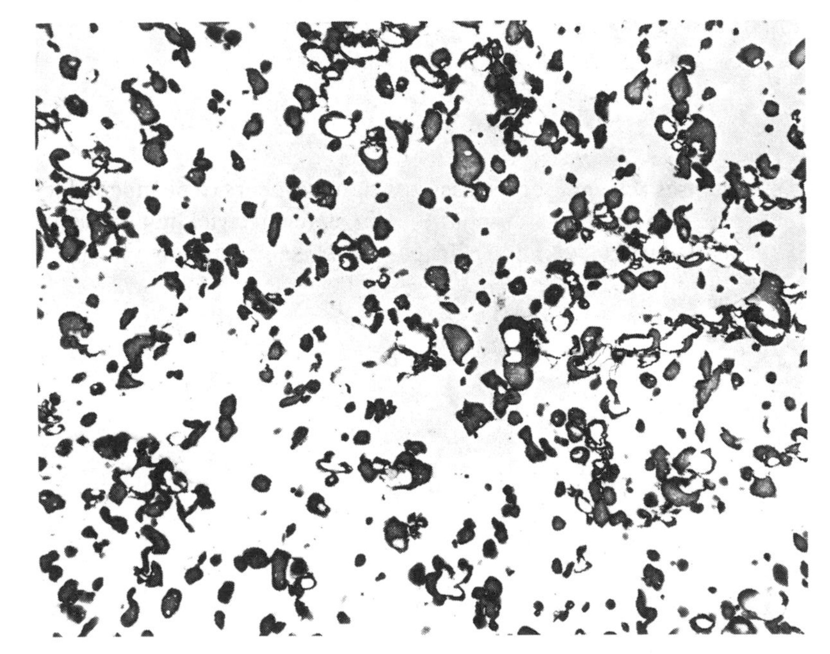

Fig. 8.3 Transmission electron micrograph taken in a plane through the process zone of a nitrile rubber–PBT blend: x plane, region 2.

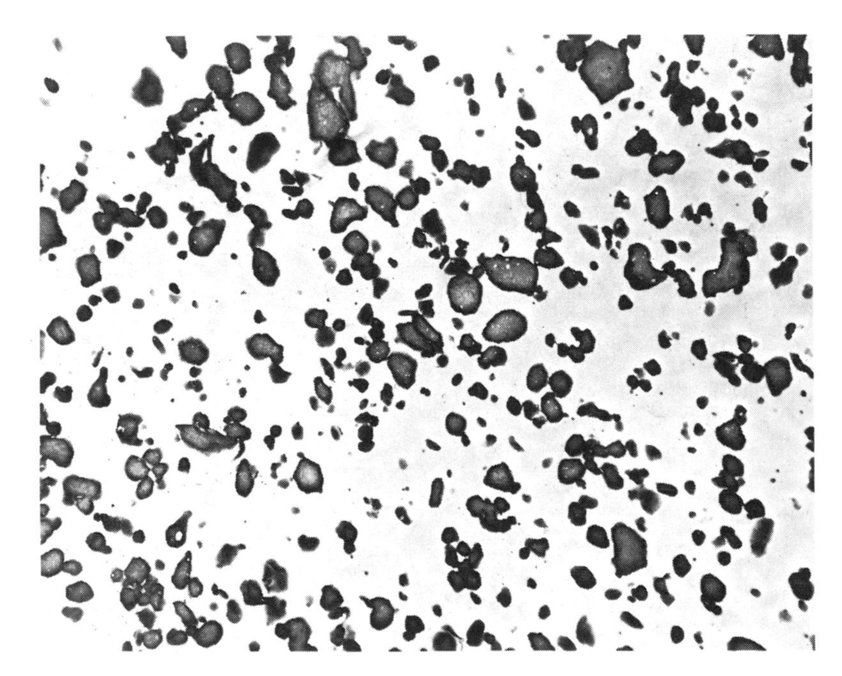

Fig. 8.4 Transmission electron micrograph taken in a plane through the process zone of a nitrile rubber–PBT blend: x plane, region 3.

Fig. 8.5 Transmission electron micrograph taken in a plane through the process zone of a nitrile rubber–PBT blend: y plane, region 1.

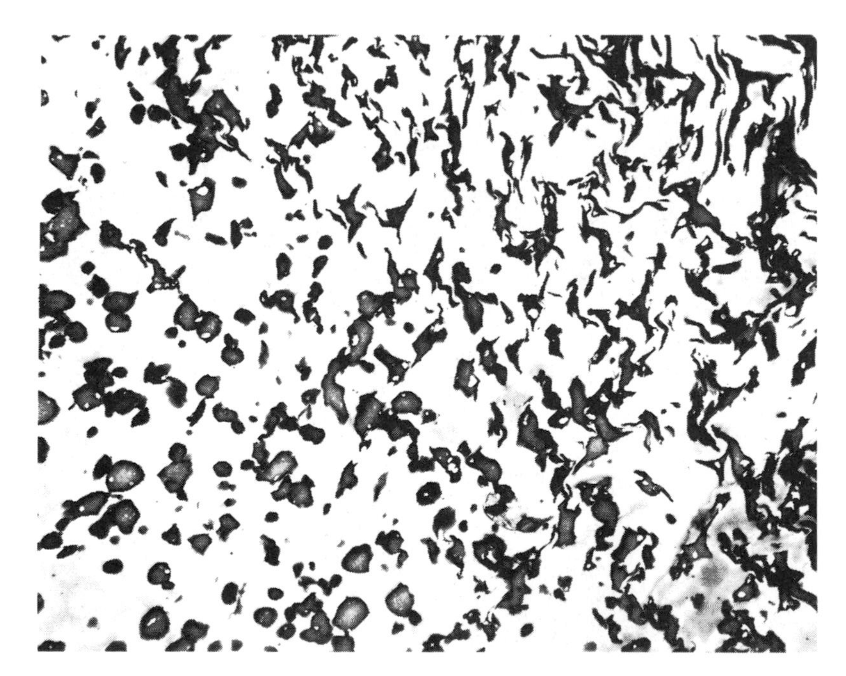

Fig. 8.6 Transmission electron micrograph taken in a plane through the process zone of a nitrile rubber–PBT blend: y plane, region 2.

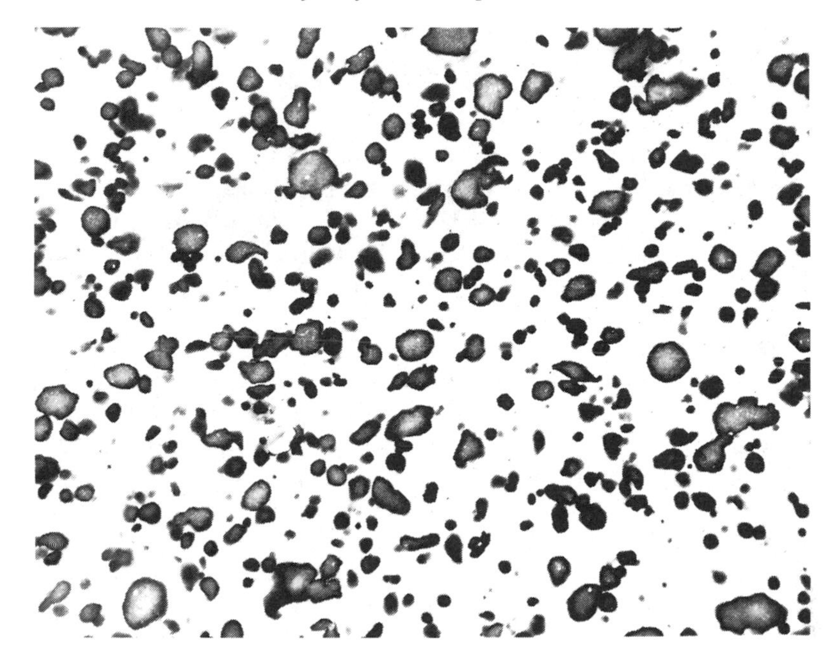

Fig. 8.7 Transmission electron micrograph taken in a plane through the process zone of a nitrile rubber–PBT blend: *y* plane, region 3.

Fig. 8.8 Transmission electron micrograph taken in a plane through the process zone of a nitrile rubber–PBT blend: *z* plane.

rubber particles and matrix without influencing the particle size of the dispersed phase. However, this was achieved with PBT blends which contained (a) an acrylonitrile–butadiene copolymer (28% nitrile content) and (b) an EPDM rubber and a small amount of the bismaleimide of diaminodiphenylmethane. In both blends, relatively large particles were produced after blending in a twin-screw extruder, and the impact strength, as determined by Charpy and Vu-Khanh [46] experiments, was low. However, when samples of these blends were post-condensed at 205 °C for 5 h before moulding, the impact properties were considerably improved despite there being no detectable difference in the morphology. This effect is not normally observed with rubber modified PBT or the pure polyester in which much smaller increases in impact strength are obtained after post-condensation. This suggests that the enhanced toughness cannot be attributed solely to the increase in molecular weight of the polyester which is known to occur during post-condensation. It is more likely to result from an increase in interfacial adhesion between the polyester and the rubber. Carboxylic acids are known to react with nitriles and it is possible that bismaleimides could react both with EPDM rubbers via a radical addition mechanism and with PBT via a condensation reaction.

These results are at variance with those of Wu [47] and others [48, 49] who believe that particle size is an overriding factor in determining impact strength. Indeed, Wu has suggested that the toughness of rubber modified thermoplastics increases as the ligament length (distance between adjacent rubber particles) is reduced, and that the interfacial adhesion has no influence above a critical value.

8.3 POLYCARBONATES

Polycarbonate (PC) polymers are used in situations where a high degree of toughness is required. The term polycarbonate, of course, covers a range of different materials, but the principal polycarbonate is that based on bisphenol A. This polymer, like both polystyrene and polymethyl methacrylate, is an amorphous, glassy material at ambient temperature. It scores over the other two materials in that it has a considerably reduced shear yield stress, with the consequence that it is a polymer with notable toughness under certain conditions, but with a tendency to brittle cracking in others. Dynamic mechanical analysis [50] and solid-state (magic angle) NMR spectroscopy [51] have been used to probe the sub T_g molecular motions in bisphenol A PC to seek the molecular origins of its outstanding toughness. References 52–56 are the fundamental studies of the fracture of bisphenol A PC. Zurimandi *et al.* [57] and others [58] have reported on the influence of internal stress on its fracture behaviour.

However, despite its inherent toughness, many successful attempts have been made to enhance its performance in this respect. Many polymers, singly and in combination, have been blended with PC and consequent property, including toughness, enhancements claimed. As the number of different polycarbonates studied is also considerable, simple classification of these materials is not possible. In this review, attention will be focused almost entirely on the polycarbonate of bisphenol A. This simplifies the data matrix, but it still remains extensive and complex. It is proposed to treat the field in the following way.

- The literature will be reviewed in terms of the principal toughening agent (polymer) added to the PC (bisphenol A PC unless otherwise stated). This will be done under the following headings:

- SBR and related materials;
- other rubbers;
- ABS;
- polyesters;
- polyamides;
- block copolymers;
- triblock copolymers;
- multicomponent copolymers;
- polyolefins;
- miscellaneous.

- The most important recent papers in this field will be reviewed.

8.3.1 SBR and related materials

Reports on the impact properties of polycarbonates containing SBR [59–64], SBS [65], hydrogenated SBS [66] and MBS [67] have been made. The final reference is to a material which did not break in a notched Izod impact test or in the unnotched double-gated Izod case.

8.3.2 Other rubbers

The toughening of polycarbonates with preformed rubbery particles has been reported by Riew and Smith [68]. They found that the polycarbonates deformed via shear yielding whether or not the rubbery particles were present, but toughness was enhanced by cavitation of the rubber in addition to the shear deformation process. Masafumi *et al.* [69] incorporated particulate rubbers (0.08–0.6 μm diameter) to achieve high notched Izod impact strengths. The rubbers were copolymers containing siloxane and acrylic components. The incorporation of acrylic core–shell latex particles has also been reported [70] to alter optical characteristics. The author reports that the modifier did not impair impact strength. The use of fluoroelastomers [71], acrylic rubbers [72, 73] and EPDM rubbers [74, 75] has been recorded.

8.3.3 ABS

Numerous reports [71, 76–83] occur of the incorporation of ABS into PC matrices and good impact strengths [78] achieved in this now commercial class of blends.

8.3.4 Polyesters

Blends of PCs with polyethylene terephthalate and other components [79, 84–92] and with polybutylene terephthalate and with polybutylene terephthalate plus other polymers [93–101] are well documented. Bertilsson, Franzen and Kubat [94] have studied the effects of aging, temperature and processing conditions on the impact properties of PC–PBT blends. Blends containing other polyesters [101–105] have also been studied.

8.3.5 Polyamides

Only very limited studies of PC–polyamide blends [106, 107] have been reported. Gambale *et al.* [106] have studied bisphenol A PC–nylon-6 blends with nylons varying in terminal NH_2 concentration. They found that lower NH_2 concentrations favoured good impact properties.

8.3.6 Block copolymers

Quite a wide range of copolymers, block, graft and random, have been compounded with polycarbonates to influence impact behaviour. Siloxane-containing copolymers [108–111] have been used with polysiloxane–polycarbonate block copolymers [109, 110] showing interesting property enhancements. Other copolymers reported are styrenic block copolymer rubbers [112], isoprene–methyl methacrylate graft copolymers [113] and maleic anhydride–styrene [114], ethylene–1-octene [115] and ethylene–ethyl acrylate [116–118] copolymers. A polyetheresteramide [119] has also been tried as a property-modifying agent.

8.3.7 Triblock copolymers

In addition to those SBS triblock materials mentioned earlier [65, 66], polyether–polycarbonate–polyether [120] and polybutadiene–polycaprolactone–polystyrene [121, 122] copolymers have been reported.

8.3.8 Multicomponent copolymers

Even more complex copolymers have been blended with PC. These include acrylonitrile–ethylene–ethylidene norbornene–styrene based graft copolymers [123] and butyl acrylate–butylene diacrylate–diallylmaleate–methyl methacrylate [124] and butadiene–divinylbenzene–methyl methacrylate–styrene [125] based copolymers.

8.3.9 Polyolefins

PC–polyolefin, especially polypropylene [99, 126–128], combinations have been investigated. In addition to polypropylene, linear low density polyethylene [129] and polyethylene [130] blends have been reported.

8.3.10 Miscellaneous

Many other homopolymers including EVA [131], polyurethane [132], phenoxy resin [133], polyetherimide (Ultem 70) [134, 135], polysulphone (Udel P1700) [136] and polystyrene [130] have been blended to varying degrees of advantage with PC. In the case of polystyrene [130], a marked drop in impact strength was reported.

8.3.11 Review of important recent papers

Gilmore and Modic [137] blended PC with 5 wt% of styrenic block copolymer rubbers to achieve improved impact resistance plus improvements in environmental stress cracking resistance and in heat-aging behaviour. They found that impact resistance was optimized when the block copolymer was present between 5 and 15 wt% and that impact behaviour improved as the polystyrene content of the block copolymer decreased. When the rubber exceeded 20 wt%, the dispersed rubbery phase changed to a semicontinuous morphology and ultimately resulted in a bicontinuous morphology.

Using notched impact tests, uniaxial tensile dilatometry [138–140] and TEM, Dekkers and coworkers [38, 39] have studied the morphology and fracture mechanisms in bisphenol A PC–PBT blends. They reported that bisphenol A PC–PBT blends can be made extremely tough by the incorporation of a core–shell impact modifier. The modifier, present in these studies at 15 wt%, consisted of latex particles of average diameter of 0.2 μm which have a rubbery core and a glassy shell. In these systems shear deformation was the major toughening mechanism and no significant degree of cavitation was observed over a wide range of temperature (−30 to 25 °C) for the blends containing only bisphenol A PC and PBT, indicating good interfacial adhesion between the two base polymers. When the modifier particles were present, extensive cavitation of the rubbery cores occurred.

Yee *et al.* [141] have reviewed the fracture behaviour of a number of two- or three-phase blends including PC–rubber and PC–ABS blends, with some success, and have applied finite element modelling to these systems.

Fujita *et al.* [142] report that both the stiffness and the toughness of ductile polymers can be enhanced by the incorporation of brittle polymer particles, citing the PC–acrylonitrile–styrene copolymer blends as an example. Their study has yielded six new ductile–brittle combinations which showed enhanced toughness. Their morphological investigations (both SEM and TEM) indicated a cold-drawing mechanism of the brittle particles which explains the improvement in toughness. Using the modified Eshelby theory [142], an analysis of stress concentration on these dispersed brittle particles was conducted. In an earlier paper [143], Koo, Inoue and Miyasaka have also reported on the PC–poly(acrylonitrile-*co*-styrene) blends which were originally discussed by Kurauchi and Ohta [144]. They [143] confirmed that toughening was achieved and proceeded to test ten other ductile–brittle combinations. These included PC–polymethyl methacrylate, PC–polystyrene and PC–polyphenylene sulphide. Only the PC–polymethyl methacrylate alloy was found to be toughened. It was concluded that the necessary conditions for toughening in such ductile–brittle blends were, firstly, significant differences in the Poisson's ratios and Young's moduli of the components which lead to high compressive stress on the occluded brittle particles. Secondly, this compressive stress should be greater than the brittle–ductile transition pressure of the brittle component in the blend and, thirdly, the occluded phases should be smaller than 1–2 μm.

Maxwell and Yee [145] have investigated PC–polyethylene (both high and low density) blends using tensile dilation measurements and transmission electron microscopy. They observed that both shear banding and voiding occurred simultaneously in these blends. The total degree of void formation increased as the strain rate increased and a sufficient degree of overlap of stress field occurred at only 4 wt% addition of the polyethylenes to modify the impact performance of the PC, i.e. to reduce its notch sensitivity.

8.4 POLYETHYLENE TEREPHTHALATE

Polyethylene terephthalate (PET) is, unlike PC, a semicrystalline thermoplastic, again, of wide utility. PET is reported [50] to have a K_{Ic} value of approximately 5 MPa m$^{\frac{1}{2}}$ at 20 °C. It is known [146, 147] that the fracture toughness of PET is improved by the incorporation of glass spheres of diameter 1–100 μm. It is believed [148] that this improvement results from a combination of crack pinning by the glass spheres and from enhanced plastic deformation of the PET matrix at the pinned crack. The chief factor controlling toughness is thought to be the distance between spheres rather than their size. Toughness appears to be at its maximum when there is poor adhesion between the glass spheres and the matrix polymer [146, 148]. This is attributed to more plastic deformation arising from increased stress concentration when voiding occurs.

For information on the mechanical properties of PET readers are directed to Refs. 149–151. There is relatively little work reported in the open literature on the toughening of PET using other polymers as the toughening agents.

There is a du Pont patent [152] outlining a process for fabricating tough PET articles with low gas and organic liquid permeabilities. The blends claimed contain 45–90% of PET and the toughening agents include ethylene–methacrylic acid (Zn or Mg salts) copolymers. Such blends have good drop impact strengths.

The most recent comprehensive account of PET toughening is by Wilfong, Hiltner and Baer [153]. They have investigated crystallization and irreversible deformation behaviour of PET blended with relatively low (1–10 wt%) concentrations of linear low density polyethylene, high density polyethylene, polypropylene and with poly(4-methylpentene-1). Three PET samples differing in molar mass were studied. The effects of these added polyolefins on nucleation, rate of crystallization and the strain rate dependence of toughening were investigated. Blending was achieved by melt extrusion. It was found that all the polyolefins were incompatible, yielding phase-separated morphologies. The polyolefin particulate phase did not nucleate crystallization of the PET matrix, but did depress the rate of crystallization. The authors observed decreases in the cold crystallization temperature of PET during heating in the blends with linear low density polyethylene and high density polyethylene which they ascribe to stress-induced crystallization created by large volume expansions associated with melting of the olefin particles. For both PET and the PET–polyolefin blends sharp transitions in fracture strain were reported with strain rate. At low strain rates fracture occurred during work hardening and as the strain rate increased the fracture event occurred during cold drawing. The transition corresponded with a decrease in draw stress, a decrease in draw ratio and an increase in the density of the neck. Wilfong, Hiltner and Baer [153] attributed the shift in the transition to higher strain rates for the blend compositions to increased rates of crystallization and orientation arising from strain-induced crystallization in the stress fields surrounding the occluded (polyolefin) phases. It is reported that, as the strain rate increased and the cold-drawing process became more adiabatic, the mechanical behaviour was governed by the kinetics of crystallization and orientation.

REFERENCES

1. Yokouchi, M., Sakakibara, Y., Chatani, Y., Tadokoro, H., Tanaka, T. and Yoda, K. (1976) *Macromolecules*, **9**, 266.
2. Ward, I. M. and Wilding, M. A. (1977) *Polymer*, **18**, 327.
3. Desborough, I. J. and Hall, I. H. (1977) *Polymer*, **18**, 825.
4. Horlbeck, G. and Bittscheidt, J. (1985) Ger. Patent 3,328,568.
5. Kasuga, T., Takahashi, K. and Tajima, Y. (1985) Eur. Patent Appl. 142,336.
6. Takahashi, K. and Tajima, Y. (1986) Jpn. Kokai Tokkyo Koho 61,174,253.
7. Inoue, K. (1986) Jpn. Kokai Tokkyo Koho 61,271,348.
8. Yonetani, K., Horiuchi, K. and Yamanaka, T. (1987) Jpn. Kokai Tokkyo Koho 62,285,947.

9. Wang, I. and Chung, W. (1988) US Patent 4,753,986.
10. Binsack, R., Rempel, D., Humme, G. and Ott, K. (1981) Ger. Patent 2,927,576.
11. Koehler, K. H., Lindner, C., Rempel, D., Weber, G., Ott, K. H. and Binsack, R. (1986) Ger. Patent 3,422,862.
12. Bier, P. and Lindner, C. (1982) Eur. Patent Appl. 63,263.
13. Blaschke, F., Klinkenberg, H. and Vollkomer, N. (1982) Ger. Patent 3,118,017.
14. Chiolle, A. and Andreoli, G. (1987) Eur. Patent Appl. 248,352.
15. Mitsubishi Rayon Company (1984) Jpn. Kokai Tokkyo Koho 59,138,256.
16. Bronstert, K., Schwaben, H. D. and Echte, A. (1988) Ger. Patent 3,706,017.
17. Lee, Y. D., Ju, S. J. and Chang, S. (1982) *J. Chin. Inst. Chem. Eng.*, **13**, 59.
18. Laurienzo, P., Malinconico, M., Martuscelli, E. and Volpe, M. G. (1989) *Polymer*, **30**, 835.
19. DiLiello, V., Laurienzo, P., Malinconico, M., Martuscelli, E., Ragosta, G. and Volpe, M. G. (1990) *Angew. Makromol. Chem.*, **174**, 141.
20. Sheer, M. L. (1982) US Patent 4,317,764.
21. Hourston, D. J., Koetsier, D. W., Lane, S. and Zhang, H. X. (1990) Proc. 33rd IUPAC Symp. on Macromolecules, Montreal.
22. Droescher, M., Gerth, C. and Bornschlegl, E. (1986) Ger. Patent 3,510,409.
23. Droescher, M., Gerth, C. and Burzin, K. (1985) Ger. Patent 3,306,008.
24. Cecere, A., Greco, R., Ragosta, G., Scarinzi, G. and Taglialatela, A. (1990) *Polymer*, **31**, 1239.
25. Atlantic Richfield Company (1986) Jpn. Kokai Tokkyo Koho 6,137,838.
26. Greco, R., Musto, P., Ragosta, G. and Scarinzi, G. (1988) *Makromol. Chem., Rapid Commun.*, **9**, 129.
27. Wefer, J. M. (1984) US Patent 4,485,212.
28. Yamanaka, T., Yonetani, K. and Inoue, S. (1987) Jpn. Kokai Tokkyo Koho 62,292,849.
29. Hourston, D. J., Lane, S. and Zhang, H. X. (1988) Proc. 20th Europhysics Conf. on Macromolecular Physics, Lausanne.
30. Hourston, D. J., Lane, S. and Zhang, H. X. (1990) Proc. 3rd Eur. Symp. on Polymer Blends, Cambridge.
31. Bootsma, J. P. C., Hourston, D. J., Koetsier, D. J., Lane, S. and Zhang, H. X. (1991) *Polymer*, **32**, 1140.
32. Pilati, F. (1983) *Polym. Eng. Sci.*, **23**, 750.
33. Pilati, F. and Pezzin, G. (1984) *Polym. Eng. Sci.*, **24**, 618.
34. Blaschke, F., Gebauer, P. and Vollkommer, N. (1982) Ger. Patent 3,150,957.
35. Mitsubishi Company (1982) Jpn. Kokai Tokkyo Koho 57,137,347.
36. Neuray, D., Nouvertne, W., Binsack, R., Rempel, D. and Mueller, P. R. (1982) Eur. Patent Appl. 64,648.
37. Verhoeven, J. J. and Pearson, R. A. (1987) Eur. Patent Appl. 239,157.
38. Hobbs, S. Y., Dekkers, M. E. J. and Watkins, V. H. (1988) *J. Mater. Sci.*, **23**, 1219.
39. Hobbs, S. Y., Dekkers, M. E. J. and Watkins, V. H. (1988) *J. Mater. Sci.*, **23**, 1225, and references therein.
40. Horlbeck, G. and Mumcu, S. (1985) Ger. Patent 3,328,567.
41. Toyobo Company (1985) Jpn. Kokai Tokkyo Koho 59,157,147.
42. Mobay Company (1981) Can. Patent 1,111,984.
43. Yates, J. B. and Ullman, T. J. (1988) US Patent 4,717,751.
44. Cimmino, S., D'Orazio, L., Greco, R., *et al.* (1984) *Polym. Eng. Sci.*, **24**, 48.
45. Hourston, D. J., Lane, S. and Zhang, H. X. (1991) *Polymer*, **32**, 2215.
46. Vu-Khanh, T. (1988) *Polymer*, **29**, 1979.
47. Wu, S. (1985) *Polymer*, **26**, 1855.
48. Borggreve, R. J. M., Gaymans, R. J., Schuijer, J. and Ingen Housz, J. F. (1987) *Polymer*, **28**, 1489.

49. Sjoerdsma, S. D. (1989) *Polymer*, **30**, 106.
50. Varadarajan, K. and Boyer, R. F. (1980) *Org. Coat. Plast. Chem.*, **42**, 689.
51. Steter, T. R., Schaefer, J., Stejskal, E. O. and McKay, R. A. (1980) *Macromolecules*, **13**, 1127.
52. Williams, J. G. (1984) *Fracture Mechanics of Polymers*, Ellis Horwood, Chichester.
53. Marshall, G. P., Culver, L. E. and Williams, J. G. (1973) *Int. J. Fract.*, **9**, 295.
54. Rikpur, K. and Williams, J. G. (1979) *J. Mater. Sci.*, **14**, 467.
55. Fraser, R. A. W. and Ward, I. M. (1978) *Polymer*, **19**, 220.
56. Newmann, L. V. and Williams, J. G. (1980) *Polym. Eng. Sci.*, **20**, 572.
57. Zurimandi, J. A., Biddleston, F., Hay, J. N. and Haward, R. N. (1982) *J. Mater. Sci.*, **17**, 199.
58. Thakkar, B. S., Barutman, L. J. and Kalpakjian, S. (1980) *Polym. Eng. Sci.*, **20**, 756.
59. Liu, P. Y. and Belfoure, E. L. (1986) US Patent 4,579,903.
60. Lee, G. F. (1985) Eur. Patent 141,197.
61. Lordi, F. E. (1984) US Patent 4,469,843.
62. Liu, P. Y. (1984) US Patent 4,424,303.
63. Liu, P. Y. (1982) Eur. Patent 59,375.
64. Liu, P. Y. (1981) UK Patent 2,057,461.
65. Lo, L. Y. and Boulier, P. R. (1989) US Patent 4,866,125.
66. Sederel, W. L. (1986) Eur. Patent 173,358.
67. Liu, P. Y. (1984) Eur. Patent 111,179.
68. Riew, C. K. and Smith, R. W. (1989) *Adv. Chem. Ser.*, **222**, 225–41.
69. Masafumi, H., Shigemitsu, H., Yamamoto, N. and Yanagase, A. (1989) Eur. Patent 307,963.
70. Crook, E. H. (1988) *Res. Discl.*, **292**, 597.
71. Choudhary, V. and Mehta, R. (1989) *Polym. Mater. Sci. Eng.*, **60**, 866.
72. Witman, M. W. and Reinert, G. E. (1986) US Patent 4,563,503.
73. O'Connell, W. J. J. (1980) PCT Int. Appl., 00,154.
74. Boutni, O. M. and Michel, R. L. (1985) Eur. Patent 131,188.
75. Wefer, J. M. (1984) Eur. Patent 107,303.
76. White, R. J. and Jrishnan, S. (1989) US Patent 4,837,243.
77. Grigo, U. R., Lazear, N. R and Witman, M. W. (1987) US Patent 4,677,162.
78. Weber, C. A. and Paige, W. P. (1986) US Patent 4,624,986.
79. Rawlings, H. L. and Reinert, G. E. (1984) US Patent 4,487,881.
80. Feay, D. C., Jeanes, T. A. and Wong, K. L. (1982) *Res. Discl.*, **217**, 146.
81. Parsons, C. F. (1989) Eur. Patent 331,970.
82. Whalen, D. and Jalbert, R. L. (1989) US Patent 4,855,357.
83. Fujimori, Y., Takabori, Y., Sakano, H. and Ito, A. (1983) Eur. Patent 74,112.
84. Wefer, J. M. (1988) Eur. Patent 287,207.
85. Belfoure, E. L. and Liu, P. Y. (1989) *Can. Chem. Abstr.*, 1,261,504.
86. Chung, J. Y. J., Neuray, D. and Witman, M. W. (1985) US Patent 4,554,314.
87. Liu, P. Y. (1985) Eur. Patent 141,268.
88. Neuray, D., Nouvertne, W., Binsack, R., Rempel, D. and Meuller, P. R. (1984) US Patent 4,482,627.
89. Endo, H. and Ishii, T. (1984) Eur. Patent 122,601.
90. Bussink, J. and Heuschen, J. M. H. (1981) US Patent 4,267,096.
91. Giles, H. F. (1984) Eur. Patent 105,244.
92. Allen, R. B., Giles, H. F., Heuschen, J. M. H. and Wiercinski, R. A. (1984) Eur. Patent 107,048.
93. Boutni, O. M. (1989) Eur. Patent 320,651.
94. Bertilsson, H., Franzen, B. and Kubat, J. (1988) *Plast. Rubber Process. Appl.*, **10**, 137.

95. Boutni, O. M. and Liu, P. Y. (1984) Eur. Patent 110,222.
96. Cohen, S. C. and Dieck, R. L. (1982) *Can. Chem. Abstr.*, 1,123,534.
97. Liu, P. Y. (1982) US Patent 4,320,212.
98. Wambach, A. D. and Dieck, R. L. (1980) PCT Int. Appl., 30.
99. McHale, A. H. and Peascoe, W. J. (1988) Eur. Patent 272,425.
100. Cartasengna, S. (1988) Eur. Patent 272,857.
101. De Rudder, J. L. (1988) Eur. Patent 273,151.
102. Boutni, O. M. (1986) US Patent 4,628,074.
103. Liu, P. Y. (1986) US Patent 4,604,423.
104. Igi, K., Okamura, T., Taniguchi, S., Ishii, M., Murata, Y., Yokota, S., Matsumoto, T., Endo, H. and Hashimato, K. (1986) Eur. Patent 186,089.
105. Luce, J. B. (1986) Eur. Patent 188,791.
106. Gambale, R. J., Maresca, L. M., Clagett, D. C. and Shafer, S. J. (1988) Eur. Patent 285,693.
107. Maresca, L. M. (1987) Eur. Patent 227,053.
108. Rock, J. A. (1989) Eur. Patent 303,843.
109. De Boer, J. and Heuschen, J. M. H. (1988) US Patent 4,788,252.
110. Marke, G. F. (1980) US Patent 4,224,215.
111. Robeson, L. M. and Matzner, M. (1983) Eur. Patent 73,067.
112. Gilmore, D. W. and Modic, M. J. (1989) *Plast. Eng.*, **45**, 51.
113. Endo, H., Hashimoto, K. and Kato, K. (1989) Eur. Patent 297,517.
114. Hansen, M. G. and Bland, D. G. (1985) *Polym. Eng. Sci.*, **25**, 896.
115. Heinert, D. H. (1981) *Res. Discl.*, **208**, 309.
116. Lui, P. Y. and Ishihira, T. (1988) US Patent 4,735,993.
117. Rosenquist, N. R. and Tyrell, J. A. (1985) US Patent 4,496,693.
118. Overton, D. E. and Liu, P. Y. (1984) Eur. Patent 119,531.
119. Witman, M. W. (1985) Eur. Patent 149,091.
120. Priddy, D. B. and Schroeder, J. R. (1989) US Patent 4,812,514.
121. Liu, P. Y. (1981) US Patent 4,255,534.
122. Liu, P. Y. (1981) US Patent 4,251,647.
123. Paddock, C. F. and Wefer, J. M. (1985) US Patent 4,550,138.
124. Liu, P. Y. and Rosenquist, N. R. (1984) US Patent 4,456,725.
125. Falk, J. C., Narducy, K. W., Cohen, M. S. and Brunner, R. (1980) *Polym. Eng. Sci.*, **20**, 763.
126. Fujita, Y., Sezume, T., Kitano, K., Narukawa, K., Mikami, T., Kawamura, T., Sato, S., Nishio, T., Yokoi, T. and Nomura, T. (1989) Eur. Patent 308,179.
127. Tergen, W. P. and Davison, S. (1981) UK Patent 1,596,711.
128. Liu, P. Y. (1981) US Patent 4,245,058.
129. Boutni, O. M. (1988) Eur. Patent 266,596.
130. Kunori, T. and Geil, P. H. (1980) *J. Macromol. Sci. Phys.*, **18**, 135.
131. Liu, P. Y. and Boutni, O. M. (1988) Eur. Patent 258,663.
132. Boutni, O. M. (1988) Eur. Patent 265,790.
133. Giles, H. F. and Liu, P. Y. (1986) US Patent 4,607,079.
134. Rock, J. A., Durfee, N. E. and Johnson, R. O. (1986) Eur. Patent 186,243.
135. Mellinger, G. A., Giles, H. F., Holub, F. F. and Schlich, W. R. (1984) PCT Int. Appl., 32.
136. Quinn, C. D. and Rosenquist, N. R. (1982) US Patent 4,358,569.
137. Gilmore, D. W. and Modic, M. J. (1989) *Plast. Eng.*, **45**, 51.
138. Heikens, D., Sjoerdsma, S. D. and Coumans, W. J. (1981) *J. Mater. Sci.*, **16**, 429.
139. Yee, A. F. and Pearson, R. A. (1986) *J. Mater. Sci.*, **21**, 4262.
140. Dekkers, M. E. J. and Heikens, D. (1985) *J. Appl. Polym. Sci.*, **30**, 3289.
141. Yee, A. F., Parker, D. S., Sue, H. J. and Huang, I. C. (1987) *Polym. Mater. Sci. Eng.*, **57**, 417.

142. Fujita, Y., Koo, K. K., Angola, J. C., Inoue, T. and Saki, T. (1986) *Kobunshi Ronbunshu*, **43**, 119.
143. Koo, K. K., Inoue, T. and Miyasaka, K. (1985) *Polym. Eng. Sci.*, **25**, 741.
144. Kurauchi, T. and Ohta, T. (1986) *J. Mater. Sci.*, **21**, 214.
145. Maxwell, M. A. and Yee, A. F. (1981) *Polym. Eng. Sci.*, **21**, 205.
146. Broutman, L. J. and Sahu, S. (1971) 26th Proc. Annu. Conf. SPI, Paper 14-C, p. 1.
147. Leider, J. and Woodhams, R. T. (1974) *J. Appl. Polym. Sci.*, **18**, 1639.
148. Brown, S. K. (1980) *Br. Polym. J.*, **12**, 24.
149. Presad, P. B. S. (1984) *Pop. Plast.*, **29**, 17.
150. Callander, D. D. (1985) *Polym. Eng. Sci.*, **25**, 453.
151. Presad, P. B. S. (1985) *Pop. Plast.*, **30**, 16.
152. Subramanian, P. M. (1987) Eur. Patent 211,649.
153. Wilfong, D. L., Hiltner, A. and Baer, E. (1986) *J. Mater. Sci.*, **21**, 2014

9

Toughened polysulphones and polyaryletherketones

G. W. Wheatley and D. Parker

9.1 INTRODUCTION

The aromatic polysulphones, bisphenol-A polysulphone (Union Carbide 'Udel Polysulphone') [1] and polyethersulphone (ICI 'Victrex PES') [2], and polyetheretherketone (ICI 'Victrex PEEK') [2, 3] are among the most commercially important examples of high performance engineering thermoplastics. The term 'engineering thermoplastic' has come to refer to those thermoplastics with a combination of good mechanical properties and excellent resistance to thermal degradation and chemical attack which allows them to compete with metals in low to medium temperature applications. Figure 9.1 illustrates the structure of the repeat unit in these polymers and Table 9.1 summarizes some key properties of Udel polysulphone and Victrex PES and PEEK.

The polysulphones and PEEK all have a linear aromatic backbone and it is this which confers excellent chemical and thermal stability on the polymers. The aliphatic side chains in Udel polysulphone do lower the thermo-oxidative stability of the material slightly with respect to the other wholly aromatic polymers but it is still possible to melt process the polymer on conventional equipment without problems [1]. The rigidity of the *para*-linked aromatic backbone coupled with strong dipolar forces is responsible for the high glass transition temperatures of the polymers. Both polysulphones exist as amorphous glasses and the upper continuous temperature limit for their use is essentially limited by the glass transition temperature (T_g) above which key mechanical properties such as modulus show a rapid decline. Commercial grades of PEEK develop significant degrees of crystallinity, typically 30–35%, owing to the combination of the polar character of the carbonyl and ether groups in the backbone and the crystallographic equivalence of these groups [4]. It is this property that is responsible for the outstanding high temperature performance of PEEK with useful mechanical properties being retained up to

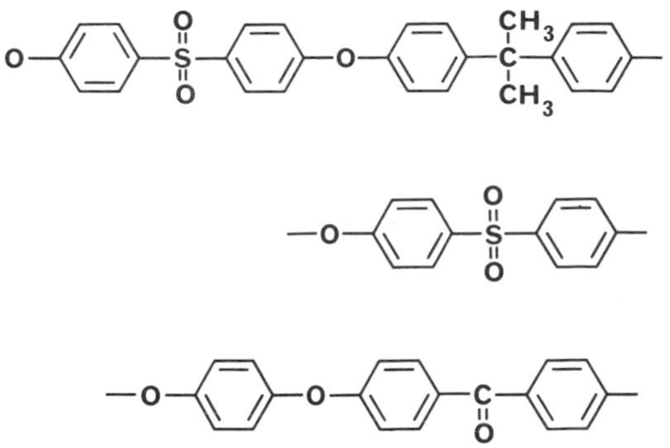

BISPHENOL - A POLYSULPHONE

PES

PEEK

Fig. 9.1 Structures of commercially important polysulphones and polyaryletherketones.

Table 9.1 Key properties of commercial polysulphones and polyaryletherketones

	Polysulphone P1700	*PES 4100G*	*PEEK 450G*
T_g (°C)	185	223	143
T_m (°C)	–	–	334
Heat distortion temperature (°C) (D648)	174	203	160
Tensile strength ($MN\,m^{-2}$) (D638)	70	84	92
Flexural modulus ($GN\,m^{-2}$) (D790)	2.7	2.6	3.7
Izod impact strength ($J\,m^{-1}$) (D256)			
Unnotched	No break	No break	No break
Notched	69	83	83

Sources: Union Carbide Co., *Udel Polysulphone: An Outstanding Engineering Polymer for Moulding and Extrusion*, 1978; R. B. Rigby, *Polymer News*, **9**, 325, 1984; C. P. Smith, *Swiss Plastics*, **4**, 37, 1981.

temperatures approaching the melting temperature of the crystallites >573 K. The crystallinity is also largely responsible for the excellent resistance to attack by solvents exhibited by PEEK [2].

The polysulphones and PEEK are all synthesized on a commercial scale by the nucleophilic attack of the anion of the appropriate bisphenol on an activated aromatic dihalide (dihalobenzophenones or dihalodiphenylsulphones) in a dipolar aprotic solvent [5–7]. Early attempts to synthesize these materials focused on electrophilic polysulphonylation [8] or polyacylation [9, 10] reactions in a strong acid but these routes were abandoned owing to the toxic and corrosive nature of the reaction medium and the introduction of structural irregularities in the backbone which had an adverse effect on properties. The latest addition to this family of thermoplastics, polyetherketoneketone (PEKK), developed by DuPont but not fully commercialized at the time of writing of this review, is thought to be synthesized by a polyacylation type route.

A major failing of these polymers is that the presence of a defect or a moulded-in sharp corner or notch dramatically reduces the impact strength of the material (Table 9.1). The nature of many of the potential applications of these materials, for example as structural components in aerospace and public transport applications, makes notch behaviour particularly damaging to their performance profile and is an important limiting factor on their further penetration of key market areas. Significant improvements in the fracture behaviour and toughness of thermoplastics can be achieved by blending with a suitable elastomer [11, 12].

A full discussion of the theories developed to explain the phenomenon of rubber toughening is covered earlier in this volume. Other excellent and comprehensive reviews are also available [13, 14]. It is valuable at this point, however, to consider briefly the relationships between a particular characteristic of a blend and the toughness as this will prove useful later in interpreting the behaviour of the blend systems examined.

There is general agreement that the rubber toughening of polymer blends is dependent on the separation of the components into a stable two-phase morphology with the elastomer dispersed as small discrete particles in a continuous thermoplastic matrix. It is thought that the absorption of impact energy that accounts for the increased toughness of the blend occurs principally in the polymer constituting the matrix phase. Energy absorbed by deformation of the rubber particle is thought to be only a minor process contributing to toughness. The matrix polymer may absorb impact energy either by the formation of crazes (Ref. 15 and Chapter 1), regions of plastically deformed highly oriented material that resemble microcracks but are in fact load bearing [16], or by shear yielding (Ref. 17 and Chapter 1). Both of these processes are believed to be initiated at the particle–matrix interface. The major role of the rubber particles is thought to be the control of these processes by the provision of a large number of stress concentrations where localized deformation is initiated and possibly also terminated. It is found that optimum particle sizes exist for the initiation of crazing and the initiation of shearing [18] and the particle size is important in determining not only the magnitude of the toughening but also which toughening process operates. The optimum size for rubber particles for the initiation of crazing (diameter >1000 nm) is typically larger than the optimum size for the initiation of shearing (diameter 200–500 nm depending on the system).

Several important considerations arise from the function of the rubber particles as stress concentrators. For stress concentration at the particle–matrix interface to be maximized there must be a large difference in the modulus of the rubber particle and the matrix. Conventional elastomers of course typically have a much lower modulus than thermoplastics, particularly the high performance thermoplastics considered in this review. The requirement for low modulus is also closely connected with another accepted requirement that the particles must possess a low glass transition temperature (T_g), i.e. they must show elastomeric behaviour [19, 20]. T_g of the elastomer composing the rubber particles must be as low as possible and not merely below the minimum temperature at which the blend is expected to be employed. The time–temperature superposition principle predicts that under the very high strain rates encountered in an impact event T_g of the particle must be as low as possible to allow it to respond as an elastomer and not a glass over the lifetime of the impact event (Ref. 14, pp. 292–3). Excessive crosslinking within the rubber particles should be avoided as this will both increase T_g and increase modulus.

Equally critical for stress concentration is good adhesion between the rubber particles and the matrix [21, 22]. As adhesion normally results from partial mixing of the phases [23] the absolute requirement for phase separation and the need for adhesion will clearly be to some extent conflicting requirements in a toughened polymer blend. It is this requirement for good adhesion between the phases in the blend that is commonly the most difficult to achieve and hence the limiting factor on blend performance. Therefore the interrelated concepts of miscibility and interfacial adhesion in polymer blends requires further discussion.

The most widely accepted theories of polymer mixing are due to Scott [23]. An expression for the entropy of mixing of two non-identical homopolymers was developed using the lattice model of Flory and Huggins:

$$-\Delta S_{\text{mix}} = (-RV/V_r)(\phi_1 \ln \phi_1/x_1 + \phi_2 \ln \phi_2/x_2)$$

where V is the volume of the mixture, ϕ_1 the volume fraction of polymer 1, x_1 the degree of polymerization of polymer 1 and V_r the volume per monomer repeat unit. This predicts that as the molecular weight of the polymers (i.e. the degree of polymerization) increases the combinatorial entropy change on mixing decreases – as a consequence of there being fewer species and hence fewer potential ways of occupying the lattice. This entropy change will always be favourable. (It should be noted that this simple treatment ignores the possibility of interactions between non-identical neighbouring groups contributing to the total entropy of mixing).

The implications of this in the context of polymer blends are that low molecular weight polymers and oligomers will show greater interfacial mixing and hence interfacial adhesion than their high molecular weight analogues as the contribution of the favourable entropy of mixing to the free energy change of mixing is maximized.

The enthalpy change on mixing for two non-identical homopolymers is basically due to the replacement of nearest-neighbour contacts with identical species in the pure homopolymer by nearest-neighbour contacts with unlike species (i.e. segments of the other polymer chain) in the mixture. This is expressed in the following equation:

$$\Delta H_{mix} = B\phi_1\phi_2$$

where ϕ_1 is the volume fraction of polymer 1 and B is the interaction energy characteristic of the interactions between polymer segments in the blend. For species that do not interact strongly this enthalpy change is predicted to be endothermic by analogy with the Van Laar expression for a two-component mixture. The magnitude of the enthalpy change will depend on how different the interactions between polymer 1 and polymer 2 are compared with the interactions between chains of the same homopolymer.

This theory was further developed by Scott in collaboration with Hildebrand [24] to produce the expression

$$\Delta H_{mix} = V(\delta_1 - \delta_2)^2\phi_1\phi_2$$

where V is the total volume of the polymer mixture. δ is the solubility parameter of the polymer, as explained in Chapter 2. It is a measure of the intermolecular cohesive forces of the polymer. Hence polymers containing highly polar groups giving rise to strong dipolar interactions between chains will be characterized by high δ values. Udel polysulphone, PES and PEEK all have high δ values. These are respectively 10.6, 12.3 and 11 $(\text{cal cm}^{-3})^{1/2}$ (21.6, 25.1 and 22.5 $(\text{MJ m}^{-3})^{1/2}$).

This theory predicts that polymers with solubility parameters that are very close will show the greatest tendency to mix, all other factors such as molecular weight (MW) of the polymers being equal. The matching of the solubility parameters of the polymers in a mixture correlates to minimizing the $\delta_1 - \delta_2$ term and hence restricting the unfavourable enthalpy of mixing to a small value which allows the favourable entropy of mixing contribution to dominate the free energy expression and hence to produce miscibility. The implications of this are as follows. When the δ values of high molecular weight polymers are very different an immiscible blend with a tendency to phase separation on a macroscopic scale will be produced on mixing while the mixing of high MW polymers with near-identical δ values will produce a single-phase miscible blend. At intermediate values of the difference in the δ values of the polymers the blend will exist as a two-phase system but some limited mixing of the phases at the interface will occur. Such a blend shows no tendency toward gross phase separation as the limited mixing produces significant interfacial adhesion and is said to be compatible. The smaller the difference in the δ values for pairs of polymers in a blend is, the greater is the tendency toward interfacial mixing.

It should be noted that this simple approach fails to take into account the possibility of a favourable enthalpy of mixing leading to miscibility. This may arise when favourable 'specific interactions' between functional groups on the two polymers, such as H bonding or dipole–dipole interactions, exist.

This treatment of the problem has given a qualitative insight into the factors which will favour phase mixing in a polymer blend. Attempts have also been made to quantify the conflicting requirements for phase separation and interfacial adhesion in a multiphase blend. Theoretical calculations by Stehling *et al.* [25] have shown that for a simple binary blend of homopolymers with molecular weights of about 10^5 the degree of interfacial mixing necessary to produce adequate adhesion between the phases when the minor component content is 5–10% results only when $\delta_1 - \delta_2 < 0.8\,(\mathrm{MJ\,m^{-3}})^{1/2}$. Under the same conditions $\delta_1 - \delta_2 > 0.4\,(\mathrm{MJ\,m^{-3}})^{1/2}$ for a two-phase system to be produced.

The highly aggressive service environments in which the polysulphones and polyetheretherketone are commonly employed and their high processing temperatures dictate that any elastomer incorporated into a rubber toughened grade of these polymers must exhibit exceptional thermo-oxidative stability and resistance to chemical attack. This basically limits the choice of a conventional elastomer to either a polysiloxane 'silicone rubber' or one of a limited number of fluoropolymer elastomers. For blending with PEEK even these materials might be insufficiently stable. Certain types of acrylate rubbers might also be suitable for incorporation in the polysulphone blends that can be processed at moderate temperatures and would be employed in less severe service environments. While not conventionally considered an elastomer polyethylene has a low T_g and a relatively low modulus coupled with reasonable thermal stability (particularly linear high density polyethylene) and good chemical resistance and therefore might be used successfully to toughen blends.

The δ values for these types of polymer are fairly low with $\delta = 15.0\,(\mathrm{MJ\,m^{-3}})^{1/2}$ for polydimethylsiloxane, the prototypic silicone rubber, $\delta = 16.4\,(\mathrm{MJ\,m^{-3}})^{1/2}$ for polyethylene and δ in the range 12.3–14.3 $(\mathrm{MJ\,m^{-3}})^{1/2}$ for fluoroelastomers depending on the percentage and type of comonomers. The acrylate rubbers tend to have higher δ values than these, particularly those containing nitrile groups, but these are still generally $< 20\,(\mathrm{MJ\,m^{-3}})^{1/2}$ for most common types. Comparison of these data with the δ values for the polysulphones and PEEK presented earlier shows the difference in δ values for combinations of thermoplastic and suitable elastomer to be large. This implies phase separation will be complete and interfacial adhesion will be minimal in simple blends of these homopolymers of moderate molecular weight and above. As will be evident from the results presented later for specific examples of blend systems this prediction is largely borne out.

The problem of poor interfacial adhesion in blends of thermoplastics and elastomers has previously been eased by the addition of a suitable block or graft copolymer [26]. These copolymers are known as compatibilizers. The addition of a compatibilizer improves the interfacial adhesion by locating preferentially at the interface with the different block types solubilized in the thermoplastic matrix and the elastomeric domains respectively, and hence anchoring together the phases [27]. The segment types of the copolymer are chosen to be chemically identical to the thermoplastic and elastomer phases to facilitate their mixing with these phases. The crosslinking between phases encouraged in the production of many commercial rubber toughened blends can be thought of to some extent as the *in situ* formation of block and graft copolymer compatibilizer.

The ultimate extension of the above approach to the preparation of toughened blends is to dispense with an individual elastomer component completely and to use only a block or graft copolymer to supply the elastomer content [28, 29]. The rubber particles are then formed by the phase separation of the copolymer and the aggregation of the elastomer segments. This should lead to excellent interfacial adhesion and hence mechanical properties. The problems of demixing of the components and excessive disruption of elastomer domains during high shear stages of melt processing should also be ameliorated.

The drawback to this approach is that the cost of a block or graft copolymer when it must be prepared separately and not *in situ* usually greatly exceeds that of a simple homopolymer elastomer. This is particularly marked in the case of block copolymers which are frequently prepared by rather complex methods. However, the synthetic techniques employed for the preparation of block copolymers, normally either a sequential anionic polymerization [30] or the linking of mutually reactive preformed oligomers [31], allow very exact control over molecular architecture and molecular weight of the blocks. These attributes are very important in determining morphology and domain dimensions, and therefore the ultimate mechanical properties of the blend, and consequently block copolymers are increasingly preferred for this application owing to the greater capability to tailor precisely the performance they afford.

This use of a block copolymer containing elastomeric segments as an 'impact modifier' of a thermoplastic is particularly suitable for the polysulphones and PEEK for which simple physical blends with elastomers might be expected to display poor interfacial adhesion for the reasons discussed previously, and for which the high value–high performance applications of the materials can tolerate the higher cost.

In addition to all the above effects the characteristics of the blend and its components can exert, it is crucial not to underestimate the effect of practical considerations concerned with the processing of the blend on the ultimate mechanical properties achieved. Even when the composition of the blend is such that potentially a fine dispersion of well-adhering rubber particles could be produced, in practice inefficient mixing of the blend components might prevent this and result in poor mechanical properties.

This review will be confined to blends based on the commercially available polysulphones and polyaryletherketones, namely Udel polysulphone, Victrex PES and Victrex PEEK. Blends based on experimental polysulphones and polyaryletherketones have been deliberately omitted. The work is not intended to be an exhaustive review of the literature but rather an analysis of the most important and most informative examples of toughened blends based on each type of thermoplastic.

Before embarking on a consideration of the results for blends based on each thermoplastic it is worth establishing some terminology conventions. All δ values for polymers are obtained either from the *Polymer Handbook* [32] or are calculated using the most commonly accepted method due to Small [33]. (These older papers quote δ in $(\mathrm{cal\,cm^{-3}})^{1/2}$. This may be converted to SI units, namely $(\mathrm{MJ\,m^{-3}})^{1/2}$, by multiplying by $\sqrt{4.18}$.) All blend compositions are expressed as per cent by weight unless otherwise specified. Impact strengths are almost invariably quoted relative to the value for an unmodified homopolymer control and therefore these have been left in the units reported by the original authors. Indeed, as impact strength is substantially affected by the test method, processing conditions and characteristics of the base polymer, such as molecular weight, which vary according to commercial grade, meaningful 'absolute' impact strength values are difficult to obtain.

9.2 TOUGHENED UDEL POLYSULPHONE BLENDS

As would be expected for a polymer with such outstanding performance apart from the tendency to exhibit notch sensitivity, many attempts have been made to remedy this failing in Udel polysulphone by blending with elastomers.

The first systems developed to display improved impact performance were the polysulphone–elastomer blends described by Union Carbide [34]. The elastomers employed in these blends were polysiloxanes, polyacrylates and vinylidene fluoride–hexafluoropropylene copolymer.

A wide range of polysiloxanes were used to prepare blends including polydimethylsiloxane homopolymers, diphenyl–dimethyl siloxane copolymers and siloxanes that had predominantly the same structure as the above but included a small amount of crosslinkable vinyl containing repeat units. Most of the siloxanes employed in the blends also contained an inorganic filler (silica). Siloxane content in the blends was typically about 3%.

It was emphasized that no special method of preparation was required for the blends. The blending procedures described as examples included a melt blending step prior to extrusion or compression moulding, and this was apparently sufficient to produce a good dispersion of rubber particles in the thermoplastic matrix as evidenced by the excellent impact strengths obtained for the blends. Test pieces for impact testing were produced by compression moulding.

All the blends reported were found to display greatly increased values for the notched Izod impact strength compared with the value obtained for an unmodified polysulphone homopolymer control (Table 9.2). For example, a polysulphone blend containing 3% polydimethylsiloxane (apparently uncrosslinked) with 0.64% silica filler showed an 850% increase in impact strength

Table 9.2 Impact strength of polysulphone–polysiloxane blends

Polysiloxane type	*Polysiloxane content (%)*	*Filler content (%)*	*Notched Izod impact strength ($J\,m^{-1}$)*
–	0	0	69.3
PDMS	3	0.64	506.3
PDMS	3	1.10	847.5
DMS–VM	2.25	0	474.4
DMS–DPS–VM	2.25	0	469.0
DMS–DPS–VM	3.25	0.9	655.6
DMS–VM	3.25	0.9	442.4
DMS–DPS–VM	3.25	0.6	634.3

PDMS, polydimethylsiloxane; DMS–VM, dimethylsiloxane–vinylmethylsiloxane copolymer; DMS–DPS–VM, dimethylsiloxane–diphenylsiloxane–vinylmethylsiloxane copolymer.
Source: Union Carbide Co., UK Patent 1,140,961, 1969.

compared to the polysulphone control. A blend also containing 3% polydimethylsiloxane but with silica filler content increased to 1.1% displayed an increase in impact strength of 1500% with respect to the polysulphone control. This superior behaviour for the polysiloxane with increased filler content could be due to the greater rigidity preventing excessive disruption of the rubber particles during processing. No data were reported for blends containing siloxanes that were neither filled nor crosslinked, implying that processing problems were encountered with these materials and substantiating the above suggestion.

The impact strength data obtained when varying the composition of the polysiloxane component were rather ambiguous. A blend containing 3.25% of a filled diphenyl–dimethyl siloxane copolymer had an impact strength 48% greater than that of a blend containing 3.25% of a polydimethylsiloxane containing an identical proportion of filler. This effect could be attributed to a greater degree of phase mixing, and hence improved interfacial adhesion between the polysulphone and the diphenyl–dimethyl siloxane, due to the higher solubility parameter of this siloxane and hence reduced difference in solubility parameter for the two components. However, for blends containing equal amounts (2.25%) of unfilled but crosslinked polydimethylsiloxane or diphenyl–dimethyl siloxane copolymer, no significant difference in impact strength was observed.

In addition to the large increases in notched impact strength produced by the addition of the siloxanes, a large reduction in the melt viscosity of the blend relative to that of the unmodified polysulphone was noted. This would normally be attributed to the migration of the lower viscosity elastomer to regions of high shear during processing and in effect lubricating the passage of the polymer melt. It was suggested that this property of the blends would be advantageous in that it would allow the more efficient reproduction of fine detail in complex mouldings.

Later workers were to find this behaviour a feature common to the majority of elastomer modified polysulphone and PES blends. However, the siloxanes employed in these blends were either filled or crosslinked and therefore unlikely to be very fluid under the processing conditions. It is possible that the high shear during processing resulted in scission of the crosslinks and disruption of the filled siloxane particles sufficiently to allow migration of a relatively fluid elastomer phase to produce the marked lowering in melt viscosity. Further, it is possible that the uniformly excellent impact strengths obtained for all the elastomer modified blends examined in this study were connected with the minimization of this migration by the fortunate choice of compression moulding as the route for producing specimens for impact testing. Later workers experienced great difficulty in producing good quality injection moulded specimens from polysulphone–elastomer and PES–elastomer blends as the shear induced migration was found to result in mouldings with pronounced surface delamination and very poor impact properties. Alternatively the excellent impact strengths could be explained by the use of filled or crosslinked elastomers which, although evidently not preventing the migration of the elastomer phase, might have restricted it to reasonable proportions.

Polysulphone blends containing acrylate rubbers with added silica filler were also examined in this study and found to display similar large increases in notched Izod impact strength to those observed for the polysiloxane containing blends (Table 9.3) [34]. A typical result was that obtained for a composition which contained 10% polyethylacrylate and 4% silica filler which had a notched Izod impact strength >1200% greater than that of the unmodified polysulphone control.

Impact strength values for these blends were more uniform than those obtained for the siloxanes. This could reflect the more ready attainment of a fine dispersion of well-adhering rubber particles due to improved interfacial mixing and lower interfacial tension. This would be expected as the solubility

Table 9.3 Impact strength of polysulphone–polyacrylate blends

Polyacrylate		*Elastomer*	*Filler*	*Notched Izod impact*
Type	*RV*	*content (%)*	*content (%)*	*strength ($J\,m^{-1}$)*
None	–	0	0	53.3
PEA	2.78	10	4	703.6
PEA	3.53	10	4	676.9
PBA	1.81	10	4	628.9
PBA	3.33	10	4	634.3
PEA	2.78	5	2	724.9
PEA	2.78	10	4	708.9
PEA	2.78	15	6	607.6

RV, reduced viscosity; PEA, polyethylacrylate; PBA, polybutylacrylate.
Source: Union Carbide Co., UK Patent 1,140,961, 1969.

parameters of the polyacrylate rubbers (δ in the range 18.0–19.1 $(\mathrm{MJ\,m^{-3}})^{1/2}$) are much closer to the solubility parameter of the polysulphone ($\delta = 21.7$ $(\mathrm{MJ\,m^{-3}})^{1/2}$) than those of the siloxanes.

The optimum level of added elastomer was found to be quite low. A series of polyethylacrylate containing blends with elastomer content being progressively increased from 5% to 15% but employing a fixed ratio of elastomer to filler were prepared. Maximum impact strength was obtained for the lowest elastomer content examined (5%) and impact strength began to decline significantly at higher loadings, probably as a result of the formation of fewer and larger elastomer domains which less efficiently initiate toughening processes.

The effect of the repeat unit of the added elastomer on blend performance was explored in these non-crosslinked systems. A series of blends holding the rubber content constant at 10% and the silica content constant at 4% while increasing the molecular weight of the polyacrylate rubber, as indicated by reduced viscosity measurements, were prepared. It might have been expected that lower molecular weight elastomers would show slightly better phase mixing and consequently improved interfacial adhesion and impact performance. The actual decrease in impact strength on significantly increasing the reduced viscosity and hence molecular weight of the rubber was, however, $> 5\%$ when polyethylacrylate rubber was used, and no significant change was observed for polybutylacrylate. This probably indicates the molecular weights of all grades of polyacrylate used were high and the entropies of mixing were intrinsically very small and therefore not significant. Alternatively any improvements in behaviour caused by greater phase mixing might have been offset by a loss in performance associated with the increased migration of the lower viscosity grades of elastomer to high shear regions, leading to disruption of the morphology of dispersed rubber particles.

The difference in impact strength for blends containing either polyethylacrylate or polybutylacrylate but with identical rubber content and filler content was slight. Assuming identical blending conditions for the blends this would be expected as the solubility parameters of the polyethylacrylate and polybutylacrylate are quite similar ($\delta = 18.0$ and 18.9 $(\mathrm{MJ\,m^{-3}})^{1/2}$ respectively) and sufficiently different from that of polysulphone ($\delta = 21.7\,(\mathrm{MJ\,m^{-3}})^{1/2}$) for interfacial mixing and adhesion to be approximately the same.

The final blend system described in the Union Carbide patent employed a vinylidene fluoride–hexafluoropropylene copolymer elastomer. Only one blend, with an elastomer content of 10% and no added inorganic filler, was prepared. A compression moulded specimen of this blend was found to have a notched Izod impact strength of $980\,\mathrm{J\,m^{-1}}$, the highest value for any blend described in the patent and a $> 1300\%$ increase with respect to the polysulphone control. This result rather confounds expectation as the δ value for the fluoropolymer is < 14.3 $(\mathrm{MJ\,m^{-3}})^{1/2}$ and would be expected to show minimal mixing and interfacial adhesion and, as the elastomer is non-crosslinked and unfilled, excessive migration and disruption of the elastomer domains during processing might also be expected. It was also reported that if moulded

Table 9.4 Properties of polysulphone–polyethylene blends

Blend composition (% polyethylene)	*Minimum moulding temperature (K)*	*Notched Izod impact strength ($J m^{-1}$)*
0	553	65.0
0.5	548	69.3
1.0	543	69.3
3.0	533	101.3
5.0	533	90.6
10.0	533	101.3

Source: A. C. Gowan, US Patent 3,472,810 (to General Electric), 1969.

specimens of this blend were irradiated before impact testing then the impact strength and elongation to break were both doubled with respect to the non-irradiated blend. This could represent the formation of grafts between the phases leading to increased adhesion and enhanced mechanical performance.

An investigation into the toughening of polysulphone by the addition of small amounts of a polyolefin formed the basis of a patent by Gowan of General Electric [35]. The blends were prepared by dry mixing the polymers, followed by melt extrusion. In this study the specimens for impact testing were prepared by injection moulding.

Significant improvements in notched Izod impact strength were observed for polysulphone–polyethylene blends containing between 0.5% and 10% polyethylene (Table 9.4). It was not stated whether branched low density or linear high density polyethylene was used to prepare these blends.

An optimum balance between improvement in impact strength and other properties was found for a blend containing 3% polyethylene, with impact strength increasing by >50% relative to the value obtained for an unmodified polysulphone control processed and tested in the same way. It was noted that for blends containing 5% polyethylene and above delamination problems were evident in the mouldings. At 10% polyethylene content the delamination was reported to be so severe as to compromise the mechanical integrity of the mouldings. This increase in the tendency for delamination to occur was accompanied by a decrease in the moulding temperature necessary to process the blend successfully. The minimum processing temperature decreased from 553 K for the pure polysulphone to 533 K for the blend containing 3% polyethylene. No further decrease in minimum processing temperature was noted as polyethylene content was increased above 3%. These effects can be attributed to migration of the lower viscosity polyethylene phase to the regions of high shear adjacent to surfaces of processing equipment.

Polysulphone based blends containing 1–5% added polypropylene were also investigated and broadly similar results to those for the polyethylene containing blends were obtained (Table 9.5).

Table 9.5 Polysulphone–polypropylene blends

Blend composition (% polypropylene)	*Notched Izod impact strength ($J\,m^{-1}$)*
0	64.0
1.0	69.3
3.0	85.3
3.0*	90.6
5.0	85.3

* Incorporating 2% mineral oil compatibilizer also.
Source: A. C. Gowan, US Patent 3,472,810 (to General Electric), 1969.

The maximum improvement in Izod impact strength for this simple binary system was observed for a blend containing 3% polypropylene. Mouldings prepared from blends containing 5% polypropylene showed signs of delamination. It was also reported that the inclusion of about 2% of a suitable mineral oil compatibilizer that was miscible with both polymers resulted in a modest increase in the impact strength of a polysulphone blend containing 3% polypropylene, possibly as a result of promoting some mixing at the interface of the blend. This increase in impact strength might simply be due to plasticization of the matrix polymer, however.

By comparison with the startling improvements in impact strength obtained by blending polysiloxanes, acrylate rubbers or fluoropolymers with polysulphones, the results for the polyolefin containing blends were disappointing. There are several potential explanations for this poor impact behaviour. The most likely of these is connected with the preparation and processing of this blend. The preparation of the blend involved only a simple extrusion process rather than a thorough melt mixing stage, and this could result in the production of a poor dispersion of large rubber particles in the matrix, which would lead to poor impact behaviour. No evidence indicating whether a fine dispersion of rubber particles was achieved was presented in the patent. An alternative explanation for the limited improvements in the impact strength of these polysulphone–polyolefin blends might be related to the actual fabrication method used to prepare specimens for impact testing. In this study the test pieces were produced by injection moulding and problems related to the shear induced segregation of the blend components were encountered, which led to delamination. The fact that the polyolefins were not filled or crosslinked would also tend to exacerbate this problem. Delamination would of course be expected to impair both the magnitude and the reproducibility of any improvements in toughness, as cracks would be readily initiated at voids in the poorly bound surface layer.

Even in the absence of these effects related to processing, the blends containing polyolefins would still be expected to display lower impact strengths than blends containing conventional elastomers and showing comparable interfacial adhesion. The modulus of the polyolefins is higher than that of uncrosslinked or lightly crosslinked elastomers and closer to the value for the polysulphone matrix. Hence stress concentration at the particle–matrix interface will be less pronounced and the initiation of the processes contributing to toughening will be more difficult.

A further Union Carbide study on simple physical blends of an elastomer with polysulphone was carried out by Barth [36]. The elastomers selected were ethylene–acrylonitrile random copolymers. The presence of the polar nitrile residues causes this elastomer to have a relatively high solubility parameter, between 8.5 and 12 $(MJ\,m^{-3})^{1/2}$ depending on acrylonitrile content, which is closer to that of the polysulphone base resin than those for the conventional elastomers incorporated in blends previously.

The example blends were prepared in a high shear melt mixer. It was emphasized that alternative blend preparation methods could be employed provided that they incorporated an efficient melt blending step which would thoroughly mix the components on a microscopic scale.

A series of blends in which the elastomer content was kept constant at 10% but the acrylonitrile content of the elastomer was varied between 17% and 30% were prepared. The notched Izod impact strengths of these blends were found to be excellent, with improvements of up to 1150% with respect to the polysulphone control being observed (Table 9.6) [36].

The effect of increasing the acrylonitrile content of the elastomer on impact strength was particularly informative. It was found that as the acrylonitrile content was increased from 17% to 30% for a constant elastomer content of 10% the improvement in impact strength almost doubled. This could be attributed to increased interfacial mixing and adhesion in the blend as the increasing nitrile content of the rubber raises the solubility parameter to a value approaching that of the polysulphone matrix. If present then this improved interfacial adhesion would probably improve the injection moulding performance of the blend and allow the production of injection moulded test pieces of high impact strength. Unfortunately no impact strength data for

Table 9.6 Impact strength of polysulphone–ethylene–acrylonitrile copolymer blends

Elastomer in blend (%)	*Acrylonitrile in elastomer (%)*	*Notched Izod impact strength* ($J\,m^{-1}$)
0	–	58.6
10	17	410.4
10	18.7	655.6
10	30	735.5
10*	30	815.5

* Blend crosslinked by irradiation.
Source: B. P. Barth, US Patent 3,510,415 (to Union Carbide), 1970.

injection moulded specimens of this blend were presented to allow this hypothesis to be confirmed.

Reference was made in this work to the incorporation of a silica filler in these blends which was found to encourage formation of a homogeneous mixture, presumably reinforcing the rubber particles and preventing their disruption during mixing, thus limiting any shear induced gross phase separation of the components. It was not stated whether filler was incorporated in the example blends for which data were given.

It was also found that the impact strength of these blends was increased by irradiation, leading to crosslinking of the elastomer. As it was not made clear whether the blend was irradiated before or after formation into test pieces it is impossible to say whether this improvement in impact behaviour results from decreased disruption of the crosslinked rubber particles during processing or from bond formation between the rubber particles and the matrix polymer increasing interfacial adhesion.

A patent filed by Uniroyal [37] describes the preparation and properties of blends of polysulphone with poly(epichlorohydrin) rubbers. Blends with both poly(epichlorohydrin) homopolymer and random copolymers of epichlorohydrin and ethylene oxide were investigated.

The blending methods for the example blends described involved mixing in a Banbury high shear melt blender at a temperature ≥ 503 K to achieve a good dispersion of rubber particles. Compression moulding was used to prepare specimens of the blend for impact testing.

The notched Izod impact strengths of a series of polysulphone based blends containing 5–20% poly(epichlorohydrin) rubber are given in Table 9.7 [37].

A very interesting effect is evident. For the blend containing 5% poly(epichlorohydrin) there is an insignificant change in the impact strength relative to that obtained for an unmodified polysulphone control. The tensile strength and modulus of the blend are also virtually the same as for the polysulphone homopolymer. When the rubber content of the blend is raised to 10%, however, a large increase in the impact strength of the blend is observed and this is accompanied by the reduction in tensile strength and modulus expected on the incorporation of dispersed rubber particles. These data seem

Table 9.7 Properties of polysulphone–poly(epichlorohydrin) blends

Poly(epichlorohydrin) content (%)	*Izod impact strength ($J\,m^{-1}$)*	*Flexural modulus ($GN\,m^{-2}$)*	*Tensile strength ($MN\,m^{-2}$)*
0	42.6	2.54	74.12
5	53.3	2.61	78.2
10	517.0	2.06	59.16
15	378.4	2.03	55.08
20	213.2	1.71	46.24

Source: Uniroyal Inc., UK Patent 1,416,643, 1973.

to indicate that the poly(epichlorohydrin) is significantly miscible with the polysulphone at low levels of incorporation and something approaching a single-phase blend is produced. The improvement in impact strength would then appear when the amount of added elastomer is increased such that it can no longer be completely solubilized by the polysulphone and a second phase of dispersed rubber particles is formed.

It could be argued that a 5% rubber content produced particles that were too small to initiate effectively the energy absorbing processes contributing to toughening. However, the effect of going from 5% to 10% rubber is so dramatic as to make this unlikely and a rubber content of 5% was found to be sufficient to produce rubber particles of the appropriate size for other polysulphone–elastomer blends [34]. If the size of the rubber particles was significantly smaller when this elastomer was used to modify polysulphone, and this was responsible for the lack of toughening at low rubber contents, even this might be explained in terms of increased solubility of the elastomer in the matrix. Improved phase mixing would lower the interfacial energy and allow interfacial area to be maximized which corresponds to a larger number of smaller particles being produced at a given rubber content.

It is significant that large increases in impact strength with respect to the homopolymer could be achieved in this blend system without the need for crosslinking of the elastomer or the use of fillers to prevent excessive disruption of the rubber particles. This again probably reflects improved interfacial mixing and adhesion and might allow the blends to be injection moulded successfully – an important advantage.

The suggestion that a significant degree of phase mixing exists in the polysulphone–poly(epichlorohydrin) blends is reasonable. The solubility parameter for the poly(epichlorohydrin) is about 19.5 $(\mathrm{MJ\,m^{-3}})^{1/2}$, higher than that of the elastomers used in the other polysulphone based blends with the exception of the ethylene–acrylonitrile copolymer, and hence fairly close to the value for the polysulphone matrix. In addition specific interactions could play a role in the limited miscibility of this blend. Blends of PES and polyethylene oxide, which are structurally very similar to the polymer in this blend, showing some miscibility have been identified [38]. PES also forms miscible blends with phenoxy resin [39]. It is thought miscibility arises in these cases as a result of the specific interaction between ether groups and aromatic rings on the two polymers. This type of specific interaction could also take place in the polysulphone–poly(epichlorohydrin) blend.

Blends containing $>10\%$ poly(epichlorohydrin) displayed a progressive decrease in the impact strength as the elastomer content was increased, but impact strengths were still superior to the impact strength of the polysulphone control. This behaviour can probably be correlated with increasing rubber particle size and a decreased number of particles, resulting in less effective initiation of toughening as elastomer content is increased. This is a fairly common observation for many rubber toughened blends.

Blends of polysulphone and epichlorohydrin–ethylene oxide copolymers were also prepared and evaluated in this study [37]. The elastomer content in these blends was again in the range 5–20%. A similar effect to that observed for the poly(epichlorohydrin) containing blends was again evident; an abrupt change in the properties of the blend occurred between 5% and 10% added elastomer (Table 9.8). This again probably indicates significant miscibility in the blend. As the comonomer residue in this elastomer is $—CH_2—CH_2—O—$, which is also capable of participating in specific interactions with the polysulphone, this is not surprising. The impact strengths obtained for the blends containing various levels of these copolymers were broadly similar to those obtained for the poly(epichlorohydrin) containing blends (Table 9.7). An elastomer content of about 10% was again found to be the optimum level to improve toughness in these blends, with impact strength gradually decreasing at elastomer contents higher than this.

A patent to Sumitomo Chem. Co. [40] describes similar polysulphone–poly(epichlorohydrin) blends to the above Uniroyal examples. Impact strengths were somewhat lower than those reported for comparable blends in the corresponding Uniroyal patent, with addition of 10% rubber increasing Charpy impact strength by approximately 300% with respect to a polysulphone control.

A Uniroyal patent [41] also describes toughened polysulphone blends containing acrylonitrile–butadiene–styrene (ABS). Blends containing the relatively large amount of 40–50% ABS were found to display the optimum balance of properties. Of particular interest was the observation that the impact strength of these blends varied linearly with the percentage of ABS added and no synergistic improvement in impact strength was observed. For these reasons it is debatable whether this blend can be considered a 'rubber toughened' system in the conventional sense but is simply an example of a compatible polyblend.

The fact that specimens displaying high impact strength could be produced by injection moulding implies that interfacial adhesion in this system is good. This is probably a consequence of the high solubility parameter of the

Table 9.8 Properties of polysulphone–poly(epichlorohydrin-*co*-ethylene oxide) blends

Copolymer content (%)	*Izod impact strength ($J\,m^{-1}$)*	*Flexural modulus ($GN\,m^{-2}$)*	*Tensile strength ($MN\,m^{-2}$)*
0	42.6	2.54	74.12
5	64.0	2.65	74.8
10	453.1	2.17	57.12
15	405.1	1.92	43.52
20	175.9	1.73	35.36

Source: Uniroyal Inc., UK Patent 1,416,643, 1973.

styrene–acrylonitrile resin constituting the continuous phase of the ABS, closely matching that of polysulphone.

The morphology of this blend was studied by electron microscopy and found to be remarkable. For the preferred blend compositions of 50–60% polysulphone, co-continuous polysulphone and biphasic ABS domains were found to exist. In the mid-range of compositions rapid changes in the heat distortion temperature occurred as blend composition was varied and alternative single continuous phases were established.

As an extension of the earlier work on simple physical blends of polysulphone with polysiloxane elastomers, workers at Union Carbide explored the possibility of using polysulphone–polydimethylsiloxane (PDMS) block copolymers to provide the rubber content [42]. The use of a block copolymer would be expected to increase the interfacial adhesion by introducing a covalent link between the elastomer domains and the matrix, assuming that some solubilization of the polysulphone blocks into the matrix occurred. Increased interfacial adhesion would of course be expected to result in greater improvements in impact strength than were observed for the simple physical blends of polysulphone and PDMS. A reduced tendency for phase separation on a macroscopic scale to occur for this blend compared with the simple physical blends should also be evident, thus overcoming the limitations in the injection moulding performance of simple physical blends of homopolymers showing poor compatibility.

For the synthesis of the polysulphone–PDMS block copolymers Noshay *et al.* selected a route based on the condensation of preformed, mutually reactive oligomers often termed 'telechelic oligomers'. The functionality of the terminal groups in such a synthesis is complementary, allowing reaction only between oligomers of different types, and therefore allowing precise control over block molecular weight and architecture in the absence of side reactions with impurities such as water.

The telechelic oligomer strategy has the advantage of being applicable to a wide range of polymers and is particularly suited to polymers produced from two mutually reactive monomers such as polysulphone (and also PES and PEEK). An excess of one monomer type in the polymerization then serves to control both the molecular weight of the oligomer and the terminal group functionality. If the terminal group functionality that can be accessed from the normal oligomers is unsuitable, then an alternative functionality for the oligomer may be produced by using an asymmetric 'blocking agent' in the synthesis. This has one functional group capable of participating in the polymerization and the other unreactive with respect to the monomers.

The chemistry chosen to form the link between the polysulphone and PDMS oligomers was the reaction of a hydroxy group with a silylamine group. This method has many advantages. The nucleophilic attack of a hydroxy group on a silylamine is typically very rapid and favourable with minimal side reactions [43, 44]. Difunctional hydroxy-terminated polysulphone oligomers can be readily prepared by employing an excess of the dihydroxy monomer (bisphenol-A) in the standard synthetic route to polysulphone. Appropriate silylamine-terminated (dimethylamino $Me_2N—Si—$) oligomers are also easily prepared from cheap and readily available starting materials [45].

There are only two significant drawbacks to this route. The silylamine-terminated oligomer is very susceptible to hydrolysis which can unbalance the stoichiometry of the terminal groups in the copolymerization and lead to an uncontrolled increase in the molecular weight of the siloxane block, and/or a poor yield. The Si–O–C linkage produced between blocks in a copolymer also is susceptible to hydrolysis [46].

Noshay *et al.* successfully prepared a series of alternating $(A-B)_n$ block copolymers from the reaction of hydroxy-terminated polysulphone and dimethylamino-terminated PDMS oligomers in chlorobenzene [42, 47] (Fig. 9.2). The molecular weights of the polysulphone blocks incorporated were in the range 4700–9300 $g\,mol^{-1}$, and the molecular weights of the PDMS blocks were in the range 350–25 000 $g\,mol^{-1}$. The PDMS content of the block copolymers varied between 10% and 79% depending on the molecular weights of the respective blocks. All the block copolymers were fully soluble in chloroform, indicating that they had an essentially linear structure and no crosslinking had occurred. High values for the reduced viscosities of 0.2% (weight per volume) chloroform solutions of the block copolymers were taken to be an indication that a large degree of polymerization had occurred to yield a high molecular weight product.

Phase separation in the polysulphone–PDMS block copolymers was studied by differential scanning calorimetry (DSC). It is of course important that phase separation occurs in a block copolymer intended for use as an impact modifier otherwise the elastomeric particles anchored to the matrix

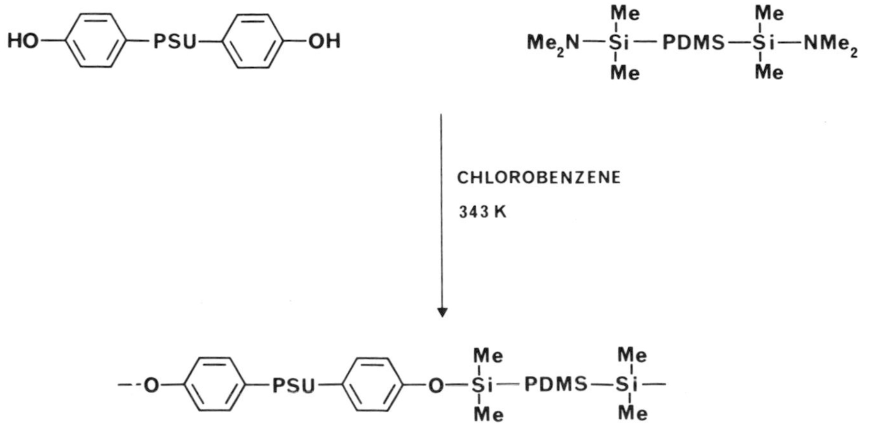

HO–C6H4–PSU–C6H4–OH + Me_2N–Si(Me)2–PDMS–Si(Me)2–NMe_2

CHLOROBENZENE
343 K

--O–C6H4–PSU–C6H4–O–Si(Me)2–PDMS–Si(Me)2–

Fig. 9.2 Synthesis of polysulphone–PDMS block copolymers.

necessary for rubber toughening would not be formed on blending with the thermoplastic. The tendency for phase separation to occur in a block copolymer will be favoured by the same factors which favour immiscibility in a simple binary polymer blend. Hence high molecular weight of the blocks favours phase separation. Calculations by Meier [48], however, have shown that the critical block molecular weight above which phase separation occurs is 2.5–5 times higher for a block copolymer compared with the corresponding physical blend owing to loss of entropy associated with the different polymer species being covalently bound together. A large difference in the solubility parameters of the block types also favours phase separation [49]; the greater the difference in solubility parameters the lower the critical molecular weight above which phase separation will occur.

It should be noted that phase separation in a block copolymer is retarded by the covalent bonding between the phases. Therefore the size of domains is very small and the situation is known as microphase separation [50] to distinguish it from the phase separation on a macroscopic scale that can occur in immiscible polymer blends.

It was found that microphase separation, as indicated by the presence of multiple glass transition temperatures characteristic of each phase, occurred at a block molecular weight of about 5000 g mol^{-1} for each segment type. When one of the block molecular weights was below this value a single glass transition temperature at a value intermediate between the values for the polysulphone and PDMS was observed, indicating a single mixed polysulphone–PDMS phase.

A series of blends of a polysulphone–PDMS block copolymer with polysulphone homopolymer were prepared by melt blending the components in a high shear mixer. Several different block copolymers were used to prepare blends. The block molecular weights of these were in the range 1700–9600 g mol^{-1} for the PDMS block and 4600–9300 g mol^{-1} for the polysulphone block. The percentage of block copolymer incorporated varied from 3% to 25%, and the PDMS content in the blends ranged from 2% to 14.25%.

In this study test pieces of the blends were produced by both injection moulding at 643 K and compression moulding at 563 K. This allowed the impact strengths of test pieces of the same blend produced by the alternative processing methods to be compared.

The notched Izod impact strengths obtained for a range of blends are given in Table 9.9 [42, 51].

The impact performance of the majority of the blends was excellent. Improvements of up to 1500% with respect to an unmodified polysulphone control were observed for compression moulded specimens. Equally important was the observation that for the majority of these blends the impact strength of the injection moulded specimens was not significantly different from that of the compression moulded specimens, being generally slightly lower. This implies the compatibility of the blend was good and rubber

Table 9.9 Properties of blends of polysulphone with polysulphone–PDMS block copolymers

Copolymer PSU–PDMS	PDMS content (%)	Copolymer in blend (%)	Total PDMS	Izod impact strength ($J\,m^{-1}$)	
				Compression moulded	Injection moulded
–	–	0	0	–	69.3
4700–5100	57.3	3	1.7	986.1	–
4700–5100	57.3	5	2.8	1125.0	1013.0
4700–5100	57.3	10	5.75	922.1	874.1
4700–5100	57.3	25	14.25	–	996.7
6600–1700	24.2	5	1.2	85.3	80.0
9300–4900	40.5	5	2	943.4	133.3
4700–9600	73.1	3	2.2	133.3	–

Sources: A. Noshay, M. Matzner, B. P. Barth and R. K. Walton, US Patent 3,536,657 (to Union Carbide), 1970; A. Noshay, M. Matzner, B. P. Barth and R. K. Walton, *ACS Organic Coatings and Plastics Division Preprints*, **34**(2), 217, 1974.

particle disruption and gross phase separation did not usually occur during processing.

Three blends deviated from the norm of a combination of greatly improved impact strength for both compression moulded and injection moulded specimens.

The blend prepared using the block copolymer with the lowest molecular weight siloxane block ($\bar{M}_n = 1700\,g\,mol^{-1}$) was found to display impact strengths for both compression moulded and injection moulded specimens that were comparable with the value for the unmodified polysulphone control. It was recognized that this result almost certainly reflects the complete solubilization of the entire block copolymer in the polysulphone matrix owing to the low molecular weight of the siloxane. This lack of phase separation prevented the formation of elastomer particles necessary for toughening. This result was not unexpected as it was noted previously that for phase separation to occur for this block copolymer the molecular weights of the blocks had to exceed $5000\,g\,mol^{-1}$. The observation that this blend was clear in contrast to the other blends was further evidence that a single-phase system had been formed.

A blend containing a block copolymer with the molecular weights of the polysulphone and PDMS blocks respectively 9300 and $4900\,g\,mol^{-1}$ was found to display an excellent Izod impact strength for compression moulded specimens ($943\,J\,m^{-1}$) but a much smaller impact strength for injection moulded specimens ($133\,J\,m^{-1}$). This is similar to the behaviour expected for a simple physical blend showing poor compatibility. The key factor in this case is probably the overall degree of polymerization of the copolymer. The 9300–4900 polysulphone–PDMS block copolymer had the lowest reduced viscosity and hence probably the lowest overall molecular weight of any of the block copolymers incorporated in the blends. Consequently the melt viscosity will be low and migration of the copolymer and disruption of the domains will occur quite readily. A similar effect has been observed in blends containing styrene–butadiene–styrene block copolymers [29]. This would account for the relatively poor impact behaviour of injection moulded specimens but good impact behaviour of compression moulded specimens of this blend. Limited interfacial adhesion in this blend might also play a role as the low molecular weight copolymer with large segment sizes would penetrate the particle–matrix interface a relatively small number of times.

A blend containing 3% of a 4700–9600 polysulphone–PDMS block copolymer was found to display an Izod impact strength of only 133 J m^{-1} for a compression moulded specimen. No data for an injection moulded specimen were quoted. No explanation for this result can be found in the molecular weight of the blocks or the overall molecular weight of the copolymer. This is worrying as it could indicate that blend preparation and processing is more critical in achieving good impact strength in these blends than at first appeared. A later publication [51] gave the impact strength of a blend containing 5% of this copolymer as being 1093 J m^{-1}, indicating a disturbing lack of reproducibility.

Although the blends prepared did not normally contain a filler, and good impact strengths were achieved without this, an example of a blend containing 5% of a 4700–5100 polysulphone–PDMS block copolymer and 2% filler was also prepared. This blend was found to display sufficient processing stability to allow recycling of material. Three injection moulding cycles were found to produce only a slight decrease in the Izod impact strength of the blend from 911 J m^{-1} to 853 J m^{-1} (Table 9.10).

Presumably the filler was added to improve rubber particle integrity during processing. Whether the fact that this blend was prepared is an indication that the unfilled blends had poor processing stability, and could not be recycled, is not clear.

A later publication contained additional data on the block copolymer modified polysulphone blends [51]. A study of the effect of the molecular weight of the matrix resin on the impact properties of the blend was conducted. The two grades of base resin employed were identified by a melt flow index

Table 9.10 Effect of recycling blend on impact strength

Injection moulding cycle	*Notched Izod impact strength (J m^{-1})*
1	911.4
2	895.4
3	852.8

Source: A. Noshay *et al.*, *ACS Organic Coatings and Plastics Division Preprints*, **34**(2), 217, 1974.

Table 9.11 Effect of rubber particle size on impact strength of block copolymer modified polysulphone blend

Maximum particle size in blend (μm)	*Notched Izod impact strength (J m⁻¹)*
<0.5	80.0
<3.0	986.1
<8.0	879.5

Source: A. Noshay *et al.*, *ACS Organic Coatings and Plastics Division Preprints*, **34**(2), 217, 1974.

rather than a molecular weight, and therefore the relationship between polysulphone base resin molecular weight and polysulphone block molecular weight is not clear. For the two grades of polysulphone resin incorporated in blends the molecular weight of the resin had a minimal effect on the impact strength of the block copolymer containing blend.

This publication also contained data describing the effect of particle size on the impact strength of the blends (Table 9.11). A series of blends in which the same block copolymer (polysulphone block $\bar{M}_n = 5000\,g\,mol^{-1}$, PDMS block $\bar{M}_n = 5000\,g\,mol^{-1}$) was employed at a constant level of 5% was prepared, and specimens for study produced by injection moulding.

For a maximum particle size <0.5 μm, improvements in impact behaviour were minimal, probably as a result of inefficient initiation of the toughening mechanisms by such a small particle. For blends with particle sizes >3 μm and <8 μm, high Izod impact strengths were observed. If this variation in rubber particle size were a random effect, or could be easily produced by incorrect processing, then this implies that problems of rubber particle integrity were present in these blends. This is probably not due primarily to poor interfacial adhesion, as the ability to produce small, well-dispersed rubber particles is an indication that this is adequate, but is rather caused by the weak interchain forces and low resistance to deformation of the non-crosslinked rubber.

The block copolymer modified polysulphone blends were the only example of a rubber toughened polysulphone, PES or PEEK blend commercialized to date, being available for a short period during the mid–late 1970s. They were, however, withdrawn apparently as a result of problems of erratic and irreproducible impact behaviour, and it is possible that this may have been connected with problems of rubber particle disruption if the material were processed incorrectly.

9.3 TOUGHENED POLYETHERSULPHONE BLENDS

The first of a series of studies conducted by ICI into the possibility of producing rubber toughened grades of PES concentrated on PES–polyolefin blends similar to those evaluated by Gowan for polysulphone. The author of this work and a subsequent patent was Hart [52, 53].

The PES–polyethylene system was studied in greatest detail. A modification in the melt flow properties of the base polymer by the addition of the polyethylene, similar to that noted for the corresponding systems based on polysulphone, was evident (Table 9.12).

A significant decrease in the melt viscosity, as measured on a ram extruder, was apparent for blends containing as little as 0.25% low density polyethylene (LDPE). The additional incremental effect declined as progressively larger amounts of LDPE were added to the blend. This is consistent with the earlier proposal based on data for the polysulphone–polyethylene blends that this phenomenon is caused by the migration of the lower viscosity polyethylene component to the high shear regions during processing. Once sufficient LDPE is added to produce a continuous surface layer of this material, additional amounts produce little extra effect.

The effect of the melt flow index (MFI), an indication of the related properties of molecular weight and melt viscosity of a polymer, of the polyethylene on the reduction in melt viscosity was studied for a series of PES–LDPE blends. For a fixed percentage of LDPE in the blend, the magnitude of the reduction in the melt viscosity was found to increase as the melt flow index of the added LDPE increased, i.e. as the melt viscosity of the LDPE decreases. This finding is again consistent with the explanation of the effect in terms of the migration of the lower viscosity component to the regions of high shear adjacent to the sides of the processing equipment and 'lubricating' the passage of the melt.

This reduction in the melt viscosity of the blend was found to correlate with a real increase in the processability of the blends compared with PES homopolymer as indicated by the torque required to mix molten blends and by spiral flow moulding trials. A similar reduction in melt viscosity was also observed for blends containing high density polyethylene (HDPE).

The notched impact strength of both LDPE and HDPE containing blends was evaluated. Specimens for impact testing were produced by injection

Table 9.12 Reduction in melt viscosity on blending LDPE with PES

LDPE added (%)	*Melt viscosity at 623 K ($kN\,s\,m^{-2}$)*
0	0.57
0.25	0.52
0.50	0.48
1.0	0.38
1.5	0.33
2.0	0.30

Sources: C. R. Hart, unpublished ICI results; C. R. Hart, Ger. Patent 2,122,754, 1971.

moulding. The method for evaluation of impact strength of test pieces probably did not conform to an internationally accepted standard method (details were not given, method described as 'standard within ICI Plastics Division') but was constant for all blends studied. For purposes of comparison, specimens of pure PES homopolymer of the same grade as that employed as the base polymer in the blends were processed in an identical manner to the blends and their impact strength evaluated by the above method.

It very rapidly became apparent that the choice of an appropriate blend preparation method that would efficiently mix the components of the blend on a microscopic scale was critical for obtaining significant and reproducible improvements in impact strength.

The initial blend preparation method employing dry mixing of the granulated polymers followed by simple melt extrusion produced PES–LDPE blends with a generally lower impact strength than the base homopolymer. A lack of reproducibility in the impact strength results was also evident. The specimens produced for impact testing from these blends had a very poor surface finish.

The effect of the melt flow index of the LDPE component on the impact strength of the blend was studied (Table 9.13).

A general trend for impact strength to increase as the MFI increased, and hence the melt viscosity and molecular weight of the LDPE decreased, was observed. The only blend to display a significantly greater impact strength than the homopolymer control at any level of LDPE was the blend containing the LDPE of highest MFI. This was attributed by the author of the report to improved interfacial mixing for this lowest molecular weight material, leading to improved adhesion between rubber particles and matrix and hence superior mechanical properties. Alternatively this result might represent the domains of the lower molecular weight, lower viscosity LDPE being broken down more effectively than those of the higher melt viscosity grades under the conditions of processing to give an improved dispersion of small rubber particles. The impact strengths of this first series of PES–polyethylene blends passed

Table 9.13 Effect of polyethylene MFI on the impact strength of PES–LDPE blends

Added LDPE (%)	*MFI of LDPE (%)*	*Impact strength ($kJ\ m^{-2}$)*
0	–	51
1.0	0.3	30
1.0	7	56
1.0	70	34
0.75	200	80

Sources: C. R. Hart, unpublished ICI results; C. R. Hart, Ger. Patent 2,122,754, 1971.

through a maximum at a level of ≤1% added LDPE and decreased markedly at higher additive levels.

Optical micrographs of these blends showed that the majority of the polyethylene particles dispersed in the matrix phase had diameters in the range 1–3 μm, but much larger particles with diameters in the range 20–30 μm were also present. Of particular interest were micrographs of fractured test pieces that had given low values for impact strength. These appeared to show cracks being initiated from the larger particles, probably indicating these particles were poorly bonded and voids were present at the particle–matrix interface.

The presence of the large rubber particles in the blends prepared simply by melt extrusion was taken to be an indication of inefficient mixing of the components during blending. Consequently a series of PES–LDPE blends were prepared using an improved blending technique that included an additional high shear melt blending step prior to extrusion and injection moulding. The MFI of the LDPE employed in the blends was held constant at the optimum value of 200, determined in the earlier work. LDPE content was in the range 1–5%. The impact strength values obtained for these blends are given in Table 9.14.

Blends containing 1% LDPE displayed an impact strength almost twice that of the unmodified PES control. At an LDPE level of 2% the impact strength of the blend was comparable with that of the PES control, while at additive levels ≥3% the impact strength was much lower than that of the homopolymer. A correlation was noted between the impact strength of the blends and the surface appearance of the injection moulded test pieces. The blends containing ≤2% LDPE and displaying good impact behaviour gave injection mouldings with an excellent surface finish and few signs of defects. The blends with ≥3% LDPE, which displayed poor impact behaviour, produced injection mouldings with poor surface quality and signs of delamination. Optical micrographs of these blends prepared using the improved blending regime showed a good dispersion of rubber particles with diameters in the range 1–2 μm and the absence of any large particles which would indicate inefficient mixing of the blend components.

A limited study of the effect of injection moulding conditions on the impact

Table 9.14 Impact strength of PES–LDPE blends

Added LDPE (%)	*Impact strength (kJ m^{-2})*
0	54
1.0	104
2.0	59
3.0	14
5.0	31

Sources: C. R. Hart, unpublished ICI results; C. R. Hart, Ger. Patent 2,122,754, 1971.

strength of test specimens produced from these well-mixed blends was made. It was found that increasing the injection speed used to prepare the test piece could increase the impact strength of a blend containing 2% LDPE by over 85%, that is the impact strength could be increased from a value comparable with that of the PES homopolymer to a value comparable with that obtained for the optimum blend composition of 1% LDPE. These data suggest that demixing of the blend occurred during processing and that this unsurprisingly had an adverse effect on the mechanical properties of the blend. These findings concerning surface delamination and the effect of injection moulding conditions on impact strength are once again consistent with the explanation of the decreased melt viscosity of this type of blend in terms of shear induced migration of the lower viscosity component to high shear regions.

A complementary study of PES–HDPE blends was carried out. The blend preparation method incorporated the refinements developed during the study on the LDPE containing blends. The results obtained for this blend system were very similar to those obtained for the PES–LDPE (Table 9.15).

Addition of 1% HDPE produced an 85% increase in the impact strength of the blend relative to that of the PES homopolymer control. A blend containing 2% HDPE showed an enormous reduction in impact strength compared with the impact strength of PES. This blend also displayed a poor surface, showing signs of delamination. The fact that the onset of delamination and an accompanying marked decrease in impact strength occur at lower additive levels in this blend system than for PES–LDPE could indicate poorer interfacial mixing than is present in the LDPE containing blend, but more likely reflects a difference in the melt viscosity of the polyethylenes used and hence a difference in ease of mixing of blend components. This conclusion is supported by the finding that the preparation of the 2% HDPE blend using a more efficient melt blending apparatus than the one normally employed increased the impact strength of specimens prepared from the blend to a value >80% of that for the PES homopolymer. This probably reflects the more efficient breaking down of a high viscosity HDPE phase to give a good dispersion of particles during mixing. No data indicating the MFI and hence melt viscosity of the HDPE used in the blend were given.

The dispersion of HDPE particles in these blends was examined by optical microscopy. In the blend containing 1% HDPE, the particles had diameters in

Table 9.15 Impact strength of PES–HDPE blends

HDPE added (%)	*Impact strength (kJ m^{-2})*
0	71
1.0	132
2.0	16

Sources: C. R. Hart, unpublished ICI results; C. R. Hart, Ger. Patent 2,122,754, 1971.

the range 1–3 μm, and in the blend containing 2% HDPE particle diameters were 2–6 μm. This difference in particle sizes could be another source of the rapid drop in impact strength for the blends when HDPE content is increased, as optimum particle sizes exist for the initiation of crazing and shear yielding. On the basis of the admittedly limited amount of data available it seems the particle sizes in this system increase quite rapidly as polyethylene content increases. This could be an artifact of the blending process and a high melt viscosity of the HDPE.

The effect of notch radius on the impact strength of PES–HDPE blends was studied in an attempt to gain some basic understanding of the mechanisms underlying the improvement in toughness of the blends (Table 9.16).

The blend containing 1% HDPE was tougher than the base homopolymer at notch radii in the range 2–0.25 mm. For a 0.25 mm notch, the impact strength of this blend was more than twice that of the unmodified PES control. Only for the sharpest notch studied, ≪0.25 mm, did the impact strength of this blend approach the value for the unmodified PES control. These data were taken as an indication that the crack initiation energy in the blend was increased but the crack propagation energy remains unaltered (as crack initiation would not be expected to be the limiting factor in a very sharply notched specimen with large stress concentrations). The increase in crack initiation energy in the blend would of course be due to energy dissipation by crazing or shearing.

The blend containing 2% HDPE had much inferior impact strength for blunt notches, ≥0.25 mm, but for the sharpest notch of ≪0.25 mm displayed a value for impact strength superior to both the PES homopolymer control and the blend containing 1% HDPE. The poor behaviour for bluntly notched specimens would be consistent with the poorly bonded HDPE particles providing many voids at the rubber–matrix interface from which cracks could propagate. This would correlate with the tendency toward delamination noted in injection mouldings produced from this blend. A poorly bonded rubber particle would also be unlikely to initiate any energy absorbing shearing or

Table 9.16 Effect of notch radius on impact strength of PES–HDPE blends

Notch radius (mm)	*Impact strength of blend (kJ m^{-2})*		
	0% HDPE	*1% HDPE*	*2% HDPE*
2	71	132	16
1	–	42	–
0.5	–	16	–
0.25	5.6	12	9.8
≪0.25	2.3	2.2	3.5

Sources: C. R. Hart, unpublished ICI results; C. R. Hart, Ger. Patent 2,122,754, 1971.

crazing that would increase crack initiation energy and offset the crack promoting effect of the voids. The apparent increase in crack propagation energy indicated by the impact strength data for the very sharp notch is rather puzzling.

Blends of PES with polymers more conventionally accepted as rubbers, ethylene–propylene (E–P) rubber and butyl rubber, were the final systems investigated in this study. Blends containing 1% ethylene–propylene rubber and 1% butyl rubber were both found to exhibit higher impact strength than the PES homopolymer control (Table 9.17).

For the butyl rubber composition the impact strength was found to be more than doubled, a greater improvement than was observed for the addition of polyethylene at the same level. The surface of the injection mouldings produced from both these blends was of good quality and showed no signs of delamination as would be expected for blends showing high impact strength. The addition of elastomer was again found to produce a drop in the melt viscosity of the blend for these systems. It is interesting to note that in these final blend systems studied there is an inverse correlation between the decrease in melt viscosity produced by the addition of the elastomer to the blend and the impact strength; the butyl rubber blend, which shows only a small decrease in melt viscosity relative to that of the unmodified PES homopolymer, has the greatest improvement in impact strength. The drop in melt viscosity is also small by comparison with the reductions in melt viscosity of other PES blends of comparable additive level that showed smaller increases in impact strength. This again emphasizes the connection between the lack of excessive migration of the added elastomer to the surface of the blend and the attainment of high impact strength. The lack of migration in this system might be explained by the crosslinking of the unsaturated units in the rubber at the high processing temperature of >573 K.

A later study into the toughening of PES by the inclusion of filled–cross-linked polysiloxanes or fluoroelastomers (specifically vinylidene fluoride–hexafluoropropene copolymer) was undertaken by May of ICI Engineering Plastics Division in 1981 [54]. The importance of efficient mixing of the blend in the molten state prior to moulding operations was recognized, and a high shear melt mixing step was incorporated in blend preparation. The moulding

Table 9.17 PES–elastomer blends

Elastomer	*Elastomer added (%)*	*Impact strength ($kJ\,m^{-2}$)*	*Melt viscosity at 623 K ($kNs\,m^{-2}$)*
–	0	41	0.41
Butyl rubber	1.0	87	0.35
E–P rubber	1.0	67	0.28

Sources: C. R. Hart, unpublished ICI results; C. R. Hart, Ger. Patent 2,122,754, 1971.

procedures proved problematical. Compression moulding of granulated blends generated test pieces of excellent surface integrity, but injection moulding under a variety of conditions persistently resulted in test pieces with excessive delamination of the surface. This clear indication of a poorly bound surface layer was attributed to thermal degradation of the polysiloxane or fluoropolymer elastomer during processing to give low molecular weight products of low viscosity, which migrate to regions of high shear during injection moulding. Alternatively the delamination might simply reflect the incompatible nature of the PES–elastomer blend and the migration of the lower viscosity rubber component to high shear regions. Similar delamination of polysulphone blends containing a high percentage of polyolefin was noted in the study of Hart [52, 53].

The impact strengths of injection moulded test pieces were studied using a calibrated drop weight technique, the exact details of which were not specified. The impact strength data for the blends are summarized in Table 9.18.

Unnotched samples of PES–PDMS blends showed much reduced failure energies with respect to PES homopolymer controls, but interestingly the PES–fluoropolymer elastomer blend had virtually unchanged failure energy. It was suggested that this decrease in failure energy for most blends was due to the rubber phase providing a large number of stress concentrating notches or voids within the matrix. Such behaviour would be typical of a very poorly adhering rubber phase. This hypothesis was supported by studies of the appearance of the fracture surfaces of some blends. The fracture surface produced by a blend containing 5% crosslinked–filled PDMS was examined by scanning electron microscopy and exhibited features which indicated that the PDMS domains could readily be pulled from the matrix implying poor interfacial adhesion. The fracture surfaces obtained for pure PES were, by contrast, relatively featureless, with evidence of slight regions of ductile deformation.

Table 9.18 Impact strength of PES–PDMS and PES–fluoropolymer blends

	Failure energy (N m)		
Added elastomer	*Unnotched*	*3 mm notch*	*1 mm notch*
None	75	4.9	3.3
3.25% PS-1	7.6	6.2	7.6
5% PS-1	4.6	6.8	11.8
5% PS-2	7.9	6.0	9.7
5% PS-3	8.8	6.6	7.6
5% Viton FP	71	–*	3.6

* Reproducible results could not be obtained for this specimen.
PS-1, PS-2 and PS-3 are siloxane rubbers containing 2% crosslinking agent and respectively 16–35%, >35% and 16% filler. Viton FP is the fluoroelastomer.
Source: R. May, unpublished ICI results.

If the above analysis of the situation is correct, then the implication is inescapable that the fluoropolymer elastomer must adhere better than PDMS to the PES matrix. If the situation in the blends is considered using the solubility parameter approach, then the poor adhesion of the PDMS to the PES matrix is not surprising as the solubility parameters are respectively 15.2 and 25.2 $(MJ\,m^{-3})^{1/2}$. However, the vinylidene fluoride–hexafluoropropylene copolymer elastomer will also have a very low solubility parameter comparable with that of PDMS, probably in the range 12.3–14.3 $(MJ\,m^{-3})^{1/2}$, depending on the ratio of comonomers, and again the large difference in solubility parameters for elastomer and thermoplastic matrix would be expected to result in very poor interfacial adhesion. It is possible that some favourable specific interaction between the fluoropolymer backbone and PES could lead to a near exothermic enthalpy of mixing, and hence promote interfacial mixing, but there is no reason to suspect this.

There is additional physical evidence that the fluoropolymer does show improved mixing with the matrix. Micrographs showed the average particle size in the fluoropolymer blends (0.5 μm) to be much smaller than the particle size in the PDMS containing blends (2–10 μm). The smaller particle size could arise from improved interfacial mixing, lowering the interfacial tension and hence allowing a finer dispersion (corresponding to a larger interfacial area) to be achieved. The interpretation of these data is not straightforward, however. It must be noted that the PDMS is both crosslinked and filled, whereas the fluoropolymer is a linear polymer, and this is likely to prevent the disruption of the rubber particles during processing and limit the minimum particle size achieved in the PDMS containing blends.

The key to the explanation of the behaviour of the PES–fluoropolymer blends is probably connected with thermal instability of the fluoropolymer. Thermal degradation of the fluoropolymer to produce low molecular weight products was suggested by the author of the work as a possible cause for the polymer flow problems encountered during injection moulding of the blends and this type of fluoropolymer is known to degrade at temperatures close to the melt processing temperature of PES. The lower molecular weight oligomers produced by the degradation of the elastomer might be expected to display greater mixing and hence improved interfacial adhesion. This would account for the small particle size and for the absence of any voids at the particle–matrix interface, lowering the impact strength of unnotched test pieces. The low melt viscosity of low molecular weight oligomers produced by degradation would also facilitate the production of a well-mixed second phase in the blend during processing.

This explanation for the unnotched behaviour of the PES–fluoropolymer blends is also consistent with the impact behaviour of notched specimens. The impact strength of the PES–fluoropolymer blends was almost unchanged with respect to values for the unmodified PES control. The well-bonded rubber domains do not offer any stress concentrating voids and impact strength is therefore not reduced. The lack of any improvement in impact strength is probably explained by the very small size of the rubber particles in this blend, which might be below the critical size necessary for the initiation of mechanisms leading to toughening.

Thermal degradation of the fluoropolymer also points to a second potential explanation of the apparently improved interfacial adhesion in the blend. If the degradation of the fluoropolymer were accompanied by the generation of radicals, then formation of, in effect, a PES–fluoropolymer graft copolymer might occur, and this would of course be expected to reduce interfacial tension and to compatibilize the blend. Again small domains with no voids at the interface would result.

The notched impact strength of the PES–PDMS blends was found to be greater than that of the homopolymer control in all cases, in contrast to the much reduced unnotched impact strengths observed for the blends. Improvements in impact strength of up to 250% with respect to PES were achieved by the incorporation of $<5\%$ PDMS (Table 9.18). In view of the imperfect adhesion of the elastomer particles to the matrix in these blends, these results are quite encouraging and imply that blends with improved interfacial adhesion might show very high impact strengths. A slightly disturbing feature of these results was the observation that the PES–PDMS blends were tougher at sharper notches, in contrast to the normal observation that impact strength is at a minimum for the sharpest notch. The difference was, however, marginal.

The fact that the siloxane elastomers employed in these blends contained varying amounts of inorganic filler complicates the interpretation of results. By analogy with the effect of rigid thermoplastic inclusions in the rubber particles in other toughened blends such as HIPS, it might be expected that a high filler content would increase particle rigidity [55] and result in decreased stress concentration, and hence limited improvements in toughness. In fact no clear trend regarding this is observed. It seems likely that the inability to injection mould test pieces with reproducibly good surfaces showing no delamination tended to mask all but the grossest effects of blend composition on mechanical properties.

The general conclusions drawn from this study were that PES–PDMS blends could potentially provide materials with excellent impact performance in both notched and unnotched situations, but the poor interfacial adhesion in simple physical blends of the polymers resulted in less than maximum improvements in impact strength being realized. It was suggested that the obvious way to improve the interfacial adhesion in this system was to prepare a PES–PDMS block copolymer, either for use as a compatibilizing agent in a simple PES–PDMS blend or as the sole source of the rubber particles themselves, by analogy with the work of Noshay *et al.* on polysulphone blends [42]. This latter possibility was subsequently explored in the work of Morris and coworkers [56, 57] at Sheffield City Polytechnic.

Morris carried out an extensive study into the use of PES–PDMS alternating block copolymers to improve the notched impact strength of a commercially available grade of Victrex PES [56, 57].

The PES–PDMS block copolymers were prepared by the same synthetic route as that used by Noshay *et al.* for the polysulphone analogues. Hydroxy-terminated PES oligomers were condensed with dimethylamino-terminated PDMS oligomers in dry 1,2-dichlorobenzene at 443 K to produce block copolymers with a variety of block sizes and PDMS contents. PES block molecular weight was varied between 1300 and 9600 $g\,mol^{-1}$ and PDMS block molecular weight was varied between 800 and 53 000 $g\,mol^{-1}$. DSC studies and transmission electron microscopy indicated that the PES–PDMS block copolymer displayed a microphase separated structure at lower block molecular weights, about 1000 $g\,mol^{-1}$ for each block type, than the polysulphone–PDMS block copolymers. This is a consequence of the significantly larger difference in the solubility parameters of the block types in this copolymer ($\delta = 15.0, 21.7$ and $24.6\,(MJ\,m^{-3})^{1/2}$ for PDMS, polysulphone and PES respectively). Thermal stability of the block copolymers was found to be adequate for melt processing with PES resin at temperatures up to 603 K.

Blends were prepared by dry mixing the components and then melt extruding at temperatures up to 603 K. Standard Izod test pieces were then produced from these blends by injection moulding.

Morris noted that the incorporation of amounts as low as 0.5% of the block copolymer resulted in a dramatic reduction in the melt viscosity of the blend relative to the value for the unmodified PES homopolymer. This is similar to the effect observed for other physical blends of PES with an elastomer and is an indication that interfacial adhesion is probably poor. Morris and coworkers also studied simple physical blends of PES and PDMS and observed comparable behaviour [58]. The behaviour in both types of blends was attributed to the migration of the low viscosity PDMS component to the high shear regions at the periphery of the melt, to produce a morphology of a PDMS-enriched sheath surrounding a depleted core. Morris also proposed the alternative explanation that the behaviour might be related to the well-established tendency for PDMS to migrate to the surface of a multicomponent system owing to its extremely low surface energy.

The study of the Izod impact strength of block copolymer modified PES blends was limited to only a small number of blends of different composition (0.5–2.5% PDMS). Unfortunately only sufficient block copolymer was available to prepare one blend from each type of block copolymer as the demanding nature of the synthesis limited the amount of block copolymer that could readily be prepared. Samples of pure 4800P grade PES, the base resin for all the blends, processed in the same way as the blends were used as the reference for all impact testing data.

The data related to the effect of PDMS content and molecular weights of the blocks on the Izod impact strength of the blend are contained in Table 9.19.

Table 9.19 Notched Izod impact strengths of blends of PES–PDMS block copolymer with commercial PES resin

Copolymer PES–PDMS	*PDMS in copolymer (%)*	*Copolymer content (%)*	*Total PDMS*	*Izod impact strength ($kJ\,m^{-2}$)*
–	–	0	0	7.8*
4900–5100	55.6	0.9	0.5	17.4
9600–5100	38.7	6.45	2.5	18.4
9600–53000	83.3	3.0	2.5	17.4
0	–	0	2.5	17.6†

* PES homopolymer control.
† Simple physical blend containing 2.5% PDMS homopolymer.
Sources: M. Morris, PhD Thesis, Sheffield City Polytechnic, 1988; A. A. Collyer, D. W. Clegg, M. Morris, D. G. Parker, G. W. Wheatley and G. C. Corfield, *Journal of Polymer Science, Part A, Polymer Chemistry*, **29**, 193–200, 1991.

Although these data are rather limited several important features are evident. The incorporation of even small quantities of copolymerized elastomer produces significant increases in the impact strength of the blends relative to that of the PES homopolymer control (the incorporation of as little as 0.5% copolymerized PDMS more than doubles the impact strength). Also, at constant PDMS content and constant PES block molecular weight, the molecular weight of the PDMS block has little effect on impact strength when it is increased from 5100 to 53 000 $g\,mol^{-1}$. Perhaps the most significant information is that a simple physical blend of PES and PDMS has almost identical impact strength to a blend containing the same amount of copolymerized PDMS. This again seems to imply that the interfacial adhesion between the PDMS domains and the PES matrix is still poor in the block copolymer modified blend. This would also explain why the improvements in impact strength relative to the homopolymer of these blends are much less significant than the improvements observed by Noshay *et al.* for the analogous block copolymer modified polysulphone blends containing similar amounts of PDMS.

The explanation favoured by Morris for this poor impact behaviour of the block copolymer modified blends was that excessive migration of the PDMS to the surface of the domains had occurred during processing, depleting the number of dispersed elastomer domains. However, the migration of the PDMS to the surface of the blends could equally likely be a manifestation of the underlying problem of poor interfacial adhesion rather than itself being the cause of the poor impact behaviour.

Incomplete or even negligible solubilization of the PES blocks in the PES matrix might be expected in these blends as the molecular weights of the blocks are significantly lower than the molecular weight of the matrix resin (the molecular weight of commercial grades of PES such as 4800P is in excess of 30 000 $g\,mol^{-1}$). Theoretical calculations by Meier [59] have indicated that only a very small amount of block copolymer segment can be solubilized by a chemically identical homopolymer of the same molecular weight or higher. On the basis of experimental evidence, Inoue *et al.* also arrived at the conclusion that unless the molecular weight of the block is greater than that of the homopolymer minimal solubilization occurs [60]. These findings seemed to be confirmed by the electron microscopy studies of Eastmond and Phillips, who invoked incomplete solubilization of like blocks into the continuous phase to explain the unusual structures they observed in blends containing block copolymers [61]. Hence the morphology that might be expected for the blends of PES homopolymer with PES–PDMS block copolymer would be three phase, with adjacent domains of PDMS and incompletely solubilized PES from the copolymer dispersed in a continuous phase of the high molecular weight PES homopolymer. This lack of solubilization would result in poor interfacial adhesion, which would account both for the impact strength comparable with that of a simple blend of the homopolymers and for the ready migration of the PDMS to the surface of the blend under high shear.

Unfortunately no copolymers containing high molecular weight PES blocks were prepared in this study. Consequently the effect of employing a PES block molecular weight comparable with that of the base resin on the impact strength of blends could not be studied. This is a great pity as the expected increase in interfacial adhesion that should result would significantly increase the impact strength of the blends and would confirm the suspicion that lack of phase mixing was the cause of the limited improvements in impact strength.

The discrepancy between the behaviour of the polysulphone blends containing copolymerized PDMS and the PES blends containing copolymerized PDMS is rather disturbing. As was noted in the discussion of the polysulphone blends the molecular weight of the polysulphone base resin was not disclosed. However, if it was a commercial grade of polysulphone, then the molecular weight might confidently be expected to be of the order of tens of thousands, significantly higher than that of the polysulphone blocks in the copolymer blended with it. Yet for the polysulphone blends interfacial adhesion seemed to be adequate and high impact strengths for injection moulded specimens were observed. Clearly this aspect of these copolymer modified blend systems requires further study, so that the real potential of such systems can be established and further unfruitful work avoided.

Morris also studied the effect of notch radius on the impact strength of PES homopolymer and the PDMS containing blends (Fig. 9.3). As expected the impact strength of the PES homopolymer was found to increase as notch radius increased from 0.1 to 1 mm, specimens with notch radii above this value failing to break below the maximum impact energy attainable in the test method of 125 kJ m^{-2}. Blends containing PES–PDMS block copolymer were found to display greater impact strengths than the homopolymer with the same notch radius for sharply notched specimens with notch radius $\leq$0.5 mm. For bluntly notched specimens with notch radius $>$0.5 mm, how-

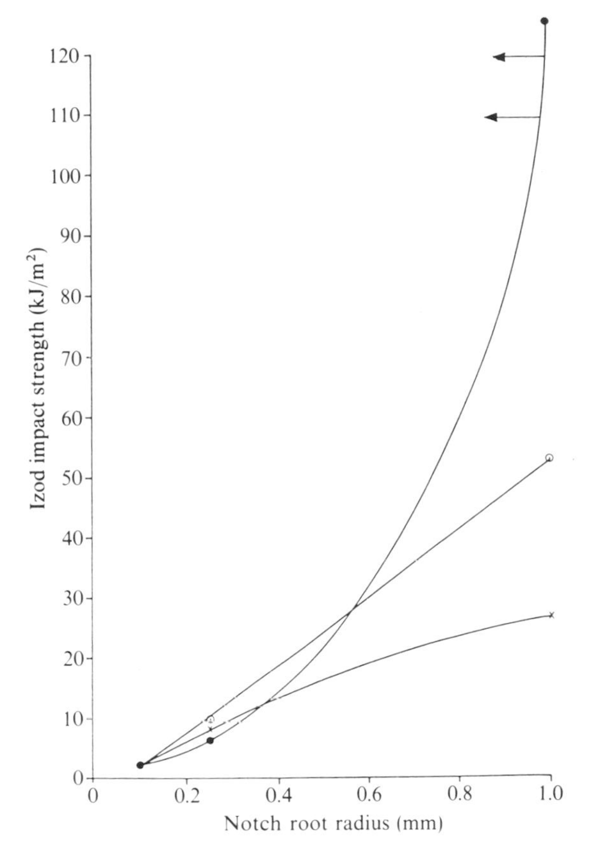

Fig. 9.3 Effect of notch radius on impact strength of PES-b-PDMS–PES blends: ●, pure PES; ×, blend 2 (2.5% copolymerized PDMS); ○, blend 9 (2.5% PDMS homopolymer). (From Ref. 56, by courtesy of M. Morris.)

ever, the copolymer modified blends had lower impact strength than the homopolymer. The extent of these modifications to impact strength was found to increase as the amount of copolymerized PDMS in the blends was increased from 0.5% to 2.5% (Fig. 9.4). The incorporation of linear PDMS homopolymer in blends once again resulted in impact strengths comparable with those obtained for the copolymer modified blends containing the same PDMS level.

The fracture surfaces of test pieces from the impact studies were examined by scanning electron microscopy (SEM). Impact tested specimens of pure PES displayed a smooth fracture surface, while the fracture surface of a copolymer modified blend tested under identical conditions showed evidence of material that had undergone plastic deformation (Fig. 9.5). A further striking feature of the fracture surfaces of copolymer modified blends was the circular depressions, which represented the sites of the PDMS domains. The surprising fact

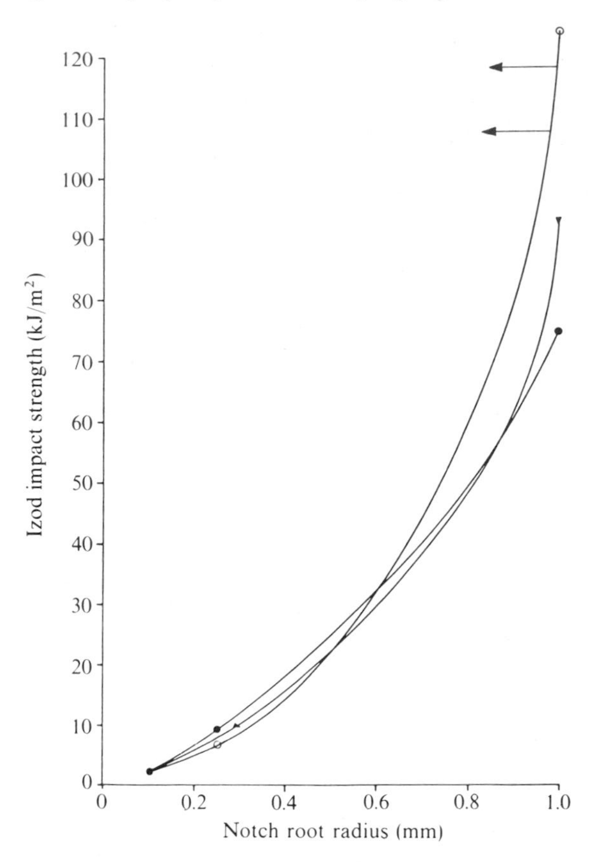

Fig. 9.4 Effect of PDMS content on impact strength of PES-b-PDMS–PES blends: ●, blend 3 (2.5% copolymerized PDMS); ○, blend 4 (0.5% copolymerized PDMS); ▼, blend 5 (1.0% copolymerized PDMS). (From Ref. 56, by courtesy of M. Morris.)

that only depressions rather than a mixture of depressions and raised domains was observed has been noted in other studies of rubber toughened polymer blends [62]. This is thought to be connected with the mismatch in the coefficients of thermal expansion of the rubber phase and the matrix, resulting in tension on cooling, which relaxes and causes contraction when a crack encounters the domain. The dimensions of these residual depressions representing the PDMS domains were typically in the range 1–5 μm. The only exception to this was the blend containing the block copolymer with the lowest molecular weight PES block ($\bar{M}_n = 1300\,\text{g mol}^{-1}$), in which the domain size was increased to a maximum of 12 μm. This probably reflects minimal solubilization of the PES blocks into the matrix, perhaps combined with incomplete phase separation in the block copolymer. This is the first direct evidence of incomplete solubilization of the PES blocks into the matrix which

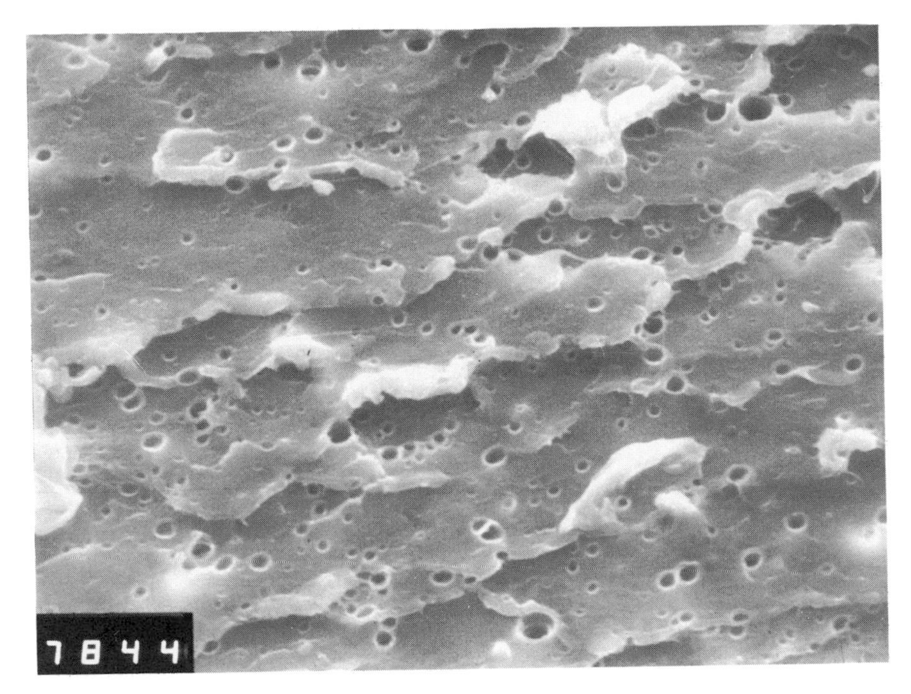

Fig. 9.5 Scanning electron micrograph of typical fracture surface of PES-b-PDMS–PES blend. (From Ref. 56, by courtesy of M. Morris.)

has been invoked as a possible explanation for the poor impact behaviour. The clean nature of the separation of the rubber particles from the matrix indicates poor interfacial adhesion, although whether this is present before the crack causes contraction of the rubber particles is uncertain.

The fracture surfaces of all blends displayed increased plastic deformation at higher temperatures. The extent of plastic deformation was found to increase with the amount of copolymerized PDMS in the blend, which correlates well with the observed effect of PDMS content on impact strength of specimens of different notch radius detailed previously.

Some transmission electron microscopy studies were conducted on microtomed sections of samples of the blends that had not been subjected to impact testing. Two micrographs in particular were very informative. An example of a rubber particle apparently undergoing contraction and leaving a void was observed, thus confirming Morris's explanation for the appearance of the fracture surfaces of impact tested blends (Fig. 9.6). More importantly perhaps, a second electron micrograph seemed to show phase separation around the site of a particle (Fig. 9.7). This micrograph is difficult to interpret but if a region of partially solubilized PES is indicated around the rubber particles, then this would confirm the poor interfacial adhesion proposed to explain the

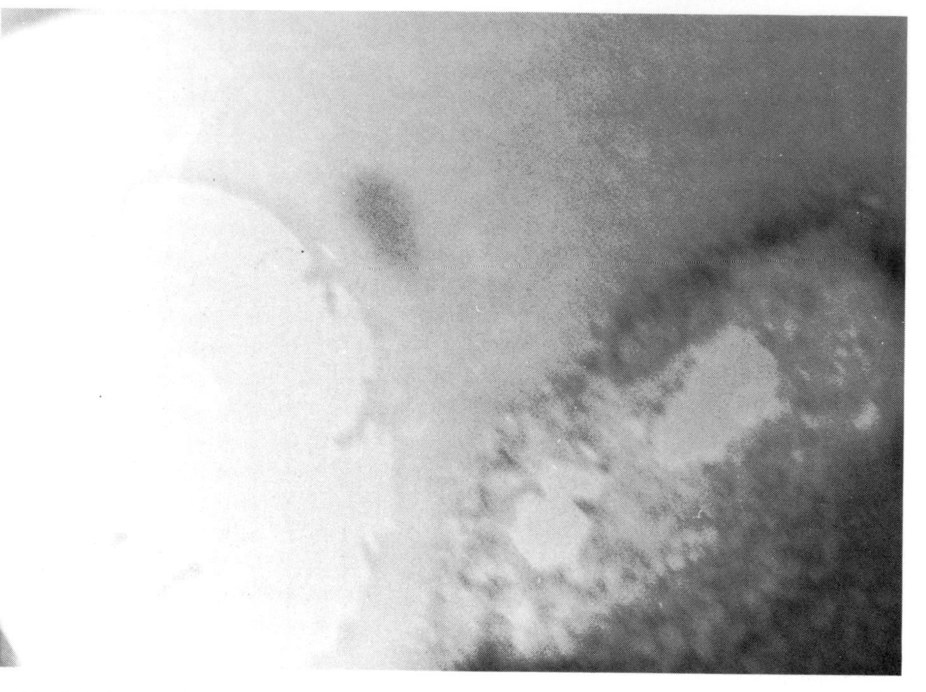

Fig. 9.6 Transmission electron micrograph of rubber particle apparently undergoing fracture and contracting to leave a void in a PES-b-PDMS–PES blend. (From Ref. 56, by courtesy of M. Morris.)

behaviour of the blends. Morris apparently considered the interpretation of this micrograph too subject to error and, perhaps wisely, did not consider this evidence in his final analysis of the mechanism of toughening in these blends.

The tensile strength of the copolymer modified blends was found to be comparable with that of the unmodified PES control and never less than 90% of this value. However, the elongation at break of the blends was significantly reduced; blends containing as little as 1% copolymerized PDMS (corresponding to about 2% copolymer in this case) suffered a reduction in the elongation at break from 24.5% for pure PES to 5% (Table 9.20). Morris interpreted this as evidence that the plastic deformation of the material was suppressed by the presence of the copolymer. The low elongation at break could be interpreted as evidence of poor bonding between the rubber particles and the continuous matrix phase.

On the basis of the body of data detailed above, Morris was able to propose a mechanism that was successful in explaining almost all the behaviour observed in the copolymer modified blends. It was suggested that two deformation processes contributed to the dissipation of energy during an impact event. The dominant process was enhanced plastic deformation (shear

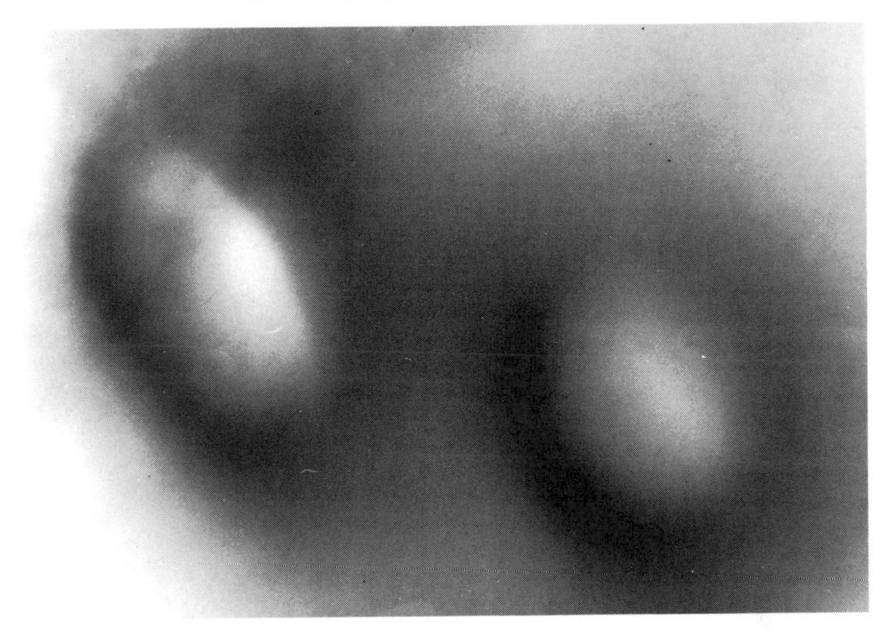

Fig. 9.7 Transmission electron micrograph showing phase separation around site of rubber particle in PES-b-PDMS–PES blend. (From Ref. 56, by courtesy of M. Morris.)

yielding) in the matrix phase triggered at the particle–matrix interface. Localized cavitation of the rubber particles at the particle–matrix interface played an important minor role in energy dissipation.

In sharply notched specimens the impact strength was significantly increased owing to shear yielding of the matrix material and deformation of the rubber particles. The magnitude of these effects was dependent on the percentage of copolymerized rubber at low copolymer contents, and hence impact strength increases as rubber content increases.

In bluntly notched or unnotched specimens of PES the absence of any stress raising features increases the amount of energy that must be supplied to initiate formation of a crack, and extensive plastic deformation is observed in these specimens before failure. When unnotched or bluntly notched specimens of block copolymer modified PES blends are subjected to impact, the loss of gross plastic deformation of the matrix exceeds the increase in energy absorbed by localized shearing initiated at the particle–matrix interface, and overall impact strength falls. Morris did not speculate on why reduced plastic deformation of the matrix in the copolymer modified blends was observed.

An alternative explanation for the behaviour of unnotched specimens, related to poor interfacial adhesion, is possible. Poorly adhering rubber particles produce voids at the particle–matrix interface and these function as stress concentrating notches and decrease the amount of energy that must be

Table 9.20 Tensile strength and elongation at break of copolymer modified PES blends

Copolymer PES–PDMS	*PDMS in blend (%)*	*Tensile strength ($MN\,m^{-2}$)*	*Elongation at break (%)*
–	0	91.9	24.4*
1300–800	2.5	83.8	2.9
1300–5100	2.5	87.0	3.8
4900–5100	2.5	89.2	5.0
4900–5100	0.5	91.3	17.0
4900–5100	1.0	91.3	4.9
9600–5100	2.5	88.9	5.5
9600–53000	2.5	87.4	5.0
Homo-PDMS	2.5	87.7	4.9

* Unmodified PES control.
Sources: M. Morris, PhD Thesis, Sheffield City Polytechnic, 1988; A. A. Collyer, D. W. Clegg, M. Morris, D. G. Parker, G. W. Wheatley and G. C. Corfield, *Journal of Polymer Science, Part A, Polymer Chemistry*, **29**, 193–200, 1991.

supplied to produce a crack. This can lead to lower impact energies in incompatible thermoplastic–elastomer blends than is observed for the unmodified thermoplastic, particularly in bluntly notched or unnotched specimens (see also the work of May [54] on physical blends of PES and PDMS presented earlier). Failure then occurs at an early stage before plastic deformation is widespread, although the reduced plastic deformation is not the cause of the failure.

A patent to Sumitomo Chem. Co. describes a PES analogue of the polysulphone–ABS blend system [63] developed by Uniroyal. Like the polysulphone system, the impact strength did not show a synergistic increase and the blend cannot be considered a rubber toughened system in the commonly accepted sense.

9.4 TOUGHENED VICTREX POLYETHERETHERKETONE BLENDS

Research into toughened PEEK blends has been limited to say the least. PEEK is the most recently introduced polymer discussed in this review, being fully commercialized in the mid 1980s. The high processing temperature of PEEK (typically 653–673 K) imposes severe restrictions on the elastomers suitable for incorporation with it to form blends. The only common elastomer likely to display sufficient thermo-oxidative stability is some type of siloxane.

The sole example in the literature to date of a rubber modified PEEK blend is a system incorporating ethene–glycidyl methacrylate elastomer described in a Sumitomo Chem. Co. patent [64]. It was reported that a 90% PEEK–10% acrylate rubber blend could be prepared at 613 K. This temperature is normally considered too low to melt process PEEK in any way and must also be close to the decomposition temperature of the acrylate rubber. The impact strength reported for this blend (8 kgf cm cm^{-1}) is comparable with that of unmodified PEEK.

The only other work related to rubber toughened PEEK blends published to date has been conducted by Wheatley *et al.* [65, 66]. The aim of the research was the preparation of PEEK–PDMS block copolymers which would then be used to prepare toughened blends with PEEK homopolymer, by analogy with the work of Noshay *et al.* on polysulphone based blends. The high solubility parameter of PEEK and semicrystalline nature imply that simple physical blends with PDMS would show poor compatibility and hencc poor impact and processing behaviour.

The synthetic route adopted for the synthesis of the PEEK–PDMS block copolymers was the same as that employed successfully for the corresponding polysulphone and PES block copolymers, namely the condensation of the appropriate pre-formed hydroxy-terminated oligomer with dimethylamino-terminated PDMS. Hydroxy-terminated PEEK oligomers were prepared by a modification of the PEEK synthesis of Attwood *et al.* with a calculated excess of the dihydroxy monomer (hydroquinone) serving both to produce the required terminal groups and to control molecular weight.

The limited number of solvents for PEEK and the very different chemical natures of PEEK and PDMS made the identification of a suitable solvent for the block copolymer synthesis problematical. The only solvent identified as being capable of dissolving both PEEK and PDMS was 1-chloronaphthalene. The synthesis of block copolymers in this solvent at 503 K gave excellent yields, in excess of 95% of theory in some cases. A series of block copolymers in which the molecular weight of the PDMS block was held constant at 4000 g mol^{-1} while that of the PDMS block was increased from 4000 to 9000 g mol^{-1} was prepared. The molecular weight of the PEEK blocks incorporated in the block copolymers was less than the molecular weight of commercial grades of PEEK resin. It was recognized that this might result in incomplete solubilization of the PEEK blocks into the matrix in blends of PEEK resin with the block copolymer and hence result in limited compatibility and poor mechanical properties. However, side reactions in the synthesis of PEEK can consume monomer and lead to loss of control over molecular weight and terminal group functionality, and 10 000 g mol^{-1} represents the maximum molecular weight of hydroxy-terminated oligomers that can be reliably synthesized.

The PEEK–PDMS block copolymers were found to be soluble only in sulpholane and 1-chloronaphthalene at high temperatures (>473 K), and this restricted the characterization of the materials. In particular an assessment of the overall molecular weight and degree of polymerization was not possible.

DSC studies showed that these block copolymers undergo phase separation into a two-phase morphology at very low block molecular weights (<4000 g mol^{-1}) as would be expected on the basis of the large difference in solubility parameters of the blocks. Crystallization of the PEEK blocks probably provided an extra driving force for phase separation.

Thermogravimetric analysis indicated that the PEEK–PDMS block copolymers were stable in an inert nitrogen atmosphere at temperatures approaching 673 K, but significant decomposition of the block copolymer, associated with degradation of the PDMS blocks, occurred above 633 K. This thermo-oxidative instability of the PDMS blocks at PEEK processing temperatures limits the preparation and processing options for PEEK blends incorporating these block copolymers.

Initial attempts to prepare PEEK blends incorporating these block copolymers in a Brabender high shear melt mixer at 633 K resulted in the apparent degradation of the block copolymer. However, these blends were prepared successfully when the Brabender blender was continuously purged with nitrogen to displace any air present. Injection moulding of the blend was not possible owing to the thermal instability of the block copolymer. Insufficient material was available for the production of the large number of test pieces necessary to achieve a valid average value for the notched Izod impact strength of the blend. Therefore samples of film for toughness testing were prepared by compression moulding the material at 633 K. The moulds used for preparing the film were quenched rapidly into water, but the crystallization rate of PEEK is so rapid that completely amorphous materials could not be produced. As the toughness of PEEK is critically dependent on the degree of crystallinity this necessitates the determination of the percentage crystallinity of the PEEK phase of the blend and the comparison of blend toughness with the toughness of a homopolymer control of comparable crystallinity.

Early indications of the toughness of PEEK–PDMS block copolymer containing blends were inconclusive and this work is still in progress.

9.5 CONCLUSIONS

The results from a wide variety of blend compositions indicate quite clearly that the attainment of stable, well-dispersed rubber domains is critical to the achievement of high impact strength.

A poor dispersion of rubber domains can arise in two ways:

- inefficient blending procedures failing to mix the components of the blend well;
- shear induced migration of the lower viscosity rubber phase resulting in phase separation of the blend components during processing.

As would be expected, shear induced phase separation is more pronounced when high shear processing operations such as injection moulding are employed.

The problem of shear induced phase separation can be minimized by the crosslinking or filling of the rubber to improve particle integrity.

Excellent interfacial adhesion can also play a role in the prevention of excessive disruption of rubber particles during processing. The polysulphone blends modified with polysulphone–PDMS block copolymers, although unfilled and not crosslinked, could display very high impact strengths for specimens fabricated by injection moulding. There were indications for this system, however, that the stability of the PDMS domains was marginal and that disruption of the relatively fluid PDMS could occur if processing conditions were not carefully controlled.

In simple physical blends interfacial adhesion often appeared adequate to produce high impact strength in compression moulded test pieces. Under the more demanding conditions of injection moulding, however, limited interfacial adhesion seemed to result in the formation of voids at the particle–matrix interface, resulting in low impact strength.

On the basis of these conclusions it is possible to outline the characteristics of the type of blend system that would display maximal impact strength.

Good interfacial adhesion would be achieved by the use of an appropriate block copolymer to provide the rubber particles. Of the suitable high performance elastomers only polysiloxanes are readily incorporated into polyarylether block copolymers, and hence this type of elastomer would be selected to provide the rubber particles. For polysulphone or PES based blends PDMS is suitable, but for PEEK based blends a more thermally stable siloxane elastomer containing diphenyl or trifluoropropyl repeat units in addition to dimethyl repeat units would be preferable. To avoid disruption of the rubber particles during high shear processing a filler should be added to the blend, or preferably the rubber phase should be crosslinked. The crosslinking of the rubber phase should be controllable to prevent excessive crosslinking, increasing T_g and the modulus of the rubber particles, which would impair the initiation of the toughening processes. This could be achieved by incorporating a small percentage of repeat units in the polysiloxane containing vinyl groups, which could be crosslinked *in situ* during blend preparation (*in situ* crosslinking would ensure that the rubber was initially fluid enough to form well-dispersed rubber domains).

An alternative to the use of a block copolymer would be to employ a conventional elastomer and to tailor the solubility parameter to fall within the narrow range of values that would produce significant interfacial mixing with the matrix polymer but still result in the necessary two-phase morphology. The critical control over solubility parameter required would probably be achieved using a random copolymer elastomer, for example a polyacrylate containing a percentage of polar repeat units such as cyano groups. The elastomer should again be filled or crosslinked for optimum results.

The use of an elastomer which contained repeat units capable of participating in specific interactions with the matrix polymer might also result in a two-phase morphology combined with some interfacial mixing.

ACKNOWLEDGEMENT

The authors extend their thanks to Dr Mick Morris for his kind permission to reproduce Figs 9.3–9.7.

REFERENCES

1. Union Carbide Co. (1978) *Udel Polysulphone: An Outstanding Engineering Polymer for Moulding and Extrusion.*
2. Rigby, R. B. (1984) *Polym. News*, **9**, 325.
3. Smith, C. P. (1981) *Swiss Plast.*, **4**, 37.
4. Dawson, P. C. and Blundell, D. J. (1980) *Polymer*, **21**, 577.
5. Clendinning, R. A., Farnham, A. G., Hall, W. F., Johnson, R. N. and Merriam, C. N. (1967) *J. Polym. Sci., Part A-1*, **5**, 2375.
6. Attwood, T. E., Barr, D. A., King, T., Newton, A. B. and Rose, J. B. (1977) *Polymer*, **18**, 359.
7. Attwood, T. E., Barr, D. A., King, T., Newton, A. B. and Rose, J. B. (1981) *Polymer*, **22**, 1096.
8. 3M (1963) UK Patent 1,060,546.
9. ICI (1964) UK Patent 971,227.
10. Raychem (1976) US Patent 3,953,400.
11. Amos, J. L. (1974) *Polym. Eng. Sci.*, **14**, 1.
12. Rosen, S. L. (1977) *Polym. Eng. Sci.*, **17**, 115.
13. Bucknall, C. B. (1977) *Toughened Plastics*, Applied Science, London.
14. Bucknall, C. B. (1977) In *Polymer Blends* (eds D. R. Paul and S. Newman), Academic Press, New York, Chap. 14.
15. Bucknall, C. B. and Smith, R. R. (1965) *Polymer*, **6**, 437.
16. Kambour, R. P. (1973) *Macromol. Rev.*, **7**, 1.
17. Newman, S. and Strella, S. (1965) *J. Appl. Polym. Sci.*, **9**, 2297.
18. Morton, M., Cizmecioglu, M. and Lhila, R. (1984) In *Polymer Blends and Composites* (ed. C. D. Han), ACS, Washington, DC.
19. Matsuo, M., Ueda, A. and Kondo, Y. (1970) *Polym. Eng. Sci.*, **10**, 253.
20. Bucknall, C. B. and Street, D. G. (1967) *Sci. Monogr.*, **26**, 272.
21. Paul, D. R. and Barlow, J. W. (1984) *Polym. Eng. Sci.*, **24**, 525.
22. Manson, J. A. and Herzberg, R. W. (1973) *J. Polym. Sci., Polym. Phys. Ed.*, **11**, 2483.
23. Scott, R. L. (1949) *J. Chem. Phys.*, **17**, 279.
24. Hildebrand, J. M. and Scott, R. L. (1950) *The Solubility of Non-Electrolytes*, Reinhold, New York.
25. Stehling, F. C., Huff, T., Speed, C. S. and Wissler, G. (1981) *J. Appl. Polym. Sci.*, **26**, 2693.
26. Paul, D. R. (1978) In *Polymer Blends* (eds D. R. Paul and S. Newman), Academic Press, New York, Chap. 12.
27. Rudin, A. (1980) *J. Macromol. Sci., Rev. Macromol. Chem. C*, **19**, 267.
28. Childers, C. W. (1969) US Patent 3,429,951.
29. Durst, R. R., Griffith, R. M., Urbanic, A. J. and Van Essen, W. J. (1974) *ACS Org. Coat. Plast. Div. Prepr.*, **34**(2), 320.
30. Morton, M. (1970) In *Block Copolymers* (ed. S. L. Aggarwal), Plenum, New York, pp. 1–10.
31. Noshay, A., Matzner, M. and Williams, T. C. (1973) *Ind. Eng. Chem. Prod. Res. Dev.*, **12**(4), 268.
32. Brandrup, J. and Immergut, E. H. (eds) (1975) *Polymer Handbook*, Wiley, New York.

33. Small, P. A. (1953) *J. Appl. Chem.*, **3**, 71.
34. Union Carbide Co. (1969) UK Patent 1,140,961.
35. Gowan, A. C. (1969) US Patent 3,472,810 (to General Electric).
36. Barth, B. P. (1970) US Patent 3,510,415 (to Union Carbide).
37. Uniroyal Inc. (1973) UK Patent 1,416,643.
38. Walsh, D. J. and Rostami, S. (1985) *Adv. Polym. Sci.*, **70**, 119.
39. Singh, V. B. and Walsh, D. J. (1986) *J. Macromol. Sci. Phys.*, **25**, 65.
40. Yamauchi *et al.* (1970) JP 7,308,860 (to Sumitomo Chem.).
41. Ingulli, A. F. and Alter, H. L. (1972) US Patent 3,636,140 (to Uniroyal).
42. Noshay, A., Matzner, M., Barth, B. P. and Walton, R. K. (1970) US Patent 3,536,657 (to Union Carbide).
43. Langer, S. H., Connell, S. and Wender, J. (1958) *J. Org. Chem.*, **23**, 50.
44. Pike, R. M. (1961) *J. Org. Chem.*, **26**, 232.
45. Nagase, Y., Masubuchi, T., Ikeda, K. and Sekine, Y. (1981) *Polymer*, **22**, 1607.
46. Steffen, K. D. (1972) *Angew. Makromol. Chem.*, **24**, 1.
47. Noshay, A., Matzner, M., Barth, B. P. and Walton, R. K. (1971) *J. Polym. Sci., Part A-1*, **9**, 3147.
48. Meier, D. J. (1969) *J. Polym. Sci., Part C*, **26**, 81.
49. Matzner, M., Noshay, A. and McGrath, J. E. (1973) *Polym. Prepr.*, **14**, 68.
50. Molau, G. E. (1971) In *Colloidal and Morphological Properties of Block and Graft Copolymers* (ed. G. E. Molau), Plenum, New York.
51. Noshay, A., Matzner, M., Barth, B. P. and Walton, R. K. (1974) *ACS Org. Coat. Plast. Div. Prepr.*, **34**(2), 217.
52. Hart, C. R. (1971) Unpublished ICI results.
53. Hart, C. R. (1971) Ger. Patent 2,122,754.
54. May, R. (1981) Unpublished ICI results.
55. Newman, S. (1977) In *Polymer Blends*, Vol. 2 (eds D. R. Paul and S. Newman), Academic Press, New York, p. 69.
56. Morris, M. (1988) PhD Thesis, Sheffield City Polytechnic.
57. Collyer, A. A., Clegg, D. W., Morris, M., Parker, D. G., Wheatley, G. W. and Corfield, G. C. (1991) *J. Polym. Sci., Part A, Polym. Chem.*, **29**, 193–200.
58. Collyer, A. A. *et al.* (1984) Proc. IXth Int. Congress on Rheology, Acapulco, Vol. 3, pp. 543–50.
59. Meier, D. J. (1977) *Polym. Prepr.*, **18**, 340.
60. Inoue, T., Soen, T., Hashimoto, T. and Kawai, H. (1970) *Macromolecules*, **3**, 87.
61. Eastmond, G. C. and Phillips, D. G. (1979) *Polymer*, **20**, 1511.
62. Kinloch, A. J., Shaw, S. J., Todd, D. A. and Hunston, D. L. (1983) *Polymer*, **24**, 1341.
63. Sumitomo Chem. Co. (1981) JP 81,107,752.
64. Sumitomo Chem. Co. (1984) JP 9,184,255.
65. Wheatley, G. W. (1988) PhD Thesis, Humberside College.
66. Corfield, G. C., Wheatley, G. W. and Parker, D. G. (1990) *J. Polym. Sci., Part A*, **28**, 2821–36.

10

Toughened polyimides

S. J. Shaw

10.1 INTRODUCTION

The use of polymeric materials in the manufacture of aerospace structures has increased markedly in recent years owing to the many desirable characteristics which they exhibit in comparison with traditional aerospace materials such as aluminium and titanium. Possibly their most desirable attribute has concerned their ability, in certain cases, to combine high strength and stiffness with low weight. Indeed, this ability can be considered as the major driving force responsible for the continually growing acceptance of these materials in various structural applications.

Two areas in which polymeric materials have been employed successfully to fulfil a structural role are as adhesives and matrix resins for fibrous reinforced composite materials. The use of such materials in aerospace structures such as aircraft and space vehicles, where weight is of substantial importance, has grown considerably in recent years and is likely to increase in the foreseeable future.

Over the past decade there has been a steadily increasing requirement for adhesives and composites, and hence the polymeric materials from which they are composed, which can endure temperatures in excess of approximately 150 °C for both short- and long-term applications. Although polymers exhibiting this capability have been available, on a commercial basis, for many years, they have all suffered to varying degrees from adverse processing characteristics which have generally hindered, or in some cases totally prevented, practical utilization. Consequently, in an attempt to overcome these problems, there has been in recent years a considerable international research effort designed principally to develop high temperature resistant polymers having considerably improved processing characteristics [1–4]. This has resulted in numerous developments some of which have produced polymers, primarily polyimides, close to or at full commercialization. Although producing greatly enhanced processability whilst retaining a high temperature capability many of these developments have yielded polymers exhibiting a high degree of brittleness. As a result many researchers have considered the technique of rubber modification, developed initially with epoxies, as a potential means of improving the toughness of cured oligomeric polyimides. This chapter will attempt to discuss these developments.

10.2 HIGH TEMPERATURE PROCESSABLE POLYMERS

Before describing some of the recent developments in high temperature processable polymers, it is perhaps advisable to consider some of the older high temperature polymers which have been traditionally employed in the past, since this will hopefully convey to the reader the inherent processability disadvantages of these systems and hence the reasons for the more recent developments in the high temperature polymer area.

10.2.1 Traditional high temperature polymers

Traditional commercially available polymers capable of operating at temperatures in excess of 150 °C for both short- and long-term applications can be divided into three broad classes, namely

- condensation polyimides,
- phenolics, and
- polybenzimidazoles.

The polymers traditionally, and indeed as employed for many current high temperature applications, are usually based on condensation polyimides. These are generally prepared by the reaction scheme outlined in Fig. 10.1, where a dianhydride is reacted with a diamine in a suitable solvent at room temperature [5]. This results in a high molecular weight polymer, known as a polyamic acid, in solution in the original solvent. The polyimide is obtained by heating the polyamic acid to approximately 200 °C with water being liberated as a byproduct. The majority of the early commercially available polyimides were derived from dianhydrides and diamines which produced polyimide structures having high degrees of intractability. Consequently these were generally processed at the intermediate polyamic acid solution stage, with conver sion to the polyimide being conducted either within the adhesive layer during the bonding process or, in the case of a composite, in the final fabrication stage under high pressure. Clearly the evolution of water during this conversion, together with the presence of the original solvent, provides a serious potential processing difficulty, producing a possibly porous, mechanically weak adhesive bond–matrix, resulting in undesirable mechanical performance. The use of high processing pressures together with precisely controlled venting techniques can reduce such problems. However, even in such circumstances severe restrictions on component size–bond area exist.

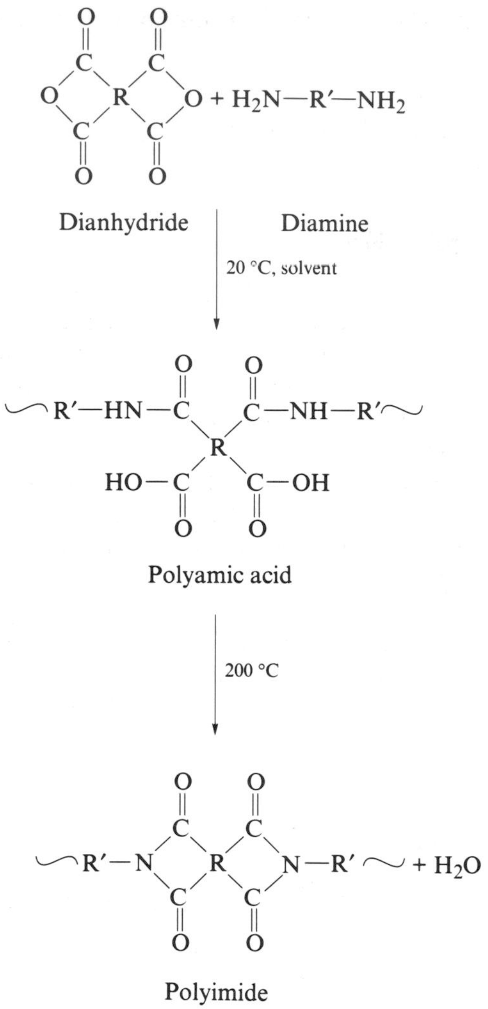

Dianhydride + $H_2N{-}R'{-}NH_2$ Diamine

20 °C, solvent

Polyamic acid

200 °C

Polyimide + H_2O

Fig. 10.1 Typical reaction scheme for a condensation polyimide.

The phenolics and polybenzimidazoles suffer from adverse processing characteristics for similar reasons to those just described [6]. Once again, successful processing requires both high processing pressures together with carefully controlled venting operations so as to alleviate effects associated with the evolution of volatiles during cure.

Such problems have resulted in the considerable international research effort mentioned above, aimed at obtaining high temperature resistant polymers with substantially improved processing characteristics. A number of interesting and significant developments have resulted from this research, some of which will now be described. For convenience they are divided into the following three principal areas:

- flexibilized polyimides;
- low molecular weight imide oligomers;
- other polymers.

10.2.2 Flexibilized polyimides

Imparting a degree of molecular flexibility to a polyimide molecule provides the possibility that the polymer may exhibit sufficient thermoplasticity above its T_g for fabrication by a 'hot melt' process. If this can be achieved, then fabrication using a fully imidized polymer is possible, thus removing the need for both solvent evaporation and polyamic acid to polyimide conversion during the final fabrication stages.

Two particularly notable examples of this approach which have resulted in commercial exploitation are LARC-TPI [7, 8] developed by the National Aeronautics and Space Administration in the USA, and polyetherimide [9, 10]. Figure 10.2 shows the molecular structures of these two polymers. In the case of LARC-TPI, an inherent degree of flexibility is achieved for two main reasons. Firstly, the presence of carbonyl groups spaced at periodic intervals along the polymer backbone induces a degree of molecular flexibility, i.e. the so-called 'hinge' effect. Secondly, the position of the nitrogen atoms

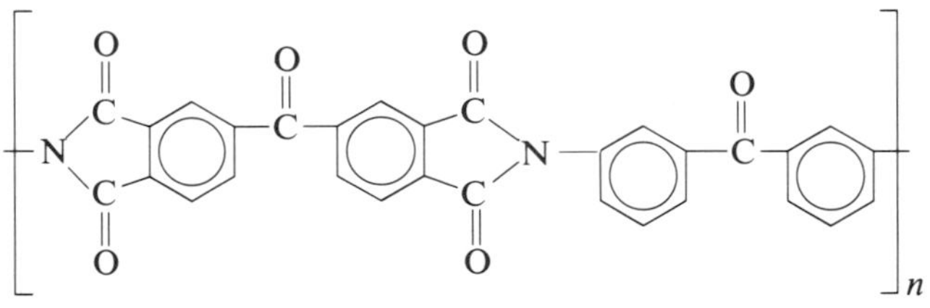

(a)

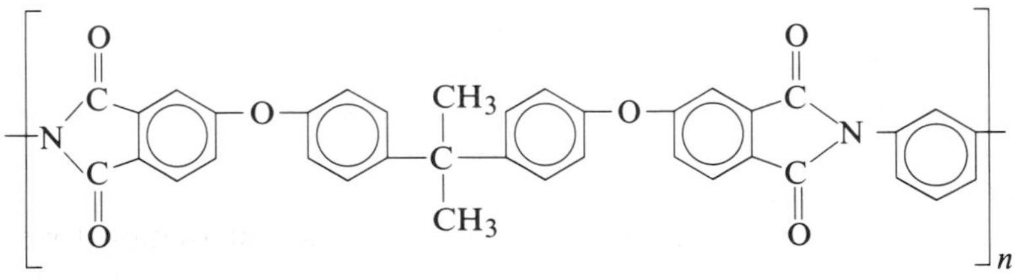

(b)

Fig. 10.2 Molecular structures of (a) LARC-TPI and (b) polyetherimide.

relative to the carbonyl group linking the two benzene rings has been shown to be critically important with a *meta* relationship allowing extensive flexibility.

It has been claimed that polyetherimides retain a high temperature capability while having melt flow characteristics sufficient for processing by injection moulding techniques [9]. This compromise has been attributed to the aromatic imide units providing stiffness and thermal resistance with the ether linkages providing the 'hinge' effect and allowing melt flow characteristics.

As would perhaps be expected, the introduction into the polyimide backbone of flexibilizing groups, as well as improving processability, also enhances polymer ductility and consequently toughness. Thus flexibilized polyimides of the types just described do not usually require toughness enhancement by any additional means such as rubber modification. However, as will be described later in this chapter, exceptions do exist and in a number of cases rubber modification of flexibilized polyimides has been attempted.

10.2.3 Low molecular weight imide oligomers

In addition to the molecular flexibilization approach the concept of low molecular weight imide oligomers has been seen as an alternative route to enhanced processability. Such systems have been developed around the following three fundamental principles. Firstly, the oligomers should be of a low molecular weight. Such a system, depending on molecular structure, could have both a low melting point and low viscosity, thus providing potential for use in 100% solids formulations (i.e. no solvents). Secondly, imide groups should be present in the oligomer, thus removing the particularly troublesome polyamic acid to polyimide process mentioned previously. Thirdly, the oligomers should have reactive terminal groups capable of reaction by an addition mechanism so as to convert the molten oligomer to a solid crosslinked polymer without the evolution of volatiles.

A wide range of materials have been developed in recent years, based essentially on this approach [7, 9, 11]. The various types differ mainly in the form of terminal reactive group employed to convert the oligomer to a crosslinked product. Although this approach has provided probably the greatest improvements in processability, i.e. greater than that achieved with the flexibilization approach discussed above, this improvement has been associated with a major difficulty, namely brittleness. Owing to the crosslinking process, some of the materials developed exhibit extremely poor toughness characteristics, thus making them a target for the principles of rubber modification. It will therefore be of no surprise to the reader to learn that, with polyimides, by far the majority of attempts at rubber modification have been directed at the low molecular weight oligomeric polyimide variants. Consequently, therefore, it is both necessary and of interest to consider these materials in greater detail prior to a consideration of rubber toughening.

Three main types of terminal reactivities have been developed, these being based on the norbornene, acetylene and maleimide chemistries. These will now be considered.

(a) Norbornene-terminated oligomers

The earliest attempt at developing low molecular weight imide oligomers was in the late 1960s when the efficacy of the so-called norbornene group as a potential crosslinking site was first realized [12]. This technology eventually became known as P13N (P for polyimide, 13 representing the average oligomer molecular weight of 1300 and N indicative of norbornene terminal functionality) [8]. Figure 10.3 shows the essential features of the P13N reaction scheme. As with conventional polyimides, a dianhydride is reacted with a diamine. However, in this case, a third reactant, 5-norbornene-2,3-dicarboxylic anhydride (commonly referred to as nadic anhydride) is also employed. As shown, the reaction is conducted at room temperature in a solvent, the end product being a low molecular weight norbornene-terminated amic acid. Heating at approximately 200 °C results in a norbornene-terminated imide which, on heating at temperatures in excess of 285 °C, undergoes polymerization and crosslinking via the norbornene end groups by an addition mechanism.

Although this general approach had potential advantages relating to enhanced processability over the more conventional condensation polyimides, the original P13N polymers did not perform particularly well. A major research programme was therefore initiated by NASA in the 1970s aimed at overcoming the deficiencies inherent in the original P13N systems. This work has culminated in a number of resin systems which have been or are close to commercialization, the most notable of these being PMR-15, LARC-160 and LARC-13.

PMR-15 was developed in the early 1970s at the NASA Lewis Research Center [13, 14]. Standing for '*in situ* polymerization of monomer reactants', its basic chemistry is similar to the P13N approach described above. Its main advantage over P13N is associated with the fact that fibre impregnation is conducted with a solution containing a mixture of monomers dissolved in a low boiling point alcohol solvent such as methanol, instead of prepregging (i.e. impregnating of fibres with resin) with norbornene end-capped amic acid oligomers employed with P13N. The monomers employed in PMR-15 are diaminodiphenylmethane, the dimethylester of benzophenone tetracarboxylic acid and the monomethylester of 5-norbornene-2,3-dicarboxylic acid (the source of the norbornene end group). Possibly the greatest advantage of the entire PMR approach concerns the ability to 'tailor make' resins having the required elements of processability and properties, this being achieved by simple alterations in monomer structure and/or stoichiometry. The most common PMR resin, PMR-15, refers to a formulation which produces an oligomer having a molecular weight of 1500.

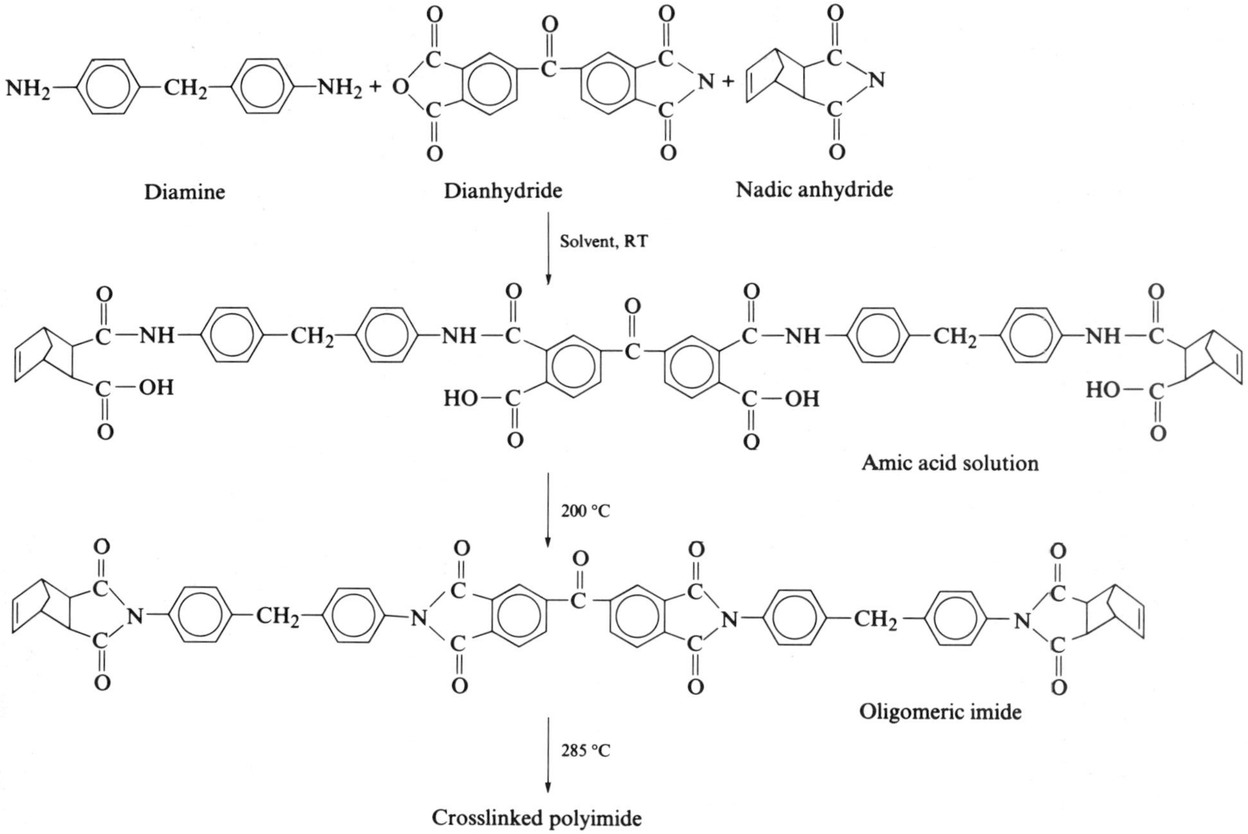

Diamine

Dianhydride

Nadic anhydride

Solvent, RT

Amic acid solution

200 °C

Oligomeric imide

285 °C

Crosslinked polyimide

Fig. 10.3 P13N reaction scheme.

As with PMR-15, LARC-160 was also developed in the early 1970s, this time at NASA Langley Research Center [15, 16]. The main aim of this research programme was to develop a polyimide resin which would confer tack and drape characteristics to a composite prepreg without the use of solvents. LARC-160 was the result. The primary difference between this resin and PMR-15 was that the former contained a commercial liquid amine mixture capable of conferring both the desired 'hot melt' processability and drape and tack characteristics without compromising high temperature capabilities. Cure mechanisms and conditions can be considered similar to those previously described for the PMR system.

While PMR-15 and LARC-160 were designed essentially as composite matrix resins, attempts were also made at NASA Langley to develop high temperature adhesive systems based on a similar approach. These efforts have culminated in the development of LARC-13 [7, 17], the preparation and structure of which are shown in Fig. 10.4.

It would seem from the published literature that LARC-13 is generally employed as a solution of norbornene-terminated amic acid in the original preparatory solvent. Thus the processing conditions which would appear necessary involve the following steps:

1. application of the solution to the substrate or carrier, followed by heating at about 100 °C to remove solvent;
2. heating at 200 °C to convert the amic acid to the imide;
3. using the resulting film or cloth and bonding at 300 °C under an applied pressure of 0.35 MPa.

The major advantage in this procedure in relation to processing of traditional condensation polyimides is that both solvent and water of imidization are removed prior to bond formation, this being possible owing to the norbornene-terminated imide having the capacity for extensive flow during cure.

Numerous experimental programmes have evaluated LARC-13, resulting in an extensive database [8, 17–23]. Within the context of this chapter it is particularly interesting to note that adhesive lap shear strength values are somewhat low relative to those obtained from some thermoplastic polyimide adhesive systems. This has been attributed to the relatively brittle nature of LARC-13 (fracture energy of approximately 100 J m^{-2}).

(b) Acetylene-terminated oligomers

An alternative route to imide oligomers using acetylenic termination was developed under US Air Force sponsorship during the early to mid 1970s culminating in a proprietary system based on the chemical structure shown in Fig. 10.5 [24, 25]. Although the initial reactants and oligomer preparation procedures employed are generally similar to those described previously for

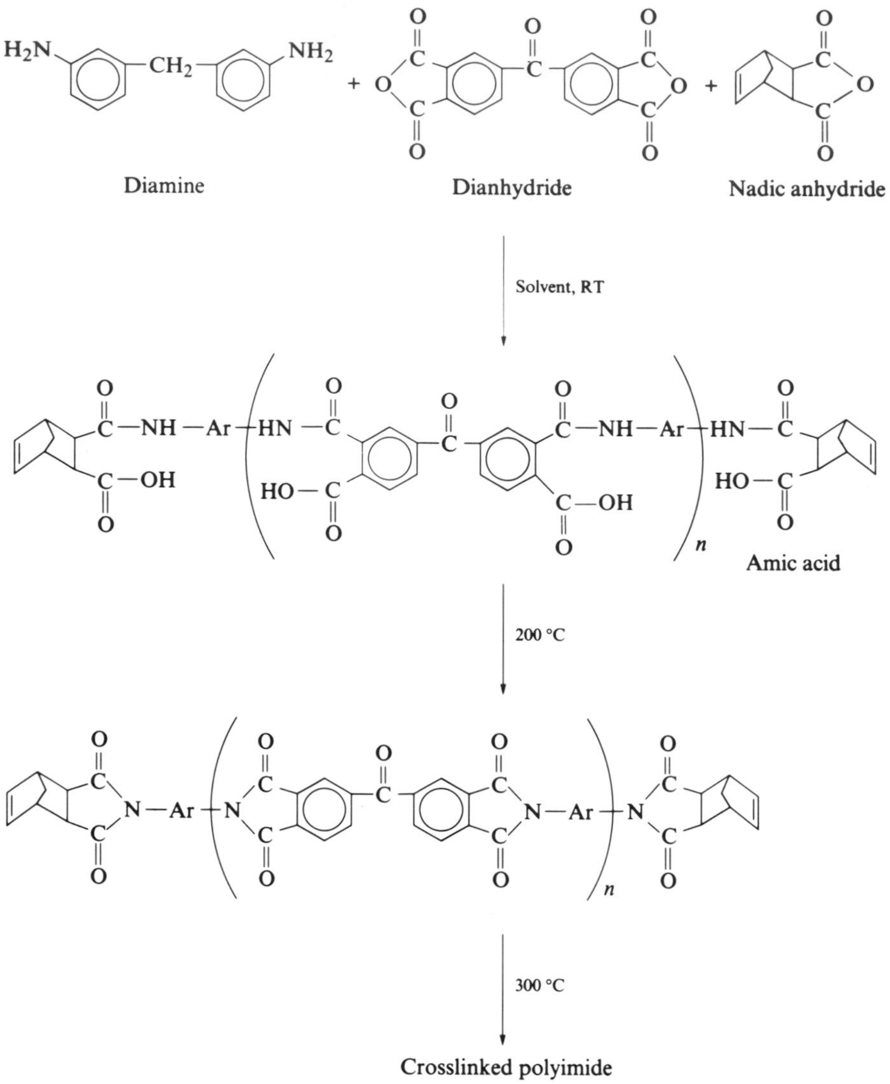

Diamine

Dianhydride

Nadic anhydride

Solvent, RT

Amic acid

200 °C

300 °C

Crosslinked polyimide

Fig. 10.4 Preparation and structure of LARC-13.

the norbornene systems, in this case terminal functionality is provided by the use of 3-aminophenylacetylene rather than a norbornene type compound.

Various publications have demonstrated the excellent properties attainable with acetylene-terminated polyimides [25–28].

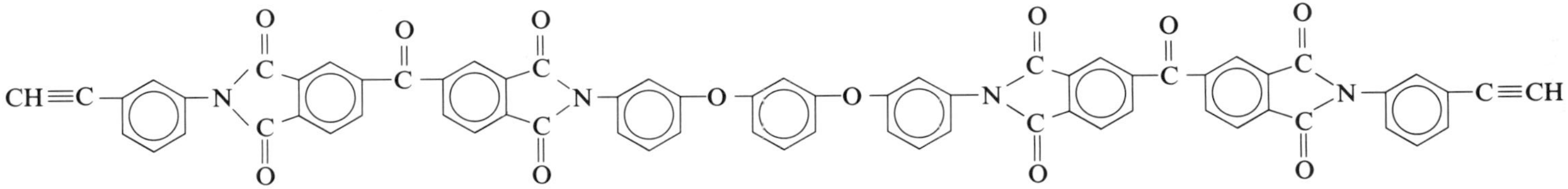

Fig. 10.5 Acetylene-terminated imide prepolymer structure.

(c) Bismaleimides

In addition to norbornene and acetylenic termination, a further way of enhancing the processability of a polyimide is to utilize the reactivity of the maleimide group, oligomers of this type being referred to as bismaleimides [11]. Bismaleimides are prepared by reacting maleic anhydride with diamines via the formation of an intermediate bismaleimic acid (Fig. 10.6). A large variety of bismaleimides can be obtained by varying the structure and molecular weight of the original diamine. Further molecular complexity can be introduced by reacting the oligomers with other compounds such as diamines [29, 30], hydrazides [11, 31, 32], epoxies [33] and bisallyl compounds [34].

Owing to their low molecular weight, bismaleimides can have, depending on molecular structure and formulation, a low melting point together with a low melt viscosity. Heating at temperatures in excess of approximately 150 °C

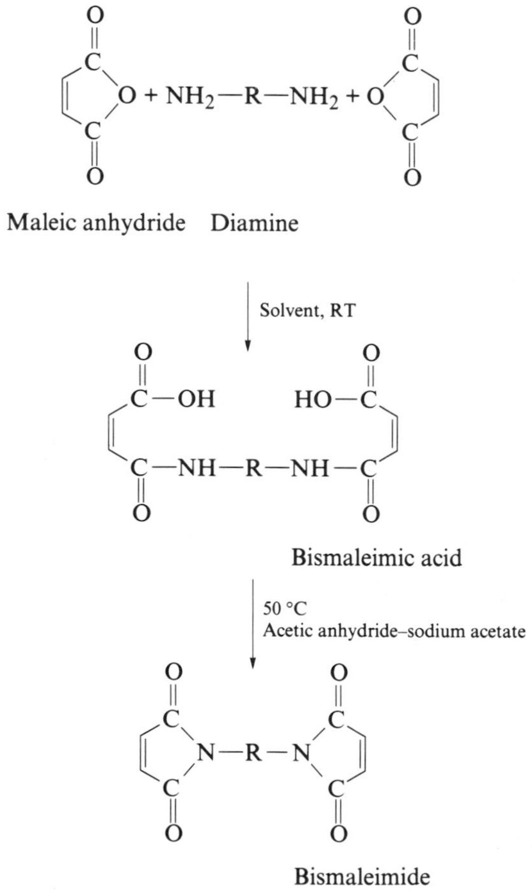

$$\text{Maleic anhydride} + NH_2\text{—R—}NH_2 + \text{Maleic anhydride}$$

Maleic anhydride Diamine

↓ Solvent, RT

$$HOOC\text{—CH=CH—C(=O)—NH—R—NH—C(=O)—CH=CH—}COOH$$

Bismaleimic acid

↓ 50 °C, Acetic anhydride–sodium acetate

$$\text{Maleimide—N—R—N—Maleimide}$$

Bismaleimide

Fig. 10.6 Bismaleimide preparation scheme and structure.

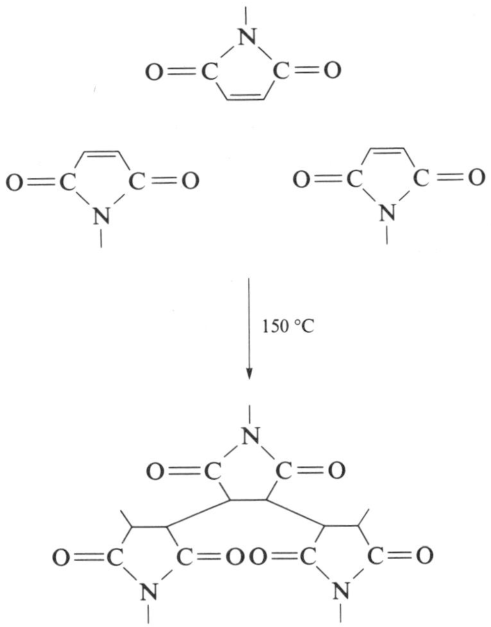

Fig. 10.7 Thermoset structure resulting from free-radical addition reaction of bismaleimides.

causes the terminal maleimide groups to react by a free-radical addition mechanism to produce a crosslinked thermoset structure (Fig. 10.7).

Although bismaleimides have the highly desirable potential processability of epoxies, they have one major disadvantage in that they depend, to a large degree, on a highly crosslinked structure for their high temperature capability. This can result in extreme brittleness and, in some cases, severe microcracking tendencies. Improvements in toughness are, for many applications, often seen as mandatory. Such improvements can be obtained by reducing crosslink density and imparting a more open, flexible molecular structure. This can be achieved by the use of high molecular weight, predominantly aliphatic, diamines in the initial reaction and/or chain extension of the bismaleimide molecule prior to crosslinking with, for example, diamines or hydrazides. Unfortunately by such an approach any resulting increase in toughness can be offset by reductions in other desirable properties such as T_g and thermal stability. As a result other means of toughness enhancement achievable without the damaging effects just mentioned are often considered necessary.

10.2.4 Other polymers

In addition to the polyimide variants discussed above other polymer types have been developed with the aim of combining a high temperature capability with acceptable processing characteristics. Notable examples include polyphenyl quinoxalines, phthalocyanines, polysulphones, polyphenylene sulphide and polyetheretherketone [2, 4]. However, since a detailed discussion of these materials is beyond the scope of this chapter, they will not be considered further.

10.3 RUBBER MODIFICATION

As discussed above, the crosslinked polyimides derived from terminally reactive oligomeric compounds can exhibit, in certain cases, extremely brittle behaviour. As discussed in Chapter 6 of this book, the modification of epoxies by elastomeric materials has been shown to produce extremely beneficial results with substantial toughness improvements being accompanied by relatively minor reductions in other important properties. As a result of this experience, a number of studies have been conducted in an attempt to determine whether a similar approach could be employed with polyimides derived from reactive oligomers. The remainder of this section, and indeed most of this chapter, will be concerned primarily with these studies. By far the majority of these investigations have been concerned with the elastomeric modification of bismaleimides and norbornene-terminated oligomers. Consequently, for convenience, these two areas will be considered separately with a third section being devoted to rubber modification studies into other polyimide types.

10.3.1 Elastomeric modification of bismaleimides

Before embarking on the main theme of this section, it is important to consider firstly some of the mechanical properties which have been measured for unmodified bismaleimides, since this will hopefully allow the reader to gain an understanding of the significance of the potential brittleness problem and hence the reasons for, and indeed the importance of, elastomeric modification.

Table 10.1 Fracture and mechanical property data for a cured unmodified bismaleimide

Property	*Value*
Fracture toughness, K_{Ic} ($MN\,m^{-3/2}$)	0.23
Fracture energy, G_{Ic} ($J\,m^{-2}$)	11
Modulus, E (GPa)	4.3
Glass transition temperature, T_g (°C)	~ 300
Lap shear strength (MPa)	4.1*

* Mild steel substrates, grit blast–degrease surface treatment, 20 °C test.
Data previously published in various forms in Refs. 35–37 by author and colleagues. © Crown Copyright.

Table 10.1 shows data obtained by Shaw and coworkers for a cured unmodified bismaleimide [35–37]. The system in question is a proprietary formulation comprising the three main molecular species shown in Fig. 10.8, which has in fact been designed specifically to allow the formation of a eutectic mixture having an advantageously low melting point of about 70 °C. Of particular relevance are the data for both fracture toughness, K_{Ic}, and fracture energy, G_{Ic}. As shown, a K_{Ic} value of 0.23 MN m$^{-3/2}$ corresponds to a G_{Ic} value of apoproximately 11 J m^{-2}, which is substantially lower than that frequently found with a range of unmodified epoxies (50–600 J m^{-2}). To put this figure in greater perspective, this fracture energy is broadly similar in magnitude to those of some of the most brittle ceramics and therefore demonstrates the improvements in toughness required if bismaleimides, particularly of the type discussed above, are to be used to their full potential. Also shown in Table 10.1 is a lap shear strength value obtained using the unmodified bismaleimide as an adhesive. A room temperature strength of 4.1 MPa can hardly be considered

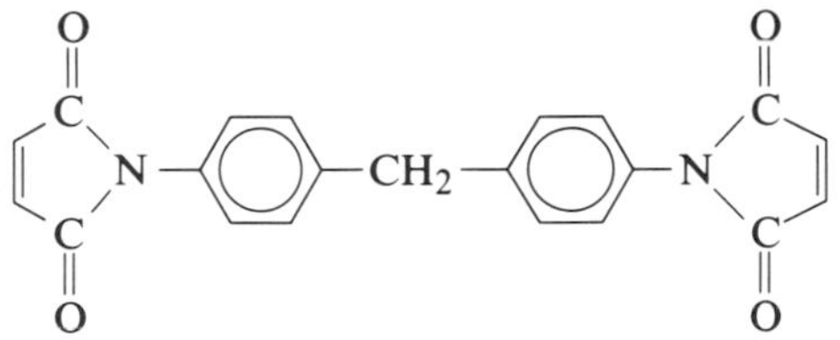

Di(4-maleimidophenyl)methane

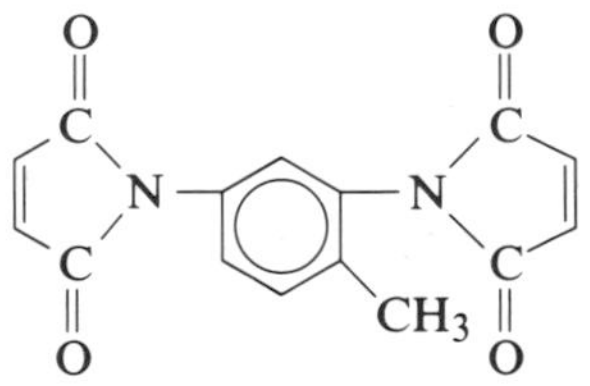

2,4-bismaleimidotoluene

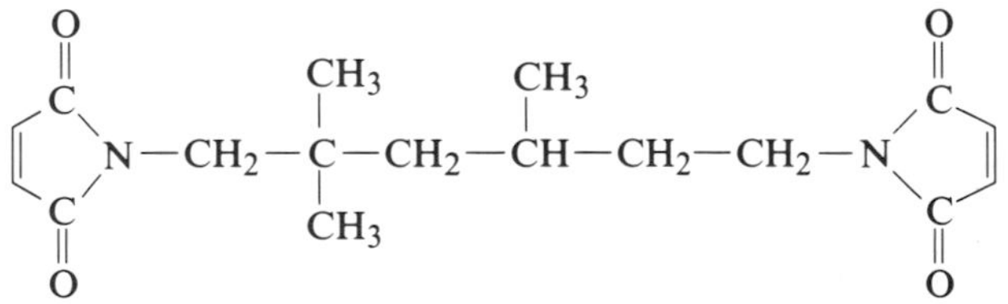

1,6-bismaleimido-2,2,4-trimethylhexane

Fig. 10.8 Constituents of the bismaleimide resin employed by Shaw and coworkers.

impressive and therefore demonstrates how extreme brittleness can influence other important properties. The high modulus and T_g values shown indicate, of course, the positive attributes of this particular bismaleimide formulation as a potential high temperature polymer.

Possibly the earliest attempt at toughness enhancement by rubber modification was conducted by McGarry and coworkers at the Massachusetts Institute of Technology in 1979, where an amine-terminated butadiene–acrylonitrile elastomer, having the structure shown in Fig. 10.9, was incorporated into a proprietary bismaleimide system [38]. Amine functionality was deliberately chosen for its potential reactivity with the bismaleimide via Michael addition. However, the toughness improvements obtained were extremely marginal.

More recently, Shaw and coworkers have conducted a series of investigations aimed at improving the toughness characteristics of the bismaleimide formulation described above [35–37]. Throughout these studies, one main type of terminally reactive elastomer was employed, namely a carboxyl-terminated butadiene–acrylonitrile elastomer having the structure shown in Fig. 10.10. In addition three variants of this elastomer type were investigated having acrylonitrile concentrations of 10%, 17% and 27% [37]. As discussed in Chapter 6, the amount of acrylonitrile contained within the elastomer has great importance in terms of polymer–elastomer compatibility and consequently the critically important phase separation process which is generally considered necessary for optimum toughness enhancement. Throughout this work, elastomer incorporation was conducted by simply adding the appropriate liquid rubber to the molten bismaleimide followed by thorough stirring. A pre-reaction period of up to 24 h at 120 °C was found to be critically important, this possibly being associated with a requirement for a certain degree of copolymer formation prior to any attempts to cure. Under the majority of circumstances curing was achieved by heating for 2 h at 170 °C followed by 5 h at 210 °C.

Bearing in mind the previous comments concerning the extremely low G_{Ic} of the unmodified bismaleimide, Fig. 10.11 demonstrates the considerable improvements in toughness which accompany CTBN addition (17% acrylonitrile) in varying concentrations. As clearly shown, the addition of up to and in excess of 100 parts per hundred (phr) of resin in polymer elastomer (equal parts bismaleimide and CTBN) results in over a fivefold increase in K_{Ic}. This corresponds to an approximately fiftyfold improvement in G_{Ic}, sufficient to raise toughness to a similar order to that obtained with the toughest unmodified epoxies. Similar observations have been found by Segal *et al.* [39] and Stenzenberger *et al.* [40] using the same system in a similar way.

Figure 10.11 further shows that variation in the amounts of acrylonitrile contained within the CTBN elastomer has a relatively minor effect on K_{Ic}, with 17% proving a marginally optimum figure. This observation can be attributed to the effect acrylonitrile will have on the phase separation process which should occur during cure. Simple visual observations of cast sheets have

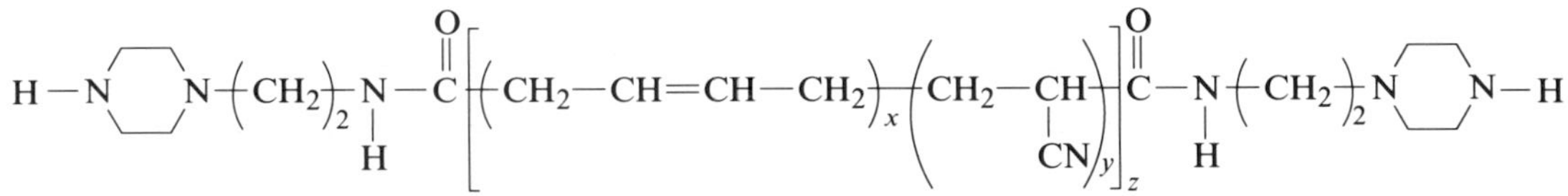

$$\text{H-N}\langle\text{piperazine}\rangle\text{N}\!-\!(\text{CH}_2)_2\!-\!\underset{\text{H}}{\text{N}}\!-\!\overset{\text{O}}{\overset{\|}{\text{C}}}\!-\!\left[(\text{CH}_2\!-\!\text{CH}{=}\text{CH}\!-\!\text{CH}_2)_x\!-\!\left(\text{CH}_2\!-\!\underset{\text{CN}}{\text{CH}}\right)_y\right]_z\!-\!\overset{\text{O}}{\overset{\|}{\text{C}}}\!-\!\underset{\text{H}}{\text{N}}\!-\!(\text{CH}_2)_2\!-\!\text{N}\langle\text{piperazine}\rangle\text{N-H}$$

Fig. 10.9 Amine-terminated butadiene–acrylonitrile rubber structure.

$$\text{HO}\!-\!\overset{\text{O}}{\overset{\|}{\text{C}}}\!-\!\left[(\text{CH}_2\!-\!\text{CH}{=}\text{CH}\!-\!\text{CH}_2)_x\!-\!\left(\text{CH}_2\!-\!\underset{\text{CN}}{\text{CH}}\right)_y\right]_z\!-\!\overset{\text{O}}{\overset{\|}{\text{C}}}\!-\!\text{OH}$$

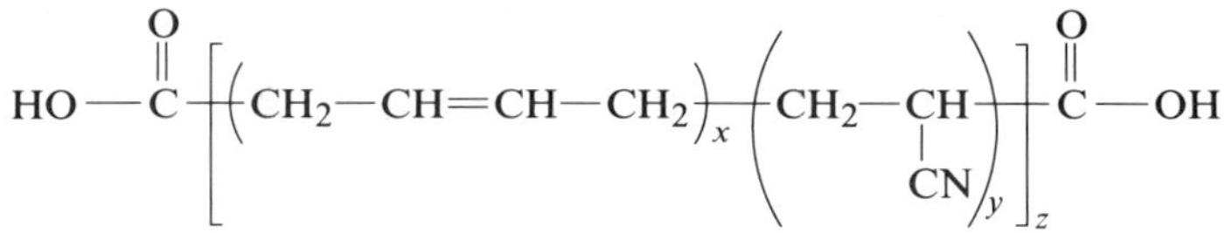

Fig. 10.10 Carboxyl-terminated butadiene–acrylonitrile rubber structure.

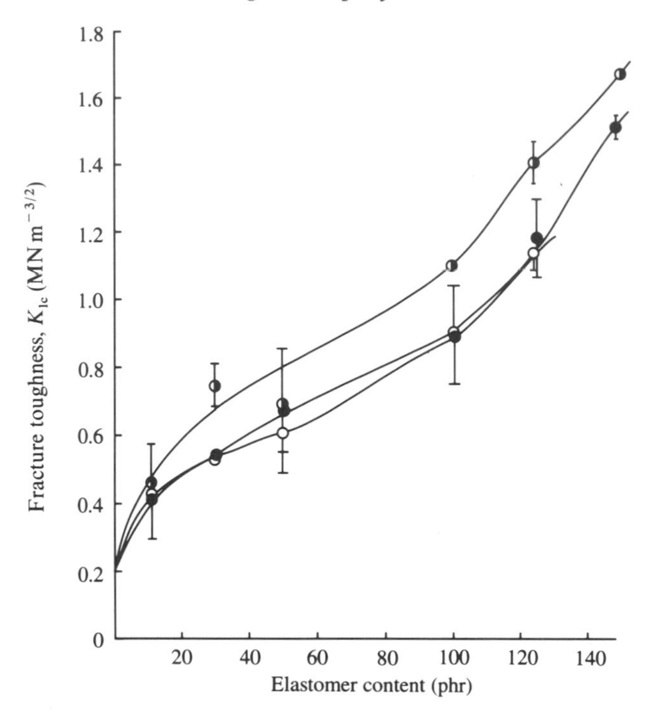

Fig. 10.11 Fracture toughness, K_{Ic}, as a function of CTBN rubber content: ○, 10% acrylonitrile rubber; ◑, 17% acrylonitrile rubber; ●, 27% acrylonitrile rubber. (Appears in paper by author in *Recent Advances in Polyimide Science and Technology* (eds M. R. Gupta and W. D. Weber), Society of Plastics Engineers, 1987. © Crown Copyright.)

shown increased casting transparency with increasing acrylonitrile level, suggesting a positive relationship between acrylonitrile concentration and bismaleimide–elastomer compatibility.

Fractographic investigations using, in particular, scanning electron microscopy have shown in many instances a similar two-phase morphology to that typically obtained with rubber modified epoxies [41].

Clearly for a high temperature application rubber modification should not result in significant deterioration in other important properties. Figure 10.12, obtained from the work of Shaw, shows the effect of both the concentration and the acrylonitrile content of CTBN elastomer on one such property, namely the modulus of the previously discussed bismaleimide system [37]. As would be expected, increasing elastomer concentration results in a modulus reduction varying from approximately 30% to 65% at an elastomer concentration of 100 phr. Interestingly, the severity of the decline depends on the nature of the rubber modifier with acrylonitrile in particular having a major influence. Clearly, if a substantial modulus reduction is to be avoided then CTBN elastomers having high acrylonitrile contents should be preferred, where, in

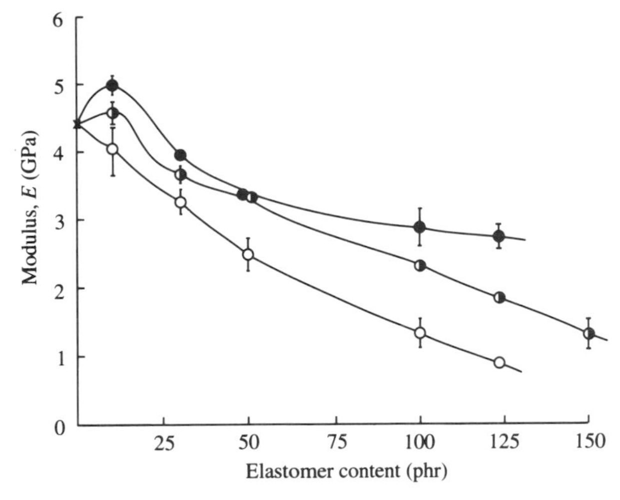

Fig. 10.12 Modulus as a function of CTBN rubber content: symbols as for Fig. 10.11. (Appears in paper by author in *Recent Advances in Polyimide Science and Technology* (eds M. R. Gupta and W. D. Weber), Society of Plastics Engineers, 1987. © Crown Copyright.)

the particular case of the CTBN containing 27% acrylonitrile, a rubber loading of 125 phr still allows a cured formulation having a room temperature modulus slightly in excess of 2.75 GPa. Although these modulus reductions can be regarded as substantial, it is of interest to note that the values obtained at high rubber loadings (50–100 phr) are roughly equivalent to those frequently found with various rubber modified epoxies [41]. The latter, however, generally have glass transition temperatures of approximately 100 °C which, as will be discussed, is some 100–200 °C lower than that usually found with the current rubber modified bismalemide systems.

One particularly important parameter that should be maintained relatively undamaged by the process of rubber modification is T_g, the glass transition temperature. This is of course of particular importance for polymers intended for use in high temperature applications since, to a first approximation at least, mechanical properties at elevated temperature are controlled by this parameter. Thus a measure of the extent of damage brought about by elastomer inclusion is critically important. This has been a factor studied in some detail by Shaw and coworkers using several different techniques [35–37]. Firstly, T_g values obtained from thermomechanical (TMA) experiments are shown in Fig. 10.13 for some of the previously described formulations. Somewhat surprisingly, addition of 10 phr CTBN increases T_g by approximately 40 °C, with further increases in rubber loading producing only a minor T_g decline from the maximum value. The precise reasons for this unusual effect are not fully understood. However, one could envisage the CTBN allowing a greater

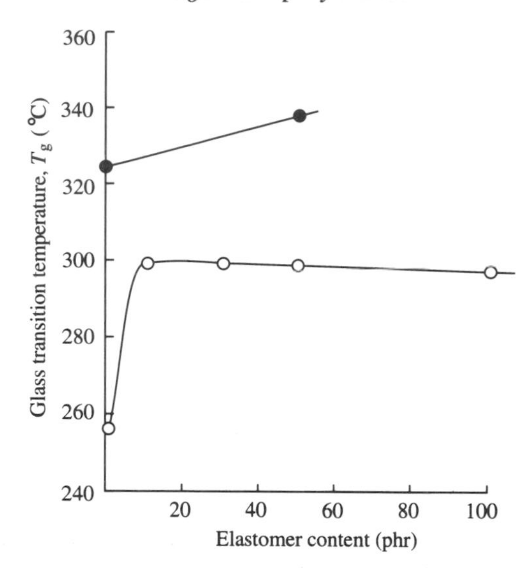

Fig. 10.13 Effect of rubber (CTBN) content on glass transition temperature, T_g, values obtained from thermomechanical analysis: ○, no post-cure; ●, post-cured at 240 °C for 16 h. (Published initially by A. J. Kinloch, S. J. Shaw and D. A. Tod in *Rubber-modified Thermoset Resins* (eds C. K. Riew and J. K. Gillham), *Advances in Chemistry Series*, **208**, American Chemical Society, 1984. © Crown Copyright.)

degree of crosslinking within the bismaleimide, thus improving T_g. A similar trend is further shown in Fig. 10.13 for two formulations which underwent a post-cure. However, as expected they exhibited greater T_g values. Similar effects have been observed by St. Clair and St. Clair whilst working with silicone rubber modified polyimides based on norbornenes (discussed later) [42]. In addition to TMA, differential scanning calorimetry (DSC) was also employed. A typical result obtained from these experiments is shown in Fig. 10.14. The trend can be regarded as anomalous in that the endothermic transition generally indicative of a glass transition was not observed with any of the formulations studied. Instead, substantial exothermic processes indicative of a post-cure phenomenon were observed. Two samples post-cured for 16 h at 240 °C which exhibited greatly reduced exotherms provided confirmatory evidence in support of this theory. Since the post-cure exotherm would be expected to swamp any endotherm occurring as a result of a glass transition, T_g could be assumed approximately equivalent to the onset of the exotherm. In all cases, this temperature was within 5–10 °C of the T_g values obtained from TMA.

Arguably one of the better techniques for studying mechanical behaviour as a function of temperature, and hence obtaining an accurate estimate of T_g, is dynamic mechanical spectroscopy. The current author has employed this technique extensively in his rubber modification studies since, in addition to

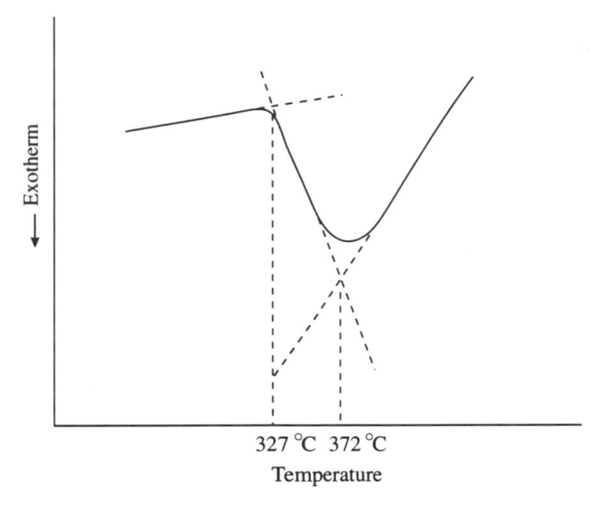

Fig. 10.14 DSC scan obtained from 30 phr CTBN (17% acrylonitrile). (Published initially by A. J. Kinloch, S. J. Shaw and D. A. Tod in *Rubber-modified Thermoset Resins* (eds C. K. Riew and J. K. Gillham), *Advances in Chemistry Series*, **208**, American Chemical Society, 1984. © Crown Copyright.)

T_g evaluation, it also allows an insight into the morphological characteristics of the polymer. Typical results obtained from some of these studies are shown in Fig. 10.15 for formulations containing 30, 50 and 100 phr of CTBN (17% acrylonitrile). For comparison, data obtained from a rubber modified epoxy are also included. The trend obtained for the epoxy can be considered typical of those generally found for a range of relatively low T_g thermosets, whereby a temperature increase eventually results in a catastrophic decline in storage shear modulus, G', which is accompanied by a pronounced loss peak at a temperature of, in this particular case, approximately 90 °C. However, the data obtained from the modified bismaleimides show substantially different trends, with the 30 and 50 phr formulations only showing a gradual decline in G' with increasing temperature, with no evidence of a catastrophic decline, at least up to the maximum temperature employed. Only at higher rubber concentrations does the G' trend begin to resemble that obtained for the modified epoxy, thus indicating in a somewhat superficial sense the high temperature superiority of the bismaleimides. Although these results initially appear to support the previously discussed T_g data obtained from TMA and DSC, an element of doubt has been expressed. Although the gradual decline in G' with increasing temperature is, of course, expected, the minor increases seen at temperatures in excess of 250 °C for the three formulations shown in Fig. 10.15 are not, and can probably be attributed to an increased crosslink density resulting from a post-cure. Clearly post-curing reactions would only be expected at temperatures sufficiently high to allow the required degree of molecular mobility.

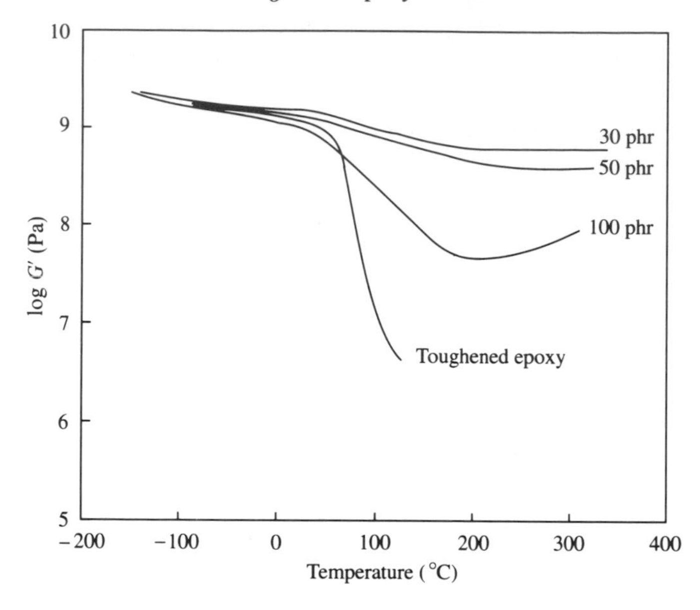

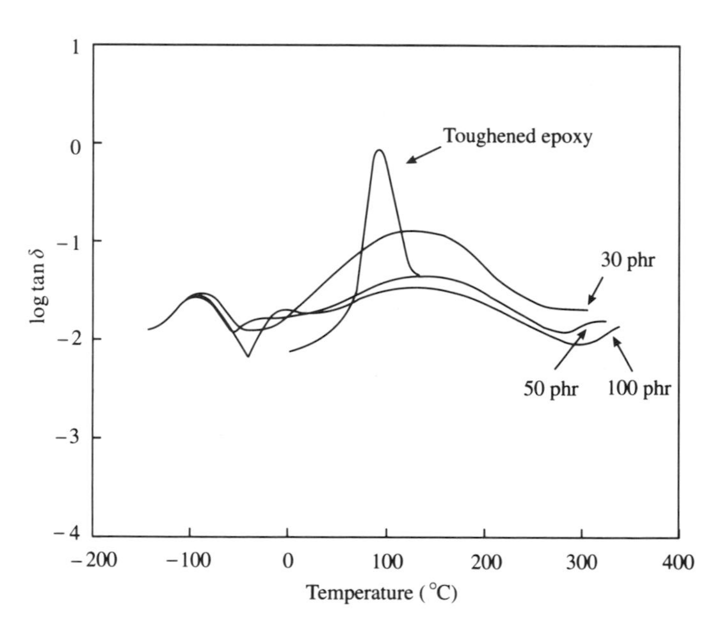

Fig. 10.15 Dynamic mechanical behaviour of toughened bismaleimide formulations containing 30, 50 and 100 phr CTBN additions (17% acrylonitrile). (Published initially by A. J. Kinloch, S. J. Shaw and D. A. Tod in *Rubber-modified Thermoset Resins* (eds C. K. Riew and J. K. Gillham), *Advances in Chemistry Series*, **208**, American Chemical Society, 1984. © Crown Copyright.)

These observations would suggest T_g values somewhat below those shown previously in Fig. 10.14, possibly in the region of 250–275 °C (lower for the 100 phr CTBN formulation). Experiments reported by Segal *et al.* [39] and Stenzenberger *et al.* [40] have essentially agreed with these findings, namely that the incorporation of relatively large concentrations of CTBN is capable of initiating significant T_g reductions.

It is important to recognize that the term 'high temperature resistance' can refer to more than one aspect of behaviour. With much of the data described so far the main theme has been concerned with maintenance of properties at elevated temperature for short periods of time. As discussed, this is primarily concerned with the glass transition temperature, T_g. However, other service requirements could involve high temperature excursions for long periods of time, i.e. many hundreds if not thousands of hours. Under these circumstances, the ability of the polymer to withstand such conditions without serious degradation would be of vital importance.

Several techniques exist for assessing the thermal stability of polymers including DSC and thermogravimetry (TG). The effects of CTBN incorporation on the thermal stability of the bismaleimide systems studied by Shaw and coworkers as assessed by TG are shown in Figs 10.16 and 10.17 for nitrogen and air environments respectively [37]. In the case of a nitrogen atmosphere, the decomposition temperature (taken as the temperature at which 10% weight loss occurs) is reduced from approximately 450 °C for the cured unmodified bismaleimide to about 400 °C for a 100 phr rubber formulation. In

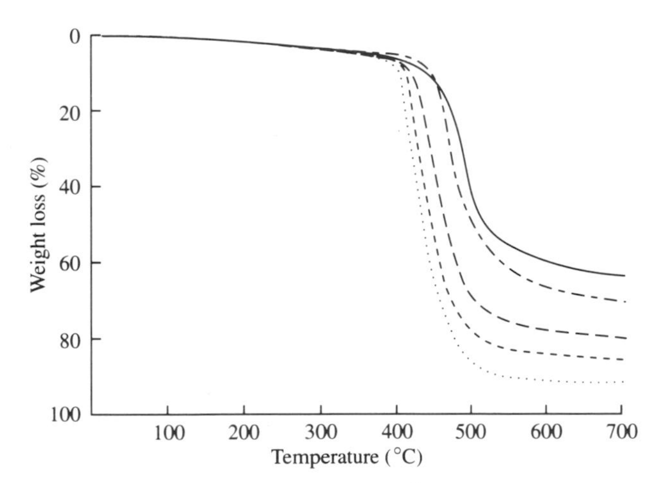

Fig. 10.16 Thermogravimetric analysis (5 °C min^{-1}) of various bismaleimide formulations in a nitrogen atmosphere: ——, 0 phr CTBN (17% ACN); -----, 10 phr CTBN (17% ACN); — —, 30 phr CTBN (17% ACN); – – –, 50 phr CTBN (17% ACN); . . . . , 100 phr CTBN (17% ACN). (Published initially by A. J. Kinloch, S. J. Shaw and D. A. Tod in *Rubber-modified Thermoset Resins* (eds C. K. Riew and J. K. Gillham), *Advances in Chemistry Series*, **208**, American Chemical Society, 1984. © Crown Copyright.)

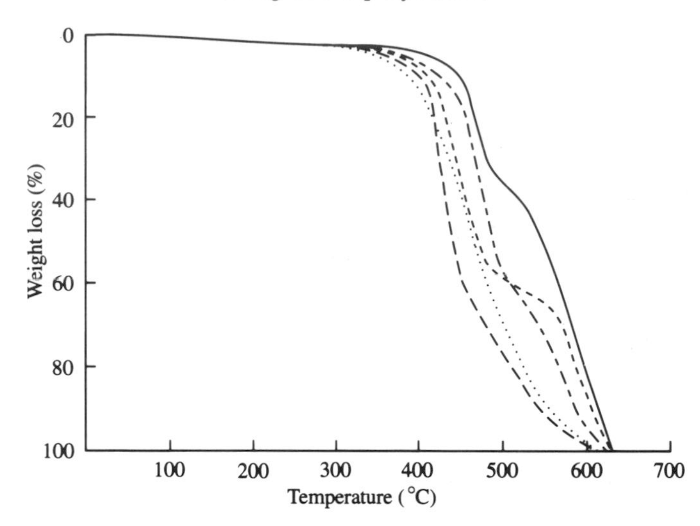

Fig. 10.17 Thermogravimetric analysis (5 °C min^{-1}) of various bismaleimide formulations in an air environment: symbols as for Fig. 10.16. (Published initially by A. J. Kinloch, S. J. Shaw and D. A. Tod in *Rubber-modified Thermoset Resins* (eds C. K. Riew and J. K. Gillham), *Advances in Chemistry Series*, **208**, American Chemical Society, 1984. © Crown Copyright.)

addition, increasing rubber concentration also increases the total weight lost by the polymer from about 60% for the unmodified system to approximately 85% for the 100 phr modified formulation.

Figure 10.17 shows similar data obtained from an air environment with decomposition temperatures only slightly lower than those obtained in nitrogen. However, in this case a two-stage degradation process occurs with an initial steep decline in weight being followed by a somewhat reduced rate of weight loss, followed in turn by a catastropic decline, resulting in complete degradation of the polymer. Similar observations have been found in numerous other investigations involving a wide range of polymers.

Although thermogravimetry can be considered a useful technique for estimating thermal stability, it is important to recognize its limitations, which have been discussed in detail by Critchley, Knight and Wright [2]. For a more accurate assessment of thermal stability it is necessary to measure mechanical performance as a function of time at the envisaged service temperatures. Shaw has employed this approach to study the influence of thermal aging on two major properties, namely fracture toughness, K_{Ic}, and modulus, using a CTBN modified bismaleimide [37]. In particular the effect of thermal aging in air at 170 °C on room temperature fracture toughness, K_{Ic}, of three bismaleimide–CTBN formulations is shown in Fig. 10.18. As indicated, the 30 phr system undergoes a reduction in K_{Ic} of approximately 30% over 1000 h. The 50 phr formulation shows signs of a peak in toughness within the first 50 h of

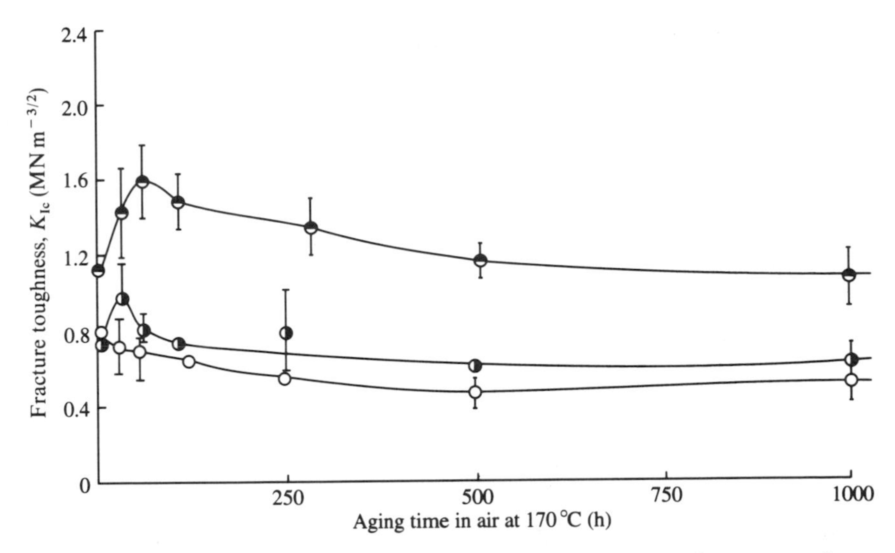

Fig. 10.18 Effect of aging in air at 170 °C on the room temperature fracture toughness of three bismaleimide–CTBN rubber (17% ACN) formulations: ◓, 100 phr CTBN; ◑, 50 phr CTBN; ○, 30 phr CTBN. (Published initially by A. J. Kinloch, S. J. Shaw and D. A. Tod in *Rubber-modified Thermoset Resins* (eds C. K. Riew and J. K. Gillham), *Advances in Chemistry Series*, **208**, American Chemical Society, 1984. © Crown Copyright.)

aging followed by a return to the original toughness value. However, statistical analysis of these data has indicated some uncertainty as to the true existence of the peak. For the 100 phr formulation, a similar but more pronounced initial rise in K_{Ic} is statistically significant, and it is interesting to note that the peak value, 1.6 MN m$^{-3/2}$, corresponds to a fracture energy, G_{Ic}, of approximately 900 J m^{-2}, which is about 90 times greater than obtained with the original cured unmodified material. The gentle decline in K_{Ic} that follows this peak results in a toughness value, after 1000 h at 170 °C, virtually identical to that of the unaged material.

Similarly the effect of thermal aging on room temperature modulus for the same formulations is shown in Fig. 10.19. As indicated, the 30 and 50 phr CTBN formulations show no significant variations in modulus over 1000 h at 170 °C. The 100 phr CTBN formulation, however, shows a substantial increase in modulus over this time period.

Thus, although 170 °C cannot be considered a particularly harsh environment, Shaw's studies have indicated that rubber modified bismaleimides can exhibit a surprising degree of thermal stability for a system containing a rubber which would not be expected to exhibit a high degree of thermal resistance. Other workers have found similar behaviour, in some cases at substantially higher temperatures, as will be discussed later.

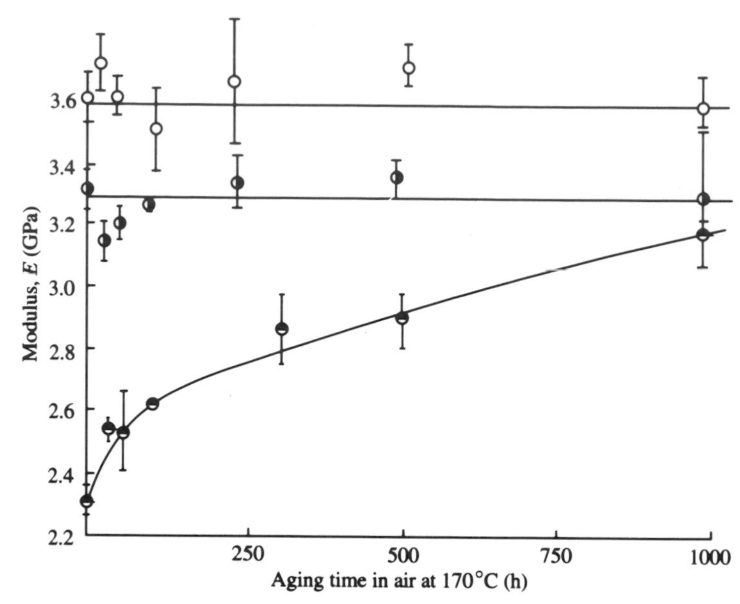

Fig. 10.19 Effect of aging in air at 170 °C on the room temperature modulus (at 20 °C) of three bismaleimide–CTBN rubber (17% ACN) formulations: symbols as for Fig. 10.18. (Published initially by A. J. Kinloch, S. J. Shaw and D. A. Tod in *Rubber-modified Thermoset Resins* (eds C. K. Riew and J. K. Gillham), *Advances in Chemistry Series*, **208**, American Chemical Society, 1984. © Crown Copyright.)

In addition to the work of Shaw and coworkers, other researchers have also attempted toughness enhancement using CTBN rubbers. For example Takeda and Kakiuchi have recently attempted the modification of bismaleimides using this approach [43]. In this work an attempt was made to relate the thermal and mechanical properties of rubber modified bismaleimides to both the quantity and the composition (acrylonitrile content) of the incorporated rubbers. Three types of bismaleimides were employed in this work, the structures of which are shown in Fig. 10.20. The first of these, 4,4′-bismaleimidodiphenylmethane (BDM) is a system which forms the basis of many proprietary bismaleimide formulations. The other two, 2,2-bis[4-(4-maleimidophenoxy)phenyl] propane (BPPP) and bis[4-(3-maleimidophenoxy)phenyl] sulphone (3,3′-BPPS) were prepared by Takeda, Akiyama and Kakiuchi [44]. These three species were studied in two combinations: a mixture of BDM 50 wt% and BPPP 50 wt%; a mixture of BDM 50 wt% and 3,3′-BPPS 50 wt%. Bismaleimide combinations such as these were found to produce eutectic mixtures, resulting in low melting points and long gel times, characteristics found by Shaw and Kinloch to be of particular importance for successful rubber modification [36, 37].

The rubber modifiers employed in this work were virtually identical to those previously discussed, namely CTBN rubbers having acrylonitrile contents of

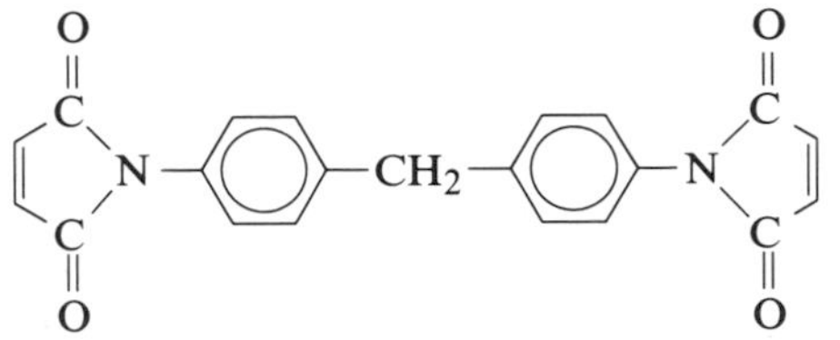

BDM

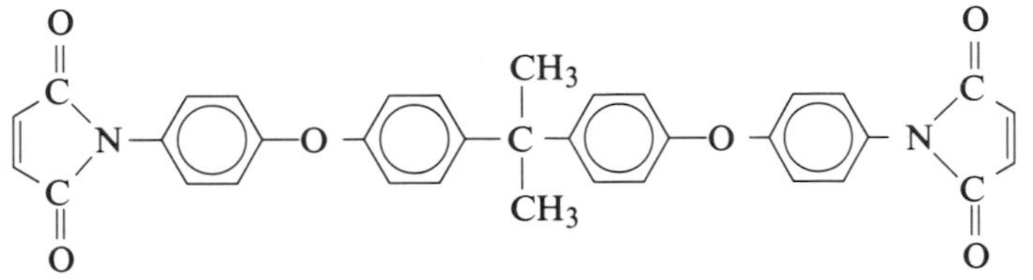

BPPP

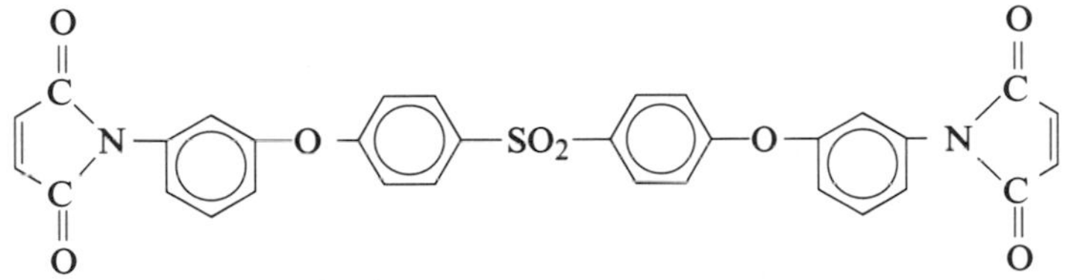

3,3′-BPPS

Fig. 10.20 Bismaleimide molecular species subjected to rubber modification studies by Takeda and Kakiuchi. (Molecular structures originally published by these authors in *Journal of Applied Polymer Science*, **35**, 1351, 1988.)

from 0% (in effect a carboxyl-terminated butadiene rubber) to 27 wt%. As in the work of Shaw and coworkers, Takeda and Kakiuchi found that pre-reaction of CTBN with the bismaleimide formulations was necessary in order to achieve significant improvements in properties and in this case employed pre-reaction conditions of 5 h at 130 °C. Interestingly, analysis of the reaction medium during the pre-reaction process indicated that the terminal carboxyl groups on the CTBN molecules did not participate in any reaction with the bismaleimide species; co-reaction almost certainly occurred between the unsaturation in the butadiene portion of the CTBN molecules and the terminal maleimide functionality of the bismaleimides, thus agreeing with the findings of Stenzenberger *et al.* [45]. Takeda and Kakiuchi [43] found that rubber modification of the above formulations resulted in quite substantial improvements in toughness, where CTBN additions to 100 phr resulted in

G_{Ic} values of approximately 300 J m^{-2}. Unfortunately G_{Ic} values for the unmodified systems were not quoted owing to experimental difficulties associated with such brittle materials. Consequently a precise indication of the degree of toughness improvement is not possible. In addition, the effect of variation in the acrylonitrile content of the CTBN rubbers on toughness was also investigated, with an inverse relationship between G_{Ic} and acrylonitrile content being demonstrated. Although this would appear, at first glance, to contradict the results of Shaw [37], where no statistically significant trend was apparent, it is important to recognize that the latter did not study such an extensive range of acrylonitrile contents; such an investigation may well have demonstrated a similar trend.

As found in numerous previous studies on both epoxies and to a far lesser extent on bismaleimides, CTBN incorporation resulted in a reduction in modulus, the severity of the decline with increasing CTBN concentration being similar in magnitude to that found by Shaw and coworkers [35–37]. However, Takeda and Kakiuchi found that the acrylonitrile content of the CTBN modifier exhibited no significant influence on modulus, a result substantially different to that previously described (Fig. 10.12), where the harmful tendencies of CTBN incorporation were dependent for their magnitude on the level of acrylonitrile present in the rubber.

Glass transition temperatures obtained by Takeda and Kakiuchi were relatively unaffected by rubber incorporation; only at CTBN concentrations greatly in excess of 50 phr were substantial reductions in T_g found. This generally agrees with the previously discussed work of Shaw and coworkers [35–37] where it was concluded that a high temperature capability would only become seriously impaired at rather extreme levels of rubber loading.

It is known that the conditions under which a thermosetting polymer is cured can have a considerable effect of properties. For example, studies conducted by Shaw and Tod on a rubber modified epoxy have shown that variations in cure temperature, in particular, can exert a considerable influence on toughness [46]. In order to determine whether similar behaviour would occur with rubber modified bismaleimides, the present author has recently conducted a similar cure investigation with these materials. Typical K_{Ic} results obtained from a formulation containing 50 phr CTBN (17% acrylonitrile), covering cure temperatures from 170 °C to 250 °C over cure times of 2–10 h, are shown in the form of a contour diagram in Fig. 10.21. The techniques employed in the construction of this diagram have been described elsewhere [46]. As indicated, variations in the cure conditions shown result in K_{Ic} changes of between 0.60 and 1.05 MN m$^{-3/2}$, i.e. a significant variation. As would perhaps be expected, the lower cure temperature regimes tended to promote the highest K_{Ic} values, the higher cure temperatures almost certainly resulting in greater crosslink densities and thus comparative brittleness.

Figure 10.22, which shows a similar contour diagram, this time constructed from modulus data, indicates definite trends regarding modulus variation as a

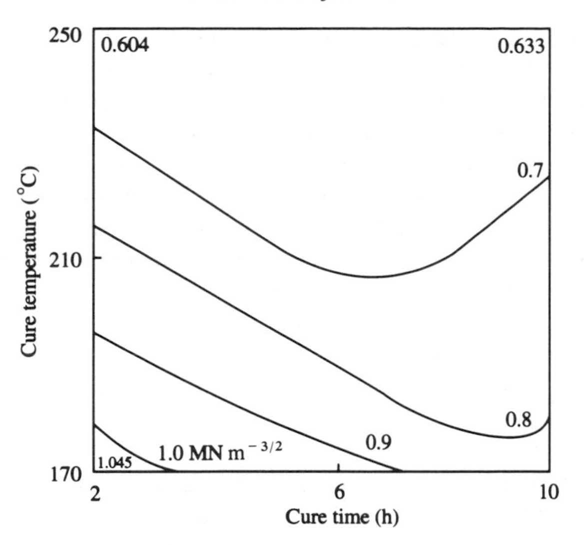

Fig. 10.21 Influence of cure conditions (temperature and time) on fracture toughness, K_{Ic} (formulation containing 50 phr of CTBN).

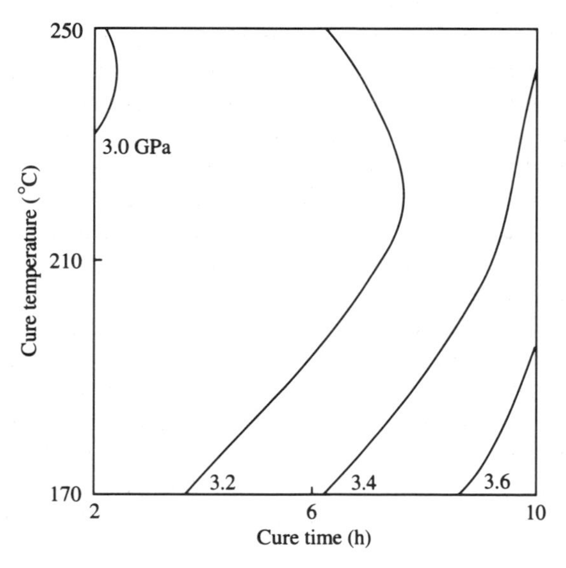

Fig. 10.22 Influence of cure conditions (temperature and time) on modulus, *E*. (Author's own data, previously unpublished.)

function of cure temperature and time. As indicated, a decrease in cure temperature and an increase in cure time are shown to increase modulus. Although the cure temperature–modulus relationship may seem intuitively strange, it is of interest to note that similar behaviour has been observed by others, where free volume related effects have been considered responsible [46–48]. Despite these trends it is of interest to note that modulus does not vary substantially with the cure conditions studied, so that the increases in toughness resulting from the decrease in cure temperature noted earlier (Fig. 10.21) do not result at the expense of modulus.

The results indicated in both Figs 10.21 and 10.22 clearly show the variation in properties which can occur by variation in cure conditions, which thus allows, in addition to variations in formulation variables, considerable scope for tailoring properties to specific requirements.

In addition to the early work of McGarry and coworkers [38] attempts at employing amine-terminated butadiene–acrylonitrile elastomers (ATBN) for toughness enhancement have been conducted by others. For example Varma *et al.* have employed a proprietary ATBN in an attempt to improve the properties of two phosphorus containing bismaleimides, the structures of which are shown in Fig. 10.23 [49]. A solution process (employing dimethyl formamide as solvent) whereby the ATBN was incorporated into the bismaleimide solution resulted in graphite–imide composites having mechanical properties which were an improvement on non-modified systems, provided that elastomer concentration was restricted to no greater than 7% by weight. A study of neat resin morphology using scanning electron microscopy indicated a distinctive two-phase morphology with wide distributions of particle size being apparent. Increasing ATBN concentration was found both to increase the size of the rubber particles and to promote a measure of irregularity. Unfortunately, quantitative assessments of toughness enhancement using fracture mechanics principles so as to yield values of fracture toughness, K_{Ic},

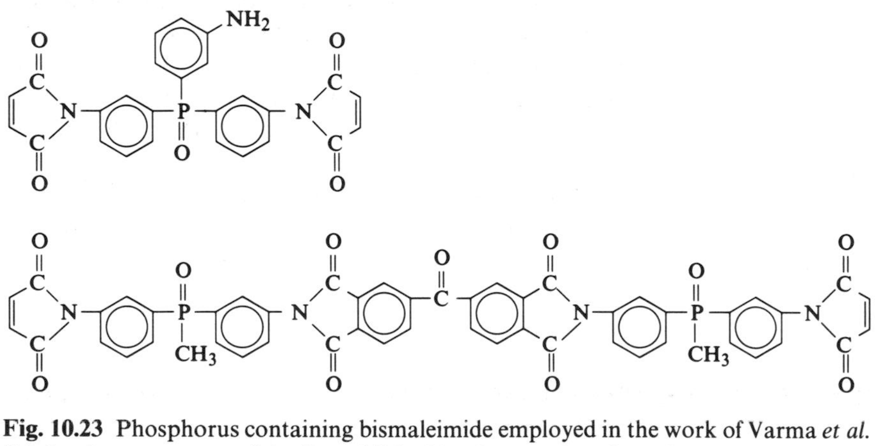

Fig. 10.23 Phosphorus containing bismaleimide employed in the work of Varma *et al.* (Molecular structures originally published by these authors in *Polyimides – Synthesis, Characterization and Applications* (ed. K. L. Mittal), Vol. 2, Plenum Press, 1984.)

and fracture energy, G_{Ic}, were not employed. It is, therefore, difficult to gauge accurately the toughness improvements obtained from this study.

Recent work by Maglio *et al.* has also investigated the incorporation of an ATBN elastomer into a bismaleimide system based on 4,4′-bismaleimidodiphenylmethane and 1,4-piperazine [50]. By incorporating ATBN at varying concentrations into the bismaleimide by Michael type addition reactions, products exhibiting only a modest sacrifice in thermal stability (at elastomer concentrations not exceeding 30 wt%), together with evidence of microphase separation, were obtained. Unfortunately, no measure of toughness enhancement was provided by the authors.

In addition to carboxyl-terminated elastomers, Shaw [37] has also conducted relatively limited investigations using a vinyl-terminated butadiene–acrylonitrile elastomer (VTBN), the structure of which is shown in Fig. 10.24.

Before discussing some of the results obtained with this elastomer it is desirable to discuss briefly the changes in chemistry which are likely to accompany modifications in elastomer terminal functionality. As previously mentioned, amine terminal groups (ATBNs) would be expected to participate in Michael type additions with terminal maleimide. Carboxyl end groups, however, would not be expected to participate in such a reaction. Indeed, as mentioned above, Takeda and Kakiuchi [43] together with Stenzenberger *et al.* [45] have shown this to be so with, in the case of bismaleimide–CTBN formulations, reactions between the two species occurring via terminal maleimides and the unsaturated butadiene portions of the elastomer. With vinyl-terminated elastomers, although reaction via the butadiene unsaturation will still be available, further reaction via a free-radical addition mechanism between the terminal vinyl groups on the elastomer and the bismaleimide would seem likely. Thus, bearing in mind this potentially more complex chemistry, it was considered worthwhile by Shaw to determine whether simple substitution of CTBN for VTBN would result in property changes. The effects on both fracture toughness, K_{Ic}, and modulus are shown in Figs 10.25 and 10.26 respectively. As indicated, the presence of either a carboxyl- or a vinyl-terminated elastomer has no major influence on either property, suggesting that any increase in the complexity of the chemistry has no major influence on mechanical behaviour.

In comparison with elastomers based on butadiene and acrylonitrile, investigations employing other elastomeric types devoted principally to bismaleimides have been far more limited in scope. Perhaps most interest has

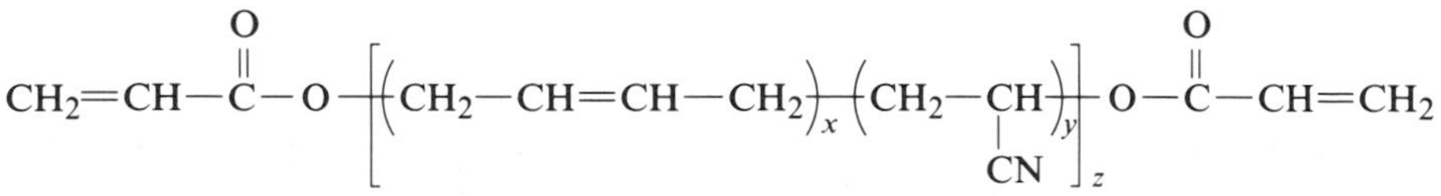

$$CH_2{=}CH{-}\overset{\overset{\displaystyle O}{\|}}{C}{-}O{-}\left[\left(CH_2{-}CH{=}CH{-}CH_2\right)_x\left(CH_2{-}\underset{\underset{\displaystyle CN}{|}}{CH}\right)_y\right]_z{-}O{-}\overset{\overset{\displaystyle O}{\|}}{C}{-}CH{=}CH_2$$

Fig. 10.24 Vinyl-terminated butadiene–acrylonitrile rubber structure.

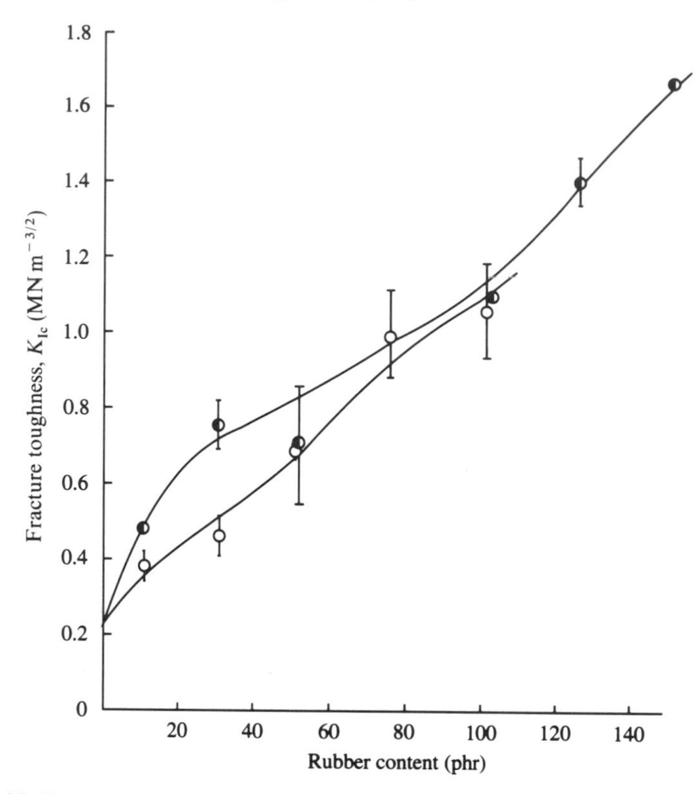

Fig. 10.25 Fracture toughness, K_{Ic}, as a function of rubber content for vinyl-terminated (○) and carboxyl-terminated (◐) rubbers. (Published initially by A. J. Kinloch, S. J. Shaw and D. A. Tod in *Rubber-Modified Thermoset Resins* (eds C. K. Riew and J. K. Gillham), *Advances in Chemistry Series*, **208**, American Chemical Society, 1984. © Crown Copyright.)

concerned studies by Maudgal and St. Clair where attempts have been made to incorporate a polysiloxane species into a bismaleimide structure [51]. The precise preparatory formulation chosen consisted of three species, namely maleic anhydride, benzophenone tetracarboxylic dianhydride and bis(γ-aminopropyl)tetramethyldisiloxane. These were co-reacted so as to yield a bismaleimide prepolymer of the type shown in Fig. 10.27, where attempts were made to exploit the properties of this siloxane containing polyimide in the form of an adhesive. In this respect successful results were obtained in that siloxane incorporation resulted in a room temperature lap shear strength of approximately 15 MPa. A number of investigations employing non-rubber-modified bismaleimide adhesive formulations have shown lap shear values of the order of 5 MPa, thus indicating the potential benefits of siloxane incorporation. In addition, the modified polymer was shown to be capable of considerably lower temperature processing in comparison with more conventional

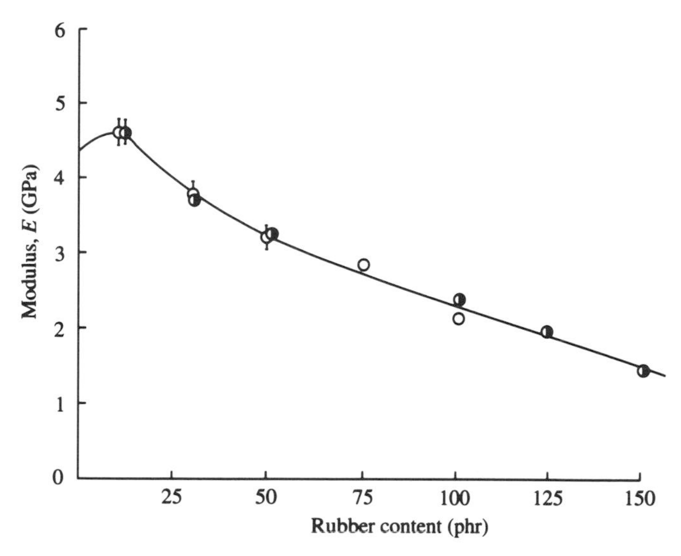

Fig. 10.26 Modulus as a function of rubber content for vinyl-terminated (○) and carboxyl-terminated (◑) rubbers. (Published initially by A. J. Kinloch, S. J. Shaw and D. A. Tod in *Rubber-Modified Thermoset Resins* (eds C. K. Riew and J. K. Gillham), *Advances in Chemistry Series*, **208**, American Chemical Society, 1984. © Crown Copyright.)

aromatic base bismaleimides, a factor which can be regarded as a significant advantage. On the negative side, Maudgal and St. Clair observed a lower thermo-oxidative stability for the siloxane containing bismaleimide than generally observed with predominantly aromatic bismaleimides. It is also likely that siloxane incorporation would result in substantial reductions in both T_g and modulus; indeed, the processing attributes mentioned above can be regarded as a direct result of the deleterious effect of siloxane incorporation on the former.

Before considering the use of elastomers in norbornene imide systems, it is of interest to discuss briefly one further advantage, associated with water absorption, which rubber modification has been shown to confer on bismaleimides. Water absorption behaviour can be a particularly important characteristic for polymers intended for adhesive and composite applications, where warm–moist atmospheres have long been recognized as potentially harmful. Perhaps of greater significance in this case is the effect absorbed water is known to have on the glass transition temperature of many polymers. A 20 °C reduction in T_g for every 1% of water absorbed is a commonly quoted phrase [52]. Thus, for hydrophilic polymers, exposure to a moist environment can have a particularly damaging effect on T_g and properties associated with it. The potential implications for polymers designed for high temperature applications can, of course, be particularly important.

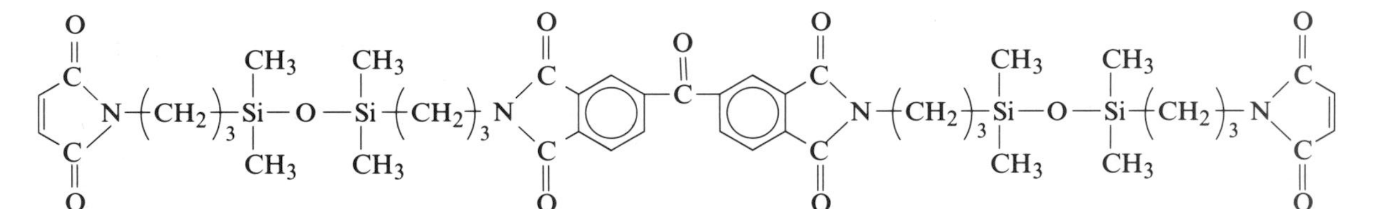

Fig. 10.27 Siloxane containing bismaleimide prepolymer species studied by Maudgal and St. Clair. (Molecular structure originally published by these authors in 29th National SAMPE Symposium, p. 437, 1984.)

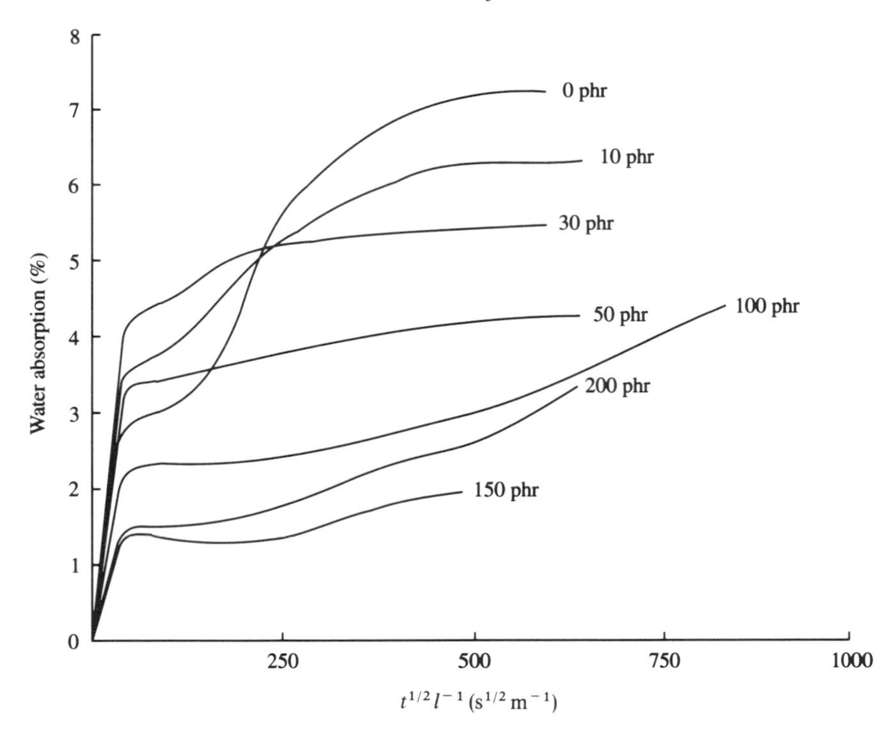

Fig. 10.28 Influence of rubber modification on the water absorption characteristics (60 °C; liquid H_2O) of a cured bismaleimide polymer. (Originally published by D. A. Tod and S. J. Shaw in *British Polymer Journal*, **20**, 397, 1988. © Crown Copyright.)

The beneficial effects of rubber modification, observed from a series of detailed studies by Tod and Shaw, are demonstrated in Fig. 10.28, which shows water uptake plotted as a function of the square root of the time normalized for specimen thickness [53]. For convenience it is necessary to divide the absorption curves into two parts, i.e. at low ($<250\,s^{1/2}\,m^{-1}$) and high ($>250\,s^{1/2}\,m^{-1}$) absorption times. At low absorption times the addition of up to 30 phr of CTBN results in a substantial increase in the amount of water absorbed. However, rather surprisingly, further increases in CTBN concentration beyond 30 phr result in a reversal of this trend. That the addition of a rubber of fairly high polarity to a relatively non-polar bismaleimide should result in reduced hydrophobicity is not unreasonable. However, the eventual reversal in behaviour for rubber concentrations in excess of 30 phr is more difficult to explain. A study of the profiles at $>250\,s^{1/2}\,m^{-1}$, however, provides a clue. For the unmodified bismaleimide a substantial change in absorption behaviour occurs beyond this absorption time where, following initial tendencies towards a plateau, water uptake rate begins to increase and continues to rise rapidly. As indicated, the 10 phr material undergoes a similar but not so severe rise, with the other formulations containing higher rubber loadings being generally unaffected. Clearly therefore, although initially reducing detrimentally the hydrophobic characteristics of the bismaleimide, the presence of rubber actually improves environmental stability. Further studies using scanning electron microscopy have indicated quite extensive cracking in both unmodified and 10 phr CTBN systems, particularly the former. As a result of these findings Tod and Shaw proposed that the two main characteristics shown in Fig. 10.28, i.e. the reversal in absorption characteristics at >30 phr CTBN and the two-stage absorption process exhibited by non- and low CTBN systems, were due primarily to crack formation within the polymer samples, caused by the generation of internal pressure by absorbed moisture. The presence of void and crack formation is generally considered to be a factor responsible for excessive water uptake by polymers and it was therefore proposed that rubber incorporation results in sufficient elevation of toughness so as to reduce substantially the extent of crack damage and hence water uptake.

10.3.2 Elastomeric modification of norbornene-terminated oligomers

Most studies concerning the elastomeric modification of norbornene-terminated imide oligomers have been conducted by St. Clair and St. Clair at NASA Langley, with most, if not all, of this work being devoted to the toughness enhancement of LARC-13 [7, 20, 42, 54].

As mentioned previously in this chapter, LARC-13 exhibits somewhat brittle characteristics exhibiting a fracture energy, G_{Ic}, of the order of $100\,J\,m^{-2}$. Although this value can be viewed in an optimistic light in relation to some of the G_{Ic} values previously quoted for unmodified bismaleimides, its brittle behaviour, relative to some thermoplastic based polyimide systems, has resulted in less than desirable adhesive joint characteristics, thus making it a ripe target for attempts at rubber modification.

Serious attempts at rubber toughening were begun in 1979. Two main approaches were employed, making use primarily of the elastomeric systems shown in Fig. 10.29. The first approach involved the physical blending of the fluorosilicone (FSE) and vinyl-terminated silicone (VTS) into LARC-13 whilst in its amic acid solution form [7]. The second approach, utilizing the two amine-terminated elastomers (ATBN and ATS), involved chemically reacting the elastomer into the LARC-13 backbone by simply replacing a portion of the original diamine, 3,3′-methylenedianiline, with the appropriate amine-terminated elastomer [7]. The main aims of these investigations were to study the effects of elastomer incorporation on properties and parameters such as T_g, thermal stability, adhesive joint strength and, of course, fracture energy, G_{Ic}.

With the exception of the FSE, elastomer incorporation at a level of 15 wt% produced a quite substantial toughening effect, particularly with the two amine-terminated rubbers (ATBN and ATS) with approximately fourfold

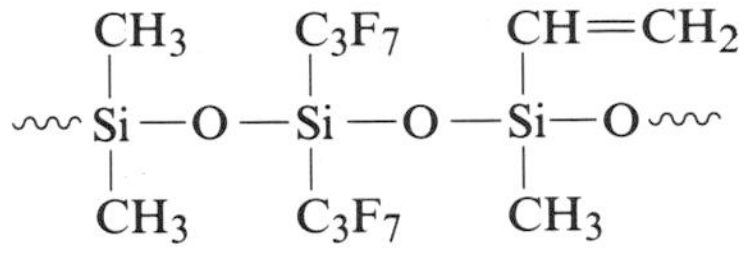

$$\sim Si(CH_3)_2-O-Si(C_3F_7)_2-O-Si(CH=CH_2)(CH_3)-O\sim$$

FSE

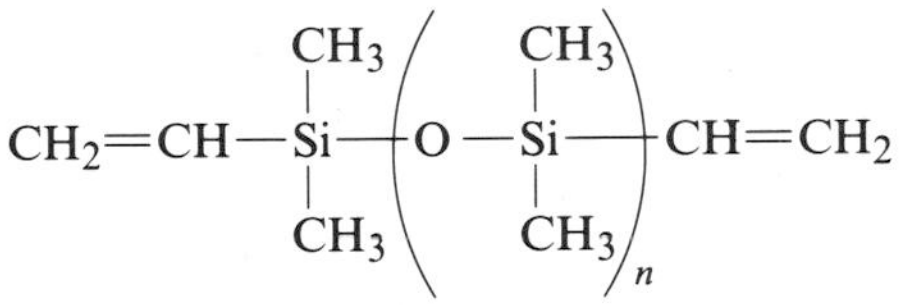

$$CH_2=CH-Si(CH_3)_2-\left(O-Si(CH_3)_2\right)_n-CH=CH_2$$

VTS

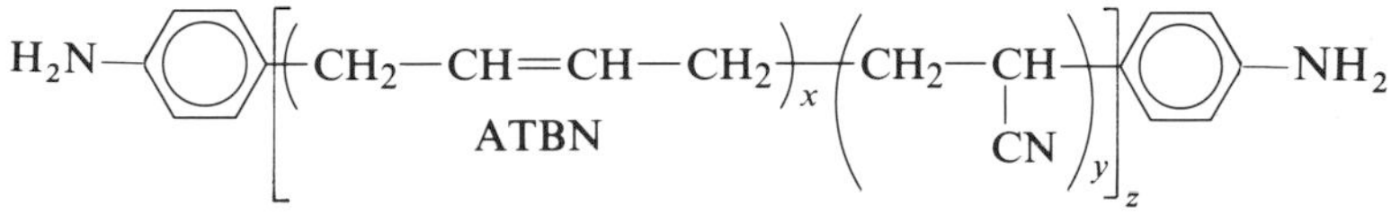

$$H_2N-C_6H_4-\left[\left(CH_2-CH=CH-CH_2\right)_x\left(CH_2-CH(CN)\right)_y\right]_z-C_6H_4-NH_2$$

ATBN

$$H_2N-C_6H_4-R-Si(CH_3)_2-\left(O-Si(CH_3)_2\right)_n-R-C_6H_4-NH_2$$

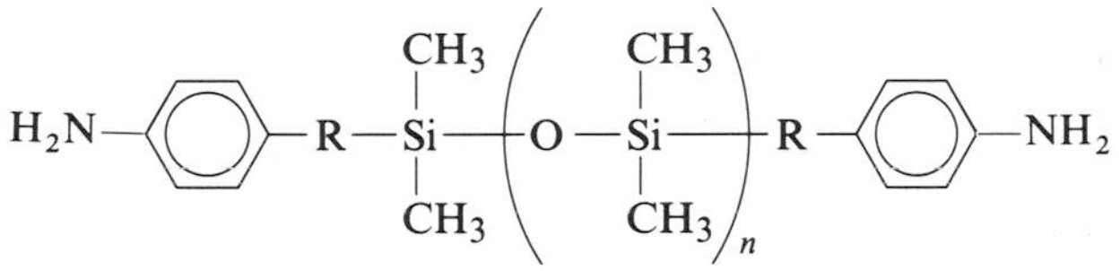

ATS

Fig. 10.29 Elastomer types employed in the toughening of LARC-13.

improvements in G_{Ic} being observed. As shown in Table 10.2 toughness enhancement was, at best, seen to provide only marginal improvements in room temperature lap shear strength. Indeed, in the majority of cases, particularly under elevated temperature test conditions, elastomer incorporation can be regarded as having a somewhat damaging effect. However, a different trend was shown to exist for T peel strength as indicated in Table 10.2. With this test geometry, elastomer incorporation produced a quite substantial reinforcement effect at both room temperature and 232 °C.

Thermomechanical properties studied using torsional braid analysis provided evidence of a two-phase morphology. For all four modified systems this technique indicated a major transition at temperatures between 277 and 300 °C together with a low temperature transition in the region from − 65 to − 115 °C. The former can be attributed to the glass transition of the cross-linked LARC-13 phase with the latter being clearly attributable to the elastomer T_g. Particularly noteworthy was the observation that elastomer incorporation at the level studied resulted, with the possible exception of the fluorosilicone elastomer, in a negligible reduction in T_g.

Table 10.2 Influence of elastomer incorporation on the adhesive joint strength characteristics of LARC-13*

	Lap shear strength (MPa)		*T peel strength ($N\,m^{-1}$)*	
Adhesive	*20 °C*	*232 °C*	*20 °C*	*232 °C*
LARC-13	20	19	230	510
LARC-13–FSE	15	12	1520	910
LARC-13–VTS	23	17	540	960
LARC-13–ATBN	25	19	960	680
LARC-13–ATS	18	12	1490	1090

* Titanium adherends.
Obtained from data previously published by St. Clair and St. Clair in *Polyimides–Synthesis, Characterization and Applications* (ed. K. L. Mittal), Vol. 2, Plenum Press, 1984.

Thermal stability, as measured by thermogravimetry, showed that, although LARC-13 was the most thermally stable resin, none of the four elastomers studied produced any major undesirable effects. Only after elevated temperature aging did some measure of undesirability emerge. For example, after aging at 232 °C followed by dynamic thermogravimetry, ATBN incorporation exhibited a reduced level of thermal stability. However, the incorporation of the vinyl-terminated silicone (VTS) resulted in an actual improvement in thermal stability.

Further studies by St. Clair and coworkers have focused attention on the ATS elastomer system [55]. In this work a new series of silicones were synthesized having the general structure shown previously in Fig. 10.29, but with varying chain lengths. In particular ATS elastomers having $n = 10$ to $n = 105$ were co-reacted into the basic LARC-13 formulation as described earlier so as to determine the effect of elastomer chain length on the final properties of the cured, modified polymer. Once again each elastomer was substituted for an appropriate portion of the original amine co-reactant so as to allow an elastomer concentration of approximately 15% by weight. Results obtained from lap shear experiments conducted using these modified LARC-13 formulations as adhesives are shown in Table 10.3. These data show that, firstly, elastomer chain length has a negligible influence on lap shear strength values and, secondly, as described above, elastomer incorporation provides no major advantage as far as lap shear strength is concerned. The best result from an elastomer modified system was obtained from a formulation comprising a 50:50 combination of ATS elastomers having repeat units of 10 and 105. Although investigations with rubber modified epoxies have shown that a similar approach regarding elastomer molecular weight can promote both an optimization in mechanical properties together with the observation of an apparent bimodal particle morphology, insufficient evidence was obtained from this study to indicate whether a similar effect would be operative with the ATS modified LARC-13 systems.

Table 10.3 Influence of elastomer chain length on the lap shear strength of LARC-13*

Adhesive	*Lap shear strength (20 °C) (MPa)*
LARC-13	20
LARC-13–ATS 105	18
LARC-13–ATS 10	20
LARC-13–ATS 10 + 105	21

* Titanium adherends.
Obtained from data previously published by St. Clair *et al.* in *NASA Tech. Memo 83172*, 1981.

From dynamic mechanical studies using TBA, St. Clair, St. Clair and Ezzell found, as in their earlier work, that ATS incorporation resulted in an apparent two-phase morphology since two major transitions were observed, one pertaining to the crosslinked LARC-13 phase, the other attributable to elastomer T_g (−113 to −119 °C). St. Clair, St. Clair and Ezzell were, as a result of this work, able to conclude that elastomer chain length exhibited very little effect on the properties of the ATS containing LARC-13 adhesive [55].

10.3.3 Elastomeric modification of other polyimide systems

Although fracture energy values obtained from some acetylenic oligomeric systems have not been quite as low as the bismaleimide and norbornene systems discussed in previous sections, it seems logical that elastomeric modification would, if successful, provide some significant benefits. It is therefore somewhat surprising to note that, as far as the author is aware, no significant investigations into the elastomeric modification of acetylene imide oligomers have been conducted, or at least published.

It seems likely that this interesting omission could be associated, partly at least, with what is generally regarded as the major deficiency with acetylenic imides. Basically acetylenics suffer from close melting and cure temperatures which result in a very narrow processing 'window' with a typical gel time of between 2 and 3 min at a typical processing temperature of 250 °C. Under such restrictive conditions it is doubtful whether highly important processes such as phase separation and morphological development could occur to the required degree. In addition, cure conditions encompassing post-cure temperatures of up to 370 °C could be regarded as somewhat extreme for some of the elastomeric materials used with bismaleimides and norbornenes.

Although linear high molecular weight polyimides such as the thermoplastic variants discussed at the beginning of this chapter generally exhibit relatively high fracture energies, this has not restricted the desirability of elastomer incorporation. In such circumstances the driving force is often associated with factors other than toughness enhancement in the conventional sense and it is of interest to consider briefly some examples of where this would appear to be the case.

Ezzell, St. Clair and Hinkley have considered the use of elastomeric modifiers as a means of imparting improved tear resistance to linear polyimide films [56]. In their work they investigated the effects of incorporating an ATBN elastomer having aromatic primary amine terminal functionality of the type previously discussed above for the work of St. Clair, St. Clair and Ezzell into a linear polyimide system based on 3,3′,4,4′-benzophenone tetracarboxylic dianhydride and 4,4′-oxydianiline. In particular, variations in the acrylonitrile content of the elastomer were studied with regard to its effects on T_g, thermal stability, mechanical properties and tear resistance of the modified polyimide films.

With ATBN elastomers containing 18% acrylonitrile, increasing elastomer concentration resulted in tear energy passing through a broad maximum with a peak at 15% elastomer by weight. However, under certain circumstances certain elastomers were shown to be capable of exhibiting a deterioration in tear resistance, thus indicating that the specific type of elastomer employed would require serious consideration.

As often found following elastomer incorporation, mechanical properties such as tensile strength and modulus generally underwent a decrease with increasing ATBN content as did T_g, although a reduction in the latter of from 281 to 264 °C on incorporation of from 0 to 20 wt% ATBN can be regarded both moderate and tolerable. Thermal stability of the polyimide films was found to be virtually unaffected by ATBN concentrations of 5% and 10% whereas 15% and 20% formulations did exhibit a reduction. This of course is not particularly surprising bearing in mind the predominantly aliphatic nature of the elastomer. However, the thermal stability decline obtained was not regarded dramatic by the authors, suggesting, as previously found by Shaw and coworkers, with rubber modified bismaleimides, that the incorporation of predominantly aliphatic elastomers into high crosslink density thermosets does not necessarily result in the major decline in thermal stability which would perhaps be envisaged.

Both torsional braid analysis and transmission electron microscopy revealed the type of two-phase morphology typically observed in many rubber modified systems with elastomeric particles dispersed in the resin matrix.

In addition to mechanical property enhancement, the incorporation of elastomeric materials into linear polyimides can give a measure of processability to what can be and often are virtually intractable materials. This can be assumed to be due to the introduction of flexibilizing segments into the main chain backbone which generally results in a reduction in T_g and enhanced solubility characteristics. In this respect the use of polysiloxane based elastomers has been proposed as a means by which such processability improvements can be achieved. In addition, siloxane incorporation would also impart further beneficial properties including reduced water sorption, good thermal and UV stability, resistance to aggressive oxygen environments as well as, of course, improved toughness.

Using this approach, Spontak and Williams have studied the microstructure together with the thermal and mechanical properties of siloxane–imide block copolymers of two types [57]. The first, a proprietary system, had the general chemical structure shown below:

$$\left[-N(CO)_2R_1(CO)_2N-R_2-\left(\underset{R}{\overset{R}{Si}}-O \right)_x-R_3-N(CO)_2R_1(CO)_2N-R_4- \right]_n$$

Both the size of the siloxane block (although of polymeric magnitude) and the nature of the groups R, R_1, R_2, R_3 and R_4 were not revealed by the manufacturer. The second siloxane–imide copolymer studied by Spontak and Williams was a proprietary prototype based on a polyetherimide, the structure of which is shown below, modified with siloxane (R_2SiO) blocks of varying size ($x = 10–15$):

Interestingly, both of the siloxane–imide copolymers revealed a two-phase morphology reminiscent of the microstructures found with both rubber modified imides of the oligomeric type and indeed epoxies. However, with the first copolyimide, a roughly bimodal size distribution of domains was found, with domain diameters of approximately 3 and 16 nm being observed. Such domain sizes are noteworthy in being approximately three orders of magnitude smaller than elastomeric domains found in many rubber modified thermoset polymers. Although particles of a similar size were observed with the polyetherimide based copolymer, the particle size range found with this siloxane–imide was substantially greater with some particles exhibiting diameters of several microns. Although it would be intuitively reasonable to assume that the particulate domains observed in the polyetherimide based copolymer were composed primarily of the siloxane component and thus elastomeric in nature, energy-dispersive X-ray microanalysis (EDX) conducted by Spontak and Williams suggested otherwise. In fact they found evidence suggesting a siloxane-rich matrix with the domains representing the imide-rich regions. This result was deemed particularly surprising in view of the fact that observations from transmission electron microscopy revealed domains relatively dark in comparison with the matrix. Regions relatively rich in siloxane would be expected to appear darker. However, Spontak and Williams found a reasonable explanation to account for this apparent paradox. The domains were found to be thicker than the surrounding matrix, thus accounting for the increased darkness for relatively silicon-lean domains. In addition, other observations, such as the presence of silicon in both domain and matrix regions, were attributed to a combination of both incomplete phase separation together with the apparent formation of a siloxane-rich layer at the copolymer–air interface. Such a low energy external layer could be viewed as forming so as to minimize the free energy of the copolymer. From these studies, Spontak and Williams concluded that the matrix was siloxane rich and thus expected to exhibit rubbery characteristics, whilst the domains, being imide rich, would be expected to behave as glassy fillers at room temperature. With regard to thermal characteristics, scanning calorimetry revealed, for the polyetherimide based copolymers, a T_g value of approximately 227 °C. A proprietary polyetherimide system has been reported as having a T_g value of 215–225 °C [9], thus suggesting that T_g of the imide component of the copolymer is virtually unaffected by the presence of siloxane. This of course implies effective microphase separation. T_g for the proprietary siloxane–imide was approximately 61 °C. Spontak and Williams attributed this to the siloxane component with a T_g value for the imide component not being observed under the appropriate test conditions, with, as suggested, T_g exceeding the temperature at which significant thermal decomposition occurs (350 °C).

Arnold *et al.* have also attempted the modification of linear polyimides using polysiloxanes [58]. In this work a series of dianhydrides and diamines were employed as intermediates for the synthesis of polysiloxane–imide copolymers together with their homopolymer counterparts. In this study, the siloxane component was provided by the use of aminopropyl polydimethylsiloxane shown below, which was incorporated into the dianhydride–diamine system in concentrations of 5–70 wt%, with molecular weights of 800–10 000 g mol^{-1}:

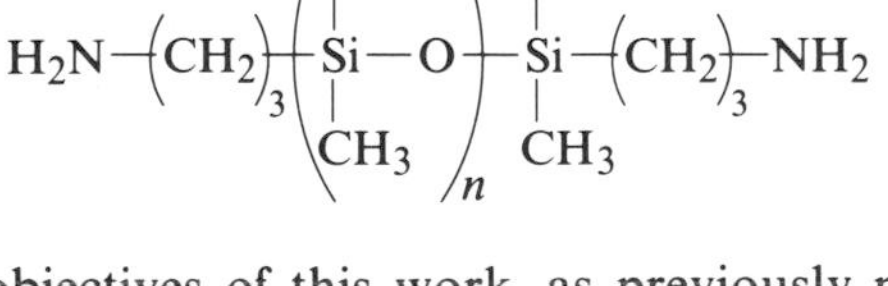

$$H_2N{-}(CH_2)_3{-}\left(\begin{array}{c}CH_3\\ |\\ Si{-}O\\ |\\ CH_3\end{array}\right)_n{-}\begin{array}{c}CH_3\\ |\\ Si\\ |\\ CH_3\end{array}{-}(CH_2)_3{-}NH_2$$

One of the main objectives of this work, as previously mentioned, was to improve the solubility of normally intractable polyimides and in this respect Arnold *et al.*'s work could be regarded successful; siloxane contents of 40 wt%

and above resulted in solubility in otherwise totally inappropriate solvents such as tetrahydrofuran and dichloromethane. In addition to this, T_g values of the siloxane modified imides were found to be a function of both the level of incorporated siloxane as well as siloxane molecular weight. Whereas increasing siloxane content resulted, as expected, in a reduction in T_g (from approximately 265 °C for the non-siloxane imide to 218 °C for a system containing 40 wt% siloxane), an increasing molecular weight for a given siloxane content had the reverse effect. Low temperature transitions within the range from −117 to −123 °C were also observed with siloxane contents in excess of 20 wt%. These, quite reasonably, were attributed to the siloxane component of the systems. The presence of both high and low temperature transitions appropriate to both the polyimide and siloxane components of the system indicated the existence of a two-phase morphology. Transmission electron microscopy conducted by Arnold *et al.* reinforced the view with evidence of siloxane microphase formation at the 5 nm level.

In addition to solubility characteristics, mechanical characterization such as stress–strain and adhesive joint behaviour together with an assessment of thermal stability by a thermogravimetric technique were also studied by Arnold *et al.* and the influence of siloxane incorporation was determined. However, of particular interest in this study was the effect of siloxane addition on hydrophobic character. As mentioned earlier, the introduction of a predominantly hydrophobic siloxane component into a linear polyimide would be expected to enhance the hydrophobic character of the resulting copolymer. Arnold *et al.* indeed found this to be the case as indicated in Table 10.4. As clearly shown, the incorporation of up to 50 wt% siloxane can dramatically reduce the quantity of water absorbed by the polymer. Improvements of this magnitude can have profound effects on the ability of a polymer to resist the particularly damaging effects associated with warm, humid conditions, possibly one of the most damaging environments that polymeric materials can encounter. Indeed, Arnold *et al.* found quite substantial improvements in the long-term durability of adhesive joints employing the siloxane modified polyimides as adhesives in comparison with their non-siloxane-modified counterparts. In addition to the undoubted improvements in processability

Table 10.4 Influence of siloxane incorporation on the hydrophobic characteristics of a linear polyimide

Siloxane content (wt %)	*Water uptake (%)*
0	2.7
30	0.8
50	0.4

Obtained from data previously published by Arnold *et al.* in *Polymer*, **30**, 986, 1989.

rendered by the use of polysiloxanes, greatly enhanced hydrophobic character will, for many applications, be seen as a major advantage.

10.4 COMPARISONS WITH OTHER TOUGHENING TECHNIQUES

Although the main theme of this chapter is toughness enhancement via rubber modification, it is of interest to consider briefly other means by which improvements in toughness can be achieved. Since improved toughness is almost universally required with oligomeric polyimides this section will be devoted to this class of material. Three main approaches will be briefly considered concerning (a) variation in initial oligomer reactants, (b) reaction of oligomers with other materials (not elastomers) and (c) modification with thermoplastics.

10.4.1 Toughness modification via oligomer reactant variations

Previous sections in this chapter have outlined the chemistry by which oligomeric imides are prepared. To recap, in the specific case of a bismaleimide, diamine is reacted under the appropriate conditions with maleic anhydride to produce the bismaleimide via an intermediate bismaleimic acid. Clearly both the structure of the bismaleimide and therefore the properties of the cured product will depend on the only formulation variable available, namely the nature of the amine. Clearly diprimary amines having a relatively high molecular weight together with a significant aliphatic composition would be expected to produce bismaleimides having substantially greater toughness characteristics than their low molecular weight, predominantly aromatic counterparts. Although this would be achieved at the expense of other equally desirable properties such as modulus and high temperature behaviour, it serves to demonstrate the potential that exists for tailor making properties to specific requirements.

A similar approach with the norbornene-terminated imide system, PMR-15, has been shown to have a similar effect, where replacement of the amine generally employed in its preparation, 4,4′-methylenedianiline, with aromatic amines containing flexible connecting groups has resulted in significant toughness improvements [59].

10.4.2 Toughness modification via reaction of oligomer with other components

Numerous investigations have indicated the ability of bismaleimides, in particular, to participate in copolymerization reactions with a number of materials. Significant amongst these are reactions with amines, hydrazides, bisallyl compounds and epoxies, as mentioned previously, all such reactions generally resulting in reductions in crosslink density on cure and thus substantial improvements in toughness. In some particularly notable cases these improvements have been achieved with very little sacrifice in other desirable properties.

10.4.3 Toughness enhancement by modification with thermoplastics

Since high molecular weight thermoplastic polymers generally exhibit substantially greater fracture energy values than crosslinked polyimides, the concept of blending thermoplastics with imide oligomers such as bismaleimides has received some consideration. One particular study conducted by Yamamoto, Satoh and Etoh has investigated the blending of a proprietary linear thermoplastic polyimide with the bismaleimide 4,4′-bismaleimidodiphenylmethane [60]. Unfortunately, precise values of fracture toughness were not apparently obtained in this work and thus a measure of the magnitude of toughness enhancement provided by the thermoplastic modification is not possible. However, it is reasonable to assume that a significant improvement in toughness would have occurred. One major disadvantage of this specific approach, highlighted by Segal *et al.* [39], has concerned the need to employ solvents such as *N*-methylpyrrolidone which are notoriously difficult to remove from the system and if retained to any significant degree can exert a harmful effect on polyimide properties. Clearly the use of high temperature thermoplastics exhibiting solubility in low boiling point solvents would be regarded as a significant advantage and clearly shows the way forward for this type of modification approach.

10.5 CONCLUDING REMARKS

The highly successful use of elastomeric modifiers as a means of enhancing the toughness characteristics of epoxy resins has resulted in attempts being made to transfer this technology to other types of thermosetting polymers, imide oligomeric compounds being no exception. Although some considerable success has been achieved in this area, particularly with bismaleimides and norbornenes, the inclusion of elastomeric compounds into these materials has posed some potentially serious problems. Obviously dramatic reductions in T_g would be regarded as unacceptable. Although many of the studies described have shown significant toughness improvements with only moderate reductions in T_g, it is likely that future developments will be based on other modification approaches, such as by thermoplastics.

REFERENCES

1. Knight, G. J. (1980) In *Developments in Reinforced Plastics – 1* (ed. G. Pritchard), Applied Science, London, pp. 145–210.

2. Critchley, J. P., Knight, G. J. and Wright, W. W. (1983) *Heat Resistant Polymers*, Plenum, New York.
3. Mittal, K. L. (ed.) (1984) *Polyimides – Synthesis, Characterization and Applications*, Vols 1 and 2, Plenum, New York.
4. Shaw, S. J. (1987) *Mater. Sci. Technol.*, **3**, 589.
5. Sroog, C. E. (1967) *J. Polym. Sci. C*, **16**, 1191.
6. Shaw, S. J. (1986) In *Adhesion–10* (ed. K. W. Allen), Applied Science, London, pp. 20–41.
7. St. Clair, A. K. and St. Clair, T. L. (1984) In *Polyimides – Synthesis, Characterization and Applications*, Vol. 2 (ed. K. L. Mittal), Plenum, New York, p. 977.
8. St. Clair, A. K. and St. Clair, T. L. (1981) *SAMPE Q.*, **13**(1), 20.
9. Johnson, R. O. and Burlhis, H. S. (1983) *J. Polym. Sci., Polym. Symp.*, **70**, 129.
10. Serfaty, I. W. (1984) In *Polyimides – Synthesis, Characterization and Applications*, Vol. 1 (ed. K. L. Mittal), Plenum, New York, p. 149.
11. Stenzenberger, H. D., Herzog, M., Romer, W., Scheiblick, R. and Reerce, N. J. (1983) *Br. Polym. J.*, **15**(2), 2.
12. Lubowitz, H. R. (1970) US Patent 3,528,980.
13. Serafini, T. T., Delvigs, P. and Lightsey, G. R. (1972) *J. Appl. Polym. Sci.*, **16**, 905.
14. Serafini, T. T. and Delvigs, P. (1973) *Appl. Polym. Symp.*, **22**, 89.
15. St. Clair, T. L. and Jewell, R. A. (1976) Proc. 8th Natl SAMPE Technical Conf., p. 82.
16. St. Clair, T. L. and Jewell, R. A. (1978) Proc. 23rd Natl SAMPE Symp., p. 520.
17. St. Clair, T. L. and Progar, D. J. (1979) Proc. 24th Natl SAMPE Symp., p. 1081.
18. Hendricks, C. L. and Hill, S. G. (1984) In *Polyimides – Synthesis, Characterization and Applications*, Vol. 2 (ed. K. L. Mittal), Plenum, New York, p. 1103.
19. Hill, S. G. and Sheppard, C. H. (1980) Proc. 12th Natl SAMPE Technical Conf., p. 1040.
20. St. Clair, A. K. and St. Clair, T. L. (1981) *NASA Tech. Memo. 83141*, July 1981.
21. Steger, V. Y. (1980) Proc. 12th Natl SAMPE Technical Conf., p. 1054.
22. Stevenson, A. A. and Wykes, D. H. (1980) Proc. 12th Natl SAMPE Technical Conf., p. 746.
23. St. Clair, A. K. and St. Clair, T. L. (1982) *Polym. Eng. Sci.*, **22**, 9.
24. Bilow, N., Landis, A. L. and Miller, L. J. (1974) US Patent 3,845,018.
25. Landis, A. L., Bilow, N., Boschan, R. H., Lawrence, R. E. and Aponyi, T. J. (1974) *Polym. Prepr.*, **15**, 537.
26. Bilow, N. and Aponyi, T. J. (1975) Proc. 20th Natl SAMPE Symp., p. 618.
27. Bilow, N. and Landis, A. L. (1976) Proc. 8th SAMPE Technical Conf., p. 94.
28. Bilow, N., Keller, L. B., Landis, A. L., Boschan, R. H. and Castillo, A. A. (1978) Proc. 23rd Natl SAMPE Symp., p. 791.
29. Bargain, M. (1971) US Patent 3,562,223.
30. Mallet, M. A. J. (1973) *Mod. Plast.*, (June), 78.
31. Stenzenberger, H. D. (1980) US Patent 4,211,861.
32. Stenzenberger, H. D. (1981) US Patent 4,303,779.
33. Jones, R. J. (1981) US Patent 4,283,521.
34. King, J. S., Choudhari, M. and Zahir, S. (1984) Proc. 29th Natl SAMPE Symp., p. 1034.
35. Kinloch, A. J., Shaw, S. J. and Tod, D. A. (1984) In *Rubber-modified Thermoset Resins* (eds C. K. Riew and J. K. Gillham), American Chemical Society, pp. 101–15.
36. Shaw, S. J. and Kinloch, A. J. (1985) *Int. J. Adhes. Adhes.*, **5**, 123.
37. Shaw, S. J. (1987) In *Recent Advances in Polyimide Science and Technology* (eds M. R. Gupta and W. D. Weber), Society of Plastics Engineers.
38. Gollob, D. S., Mandell, J. F. and McGarry, F. J. (1979) Toughening of thermoset-

ting polyimides. *Massachusetts Institute of Technology Research Report R 79-1*, August 1979.
39. Segal, C. L., Stenzenberger, H. D., Herzog, M., Romer, W., Pierce, S. and Canning, M. S. (1985) Proc. 17th Natl SAMPE Technical Conf., p. 147.
40. Stenzenberger, H. D., Romer, W., Herzog, M., Pierce, S., Canning, M. and Fear, K. (1986) Proc. 31st Int. SAMPE Symp., p. 920.
41. Shaw, S. J. (1994) In *Rubber Toughened Engineering Plastics* (ed. A. A. Collyer), Chapman & Hall, London, Chap. 6.
42. St. Clair, A. K. and St. Clair, T. L. (1981) *Int. J. Adhes.*, **2**, 249.
43. Takeda, S. and Kakiuchi, H. (1988) *J. Appl. Polym. Sci.*, **35**, 1351.
44. Takeda, S., Akiyama, H. and Kakiuchi, H. (1988) *J. Appl. Polym. Sci.*, **35**, 1341.
45. Stenzenberger, H. D., Konig, P., Herzog, M., Romer, W., Pierce, S., Fear, K. and Canning, M. S. (1987) Proc. 19th Int. SAMPE Technical Conf., p. 372.
46. Shaw, S. J. and Tod, D. A. (1989) *J. Adhes.*, **28**, 231.
47. Enns, J. B. and Gillham, J. K. (1983) *J. Appl. Polym. Sci.*, **28**, 2567.
48. Gupta, V. B., Drzal, L. T., Lee, C. Y.-C. and Rich, M. J. (1985) *Polym. Eng. Sci.*, **25**, 812.
49. Varma, I. K., Fohlen, G. M., Parker, J. A. and Varma, D. S. (1984) In *Polyimides – Synthesis, Characterization and Applications*, Vol. 2 (ed. K. L. Mittal), Plenum, New York, p. 683.
50. Maglio, G., Palumbo, R. and Vitagliano, V. M. (1989) *Polymer*, **30**, 1175.
51. Maudgal, S. and St. Clair, T. L. (1984) Proc. 29th Natl SAMPE Symp., p. 437.
52. Wright, W. W. (1981) *Composites*, **12**, 201.
53. Tod, D. A. and Shaw, S. J. (1988) *Br. Polym. J.*, **20**, 397.
54. St. Clair, A. K. and St. Clair, T. L. (1980) Proc. 12th Natl SAMPE Technical Conf., p. 729.
55. St. Clair, A. K., St. Clair, T. L. and Ezzell, S. A. (1981) *NASA Tech. Memo. 83172*.
56. Ezzell, S. A., St. Clair, A. K. and Hinkley, J. A. (1987) *Polymer*, **28**, 1779.
57. Spontak, R. J. and Williams, M. C. (1989) *J. Appl. Polym. Sci.*, **38**, 1607.
58. Arnold, C. A., Summers, J. D., Chen, Y. P., Bott, R. H., Chen, D. and McGrath, J. E. (1989) *Polymer*, **30**, 986.
59. Delvigs, P. (1989) *Polym. Compds.*, **10**, 134.
60. Yamamoto, Y., Satoh, S. and Etoh, S. (1985) *SAMPE J.* (July–August), 6.

Index